Indigenous Knowledge

Enhancing its Contribution to Natural Resources Management

Indigenous Knowledge

Enhancing its Contribution to Natural Resources Management

Edited by

Paul Sillitoe

Durham University

CABI

CABI is a trading name of CAB International

CABI
Nosworthy Way
Wallingford
Oxfordshire OX10 8DE
UK

Tel: +44 (0)1491 832111
Fax: +44 (0)1491 833508
E-mail: info@cabi.org
Website: www.cabi.org

CABI
745 Atlantic Avenue
8th Floor
Boston, MA 02111
USA

Tel: +1 (617)682 9015
E-mail: cabi-nao@cabi.org

A catalogue record for this book is available from the British Library, London, UK.

Library of Congress Cataloging-in-Publication Data

Names: Sillitoe, Paul, 1949- editor.
Title: Indigenous knowledge : enhancing its contribution to natural resources management / edited by Paul Sillitoe.
Other titles: Indigenous knowledge (C.A.B. International)
Description: Wallingford, Oxfordshire : CABI, 2017. | Includes bibliographical references and index.
Identifiers: LCCN 2017016224 (print) | LCCN 2017039849 (ebook) | ISBN 9781780648118 (ePDF) | ISBN 9781780647074 (ePub) | ISBN 9781780647050 (hbk : alk. paper)
Subjects: LCSH: Indigenous peoples--Ecology--Case studies. | Traditional ecological knowledge--Case studies. | Natural resources--Management--Developing counries--Case studies. | Conservation of natural resources--Developing countries--Case studies.
Classification: LCC GF50 (ebook) | LCC GF50 .I527 2017 (print) | DDC 304.2--dc23
LC record available at https://lccn.loc.gov/2017016224

ISBN-13: 978 1 78064 705 0

Commissioning editor: David Hemming
Editorial assistant: Emma McCann
Production editor: Marta Patiño

Typeset by SPi, Pondicherry, India
Printed and bound in the UK by CPI Group (UK) Ltd, Croydon, CR0 4YY

Contents

List of Figures

List of Tables

Contributors

Andrew Ainslie is currently a lecturer in International Rural Development in the School of Agriculture, Policy and Development at the University of Reading, Reading, RG6 6AR, UK (e-mail: a.m.ainslie@reading.ac.uk). His research interests include the study of system-wide drivers of agrarian change and ecosystem resilience, particularly in Sub-Saharan Africa. His research considers traditional/hybrid/modern innovations in resource governance; the food system, particularly in relation to global/local beef value chains; and contestations of value in knowledge systems.

Kojo Amanor is Professor at the Institute of African Studies, University of Ghana, PO Box LG 73, Legon, Ghana (e-mail: kojoamanor@gmail.com). His research interests are in the land question, smallholder agriculture, agribusiness food chains, forestry policy, environment, and participatory methodologies for rural development.

Jeffery Bentley is an agricultural anthropologist at Agro-Insight, Casilla 2695, Cochabamba, Bolivia (e-mail: jeff@agroinsight.com). He is interested in local knowledge, language, farmer experiments, and video in agricultural extension.

Samara Brock is a doctoral student in the School of Forestry and Environmental Studies, Yale University, 195 Prospect Street, New Haven, CT 06511 (e-mail: samara.brock@yale.edu). Her research interests include food systems sustainability, globalized environmental governance, and the construction of systems of knowledge and expertise.

Stephen B. Brush is Professor Emeritus in the Department of Human Ecology, University of California, Davis CA 95616 USA (e-mail: sbbrush@ucdavis.edu). His research interests focus on the indigenous knowledge of crop diversity in centres of crop origins and how culture structures crop evolution and conservation.

David A. Cleveland is Research Professor in the Environmental Studies Program and the Department of Geography, University of California, Santa Barbara, CA 93106-4160, USA (e-mail: cleveland@es.ucsb.edu). He is currently researching the potential for agrifood system localization and diet change to improve nutrition, reduce greenhouse gas emissions and promote food and climate justice.

Marion Comptour is a doctoral student at the Centre d'Ecologie Fonctionnelle et Evolutive, CNRS University of Montpellier, 1919 route de Mende, 34293 Montpellier cedex 5, France. She is working on multi-activity subsistence strategies of inhabitants of the floodplains of the *Cuvette congolaise* and on the history and dynamics of floodplain socio-ecosystems in this region since the pre-colonial period.

Luisa Cortesi is a doctoral student in the School of Forestry and Environmental Studies, Yale University, 195 Prospect Street, New Haven, CT 06511, USA (e-mail: luisa.cortesi@yale.edu). Her ethnographic work, originating from applied anthropology practice in development, embraces the environmental humanities as well as the natural sciences for an interdisciplinary outlook on water, environmental knowledge, technologies and disasters.

Isabel Díaz Reviriego is a research associate at the the Institut de Ciència i Tecnologia Ambientals, Universitat Autònoma de Barcelona, Calle de les columnes, UAB, E-08193 Cerdanyola del Vallès, Barcelona, Spain and a Researcher at the Internet Interdisciplinary Institute at the Universitat Oberta de Catalunya, Spain (e-mail: isabel.diaz@uab.cat). Her research interests include ethnoecology, biocultural diversity, local ecological knowledge, global environmental change, gender studies and feminist political ecology.

Michael R. Dove is Margaret K. Musser Professor of Social Ecology, School of Forestry and Environmental Studies, Yale University, 195 Prospect Street, New Haven, CT 06511, USA (e-mail: michael.dove@yale.edu). His current research interests include the anthropology of climate change, natural hazards, disasters and resource degradation; indigenous environmental knowledge; development and environmental institutions, discourses and movements; the sociology of the environmental sciences; and post-humanist study of environmental relations.

Romain Duda is a PhD student at Institut de Ciència i Tecnologia Ambientals, Universitat Autònoma de Barcelona, Calle de les columnes, UAB, E-08193 Cerdanyola del Vallès, Barcelona, Spain (e-mail : romainduda@gmail.com). He is associated with the Eco-anthropology and Ethnobiology Laboratory of the Musée de l'Homme, Paris. His interests lie in ethnoecology and specifically in human–animal relations and wildlife conservation. His current research focuses on the changes affecting the socio-cultural aspects of hunting among the Baka (Cameroon) in the context of wildlife decline and the bushmeat trade.

Roy Ellen is Emeritus Professor of Anthropology and Human Ecology in the School of Anthropology and Conservation at the University of Kent, Marlowe Building, Canterbury, Kent CT2 7NR, UK (e-mail: R.F.Ellen@kent.ac.uk). His current research interests include cultural cognition, ethnobiological knowledge systems, and the resilience of traditional modes of exchange and social reproduction in the Moluccas.

James Fairhead is Professor of Anthropology at the University of Sussex, Arts Building C, Arts Road, University of Sussex, Falmer, Brighton, BN1 9SJ, UK (e-mail: j.r.fairhead@sussex.ac.uk). His research examines existing agro-ecological and medical practices in Central and West Africa and their interface with global scientific, policy and international development communities.

Álvaro Fernández-Llamazares is a research associate at the Institut de Ciència i Tecnologia Ambientals, Universitat Autònoma de Barcelona, C/ de les columnes, UAB, E-08193 Cerdanyola del Vallès, Barcelona, Spain and postdoctoral fellow at the Metapopulation Research Centre, Department of Biosciences, University of Helsinki, PO Box 65, Viikinkaari 1, FI-00014, Helsinki, Finland (e-mail: alvaro.fernandez-llamazares@helsinki.fi). His current research interests include biocultural conservation approaches and the role of indigenous lands in buffering biodiversity loss.

James Fraser is a lecturer in the Environment Centre at the University of Lancaster, Lancaster University, United Kingdom LA1 4YQ, UK (e-mail: james.angus.fraser@gmail.com). His research has examined how local people in the Afro- and Neo-tropics (Brazilian Central Amazonia and the Liberian Upper Guinea Forest of West Africa) manage soils, plants, trees and landscapes in their regions, and the social, natural and normative dimensions.

Sandrine Gallois is a research associate at the Institut de Ciència i Tecnologia Ambientals, Universitat Autònoma de Barcelona, Calle de les columnes, UAB, E-08193 Cerdanyola del Vallès, Barcelona, Spain and at the French National Museum of Natural History, Paris (e-mail: galloissandrine.ethnoeco@gmail.com). Her research areas are ethnoecology and biocultural diversity, focusing mainly on the acquisition and transmission of local ecological knowledge of small-scale societies. Her PhD work was on the dynamics of local ecological knowledge among a hunter-gatherer society from the Congo Basin (the Baka).

Samuel Guindo is an agronomist and is Coordinator of the Regional Rice Program of the Syngenta Foundation for Sustainable Agriculture in Mali (e-mail: samuelamaibe@gmail.com). He is interested in disseminating agricultural technologies and in agricultural extension through training videos and in networking farmers to make better use of their local knowledge.

Chris Hebdon is a doctoral student in the School of Forestry and Environmental Studies, Yale University, 195 Prospect Street, New Haven, CT 06511, USA (e-mail: chris.hebdon@yale.edu). His dissertation research focuses on Kichwa-speaking communities in the Ecuadorian Amazon.

Patricia L. Howard is a professor in the Department of Social Sciences, Wageningen University, The Netherlands, and Professor in Ethnobotany, School of Anthropology and Conservation at the University of Kent, Marlowe Building, Canterbury, Kent CT2 7NR, UK (e-mail: patricia.howard@wur.nl). She focuses on contemporary change in food systems, food security, and the relations between agrobiodiversity (wild and cultivated) and human well-being, especially in the domestic sphere: home gardening, post-harvest processing and culinary traditions.

Sascha Huditz is a human geographer (TU Dresden, Germany) who collaborated as a research assistant in the project 'The Adaptive Nature of Culture: a Cross-cultural Analysis of the Returns of Local Environmental Knowledge in Three Indigenous Societies', among the Tsimane' in Bolivia.

Amy Johnson is a doctoral student in the School of Forestry and Environmental Studies, Yale University, 195 Prospect Street, New Haven, CT 06511, USA (e-mail: amy.l.johnson@yale.edu). Her work focuses on federalism and environmental sentiment in Nepal.

Alder Keleman-Saxena is a doctoral student affiliated with the School of Forestry and Environmental Studies, Yale University, 195 Prospect Street, New Haven, CT 06511, USA; and Anthropology Department, School of Culture and Society, Aarhus University, Aarhus, Denmark (e-mail: alder.keleman@gmail.com). Her research uses mixed-method approaches to explore the role of agrobiodiversity in household food security in the Bolivian Andes, incorporating the ethnobotanical, nutritional and cultural values of Andean crops.

J. Stephen Lansing is Director of the Complexity Institute and Professor in the Asian School of Environment at Nanyang Technological University in Singapore (e-mail: jlansing@ntu.edu.sg). He is an external professor at the Santa Fe Institute, emeritus professor of anthropology at the University of Arizona, and senior research fellow at the Stockholm Resilience Center. Website: www.slansing.org. His research has to do with anthropology and the environment in Indonesia.

Melissa Leach is Professorial Fellow and Director of the Institute of Development Studies at the University of Sussex, Library Road, Brighton, BN1 9RE, UK (e-mail: m.leach@ids.ac.uk). She is a geographer and social anthropologist. Her interdisciplinary, policy-engaged research in Africa and beyond links environment, agriculture, health, technology and gender, with particular interests in knowledge, power and the politics of science and policy processes.

Rosalinda Hidalgo Ledesma is with the Manejo Integral de los Montes de la Sierra de Zongolica, Veracruz/People and Plants International, Inc. (email: rosalindahidalgo@gmail.com). She is a social anthropologist and specializes in rural development, with more than 10 years of experience working on research action projects with indigenous and farmer organizations. She is actively involved with the environmental justice movement surrounding the protection of river systems and people's territorial rights in Mexico.

Johannes Lehmann is Professor of Soil Science at Cornell University, School of Integrative Plant Science, College of Agriculture and Life Sciences, 232 Emerson Hall, Cornell University, Ithaca, NY 14853, USA (e-mail: cl273@cornell.edu). His research focuses on soil biogeochemistry and soil fertility management. His specialization is in soil organic matter and nutrient studies of managed and natural ecosystems with a focus on soil degradation and sustainable agriculture in the tropics (especially Africa), bio-energy, greenhouse gas emissions from soil and headwaters, and synchrotron-based methods for soil research.

Francis Michael Ludlow is Marie Skłodowska-Curie Individual Fellow 2016–2018, in the Department of History, School of Histories and Humanities, Trinity College, Dublin 2, Ireland (e-mail: ludlowf@tcd.ie). He is a climate historian who specializes in reconstructing past climatic conditions from documentary evidence, and in understanding the impacts of abrupt climatic changes and extreme weather on past societies.

Doyle McKey is a professor at the Centre d'Ecologie Fonctionnelle et Evolutive, CNRS University of Montpellier, 1919 route de Mende, 34293 Montpellier cedex 5, France (e-mail: doyle.mckey@cefe.cnrs.fr). He studies interactions between plants and animals, between plants and humans and between humans and ecosystems in tropical environments.

Citlalli López Binnqüist is an ethnoecologist, researcher and teacher-facilitator at Centro de Investigaciones Tropicales, Universidad Veracruzana, Calle J.M. Morelos 44, Centro Histórico, CP 91000, Xalapa, Veracruz, México (e-mail: cilopez@uv.mx). Her work focuses on the study and revitalization of traditional ecological knowledge in mountainous regions, specializing in the socio-cultural, organizational and institutional aspects relating to the use and management of forest and agroforestry systems.

Paul Van Mele is an agricultural scientist at Agro-Insight, Grote Baan 24, 3670 Meeuwen-Gruitrode, Belgium (e-mail: paul@agroinsight.com). He has a special interest in producing farmer training videos as a way of merging scientific with local knowledge.

Tom Van Mourik is an agronomist and nutrition-focused agriculture specialist at Helen Keller International, Yoff, Toundoup Rya Lot 122, Route de l'aeroport, Dakar, Senegal (e-mail: tvanmourik@hki.org). He is interested in the development of tools for communicating and promoting good practices in agriculture, nutrition, and water and sanitation hygiene.

Lars Otto Naess is a research fellow with the Resource Politics Cluster, Institute for Development Studies, Brighton, UK (e-mail: l.naess@ids.ac.uk). His research interests centre on the social, political and institutional dimensions of adaptation to climate change in the context of international development.

Florencia G. Palis is Professor of Anthropology at the University of the Philippines at Los Baños College, Laguna, 4031, The Philippines (e-mail: florenciagpalis@gmail.com). Her research interests include natural resources management, agriculture and health, indigenous knowledge, biodiversity conservation, climate change, agricultural innovation systems, multi-stakeholder partnerships, impact assessment and social enterprises, with research experience in Asia.

Fortunata Panzo Panzo works with Manejo Integral de los Montes de la Sierra de Zongolica, Veracruz/ People and Plants International, Inc. (email: fortupanzo@gmail.com). Graduated from the Universidad Veracruzana Intercultural with a BA in Intercultural Management for Development with a specialization in Sustainability. Her work relates to the traditional environmental knowledge of the Sierra de Zongolica, actively promoting food sovereignty and biodiversity conservation.

Victoria Reyes García is ICREA Research Professor at the Institut de Ciència i Tecnologia Ambientals, Edifici Z ICTA-ICP Carrer de les columnes, Universitat Autònoma de Barcelona 08193, Bellaterra, Cerdanyola del Vallès-Barcelona, Spain (e-mail: victoria.reyes@uab.cat). She coordinates the Ethnoecology Laboratory at ICTA-UAB (http://icta.uab.es/Etnoecologia/mission.htm). Her research addresses the benefits generated by local ecological knowledge among indigenous populations.

Delphine Renard is a postdoctoral researcher in ecology at the Bren School of Environmental Science and Management, University of California Santa Barbara, Santa Barbara, California, 93106–5131, USA (e-mail: delphinerenard39@gmail.com). Her research interests focus on regularly patterned landscapes of both natural and cultural origin, and their interactions. She has studied earth-mound landscapes in seasonally flooded savannas in French Guiana, Colombia and Bolivia. She is also interested in the ecology of agroecosystems.

Gede Sedana is Dean of the Faculty of Agriculture, University of Dwijendra, Bali, Indonesia. He is an expert on Balinese *subaks* and a leading Balinese public intellectual.

Christopher Shepherd is a research fellow in the Department of Anthropology at the School of Culture, History, and Language, Australian National University, Acton, ACT, 2601, Australia (e-mail: chris.shepherd@anu.edu.au). He conducts research on development, natural resource management and mining in Latin America and Southeast Asia.

Paul Sillitoe is Professor of Anthropology, University of Durham, Dawson Building, South Road, DH1 3LE, UK (e-mail: paul.sillitoe@durham.ac.uk). His interests include natural resources management, technology, sustainable development and social change, and livelihoods, human ecology and ethno-science, with research experience in the Pacific, South Asia and Gulf regions. He has made several contributions to indigenous knowledge in development debates.

Daniela Soleri is an associate research scientist in the Department of Geography, University of California, Santa Barbara, CA 93106-4060, USA (e-mail: soleri@geog.ucsb.edu). An ethnoecologist, her current research quantifies practices and investigates informal knowledge in small-scale agriculture, seed and food systems, and how scientists can partner with these to respond to climate, environmental and social changes effectively and with equity.

Dawit Solomon is a senior research associate, Atkinson Center for Sustainable Future, Department of Crop and Soil Sciences, College of Agriculture and Life Sciences, 920 Bradfield Hall, Cornell University, Ithaca, NY 14853, USA (e-mail: ds278@cornell.edu). He is a soil scientist who works on soil fertility and land degradation, sustainable agriculture, biogeochemistry, and instrumental chemical analysis, with a strong interest in development and outreach activities geared towards increasing agricultural productivity and food security in a sustainable manner based on either using locally available techniques or through the development and transfers of effective and environmentally sound technology.

Sidi Toure is a research technician (zootechnician, senior facilitator GISF/CEPGV) with the International Crops Research Institute for the Semi-Arid Tropics (ICRISAT) in Mali (e-mail: sidito6@yahoo.fr). He is interested in the development of tools for communicating and promoting good practices in agriculture.

Thérèse de Vet taught anthropology at the School of Anthropology, University of Arizona, Tucson, Arizona, USA. She has written widely on Balinese culture, history and society. A key focus of her research was the relationship between orality and literacy.

Wayan Windia is Distinguished Professor of Agriculture at Udayana University, Bali, Indonesia. He was a leader of the team that drafted the proposal for UNESCO to designate the *Subak* in Bali as a World Cultural Heritage. Besides his researches on *subaks* and associated water-management system for irrigation, he has also researched community empowerment through agroecotourism.

Gérard Zoundji is a PhD candidate at the University of Abomey-Calavi, Faculty of Agronomic Sciences, 081 BP 7145 Cotonou, Benin (e-mail: gezoundji@yahoo.fr). He has a keen interest in farmer learning and innovation.

Preface

It is increasingly recognized that indigenous knowledge (IK), which has featured centrally in the management of natural resources for millennia, should play a significant part in programmes that seek to increase land productivity, food security and environmental conservation. This is evident in the relatively new 'bio-cultural' approach to environmental stewardship, which underlines the connection between biological and cultural diversity. It acknowledges that indigenous knowledge and management of natural environments – the soils, plants and animals – feature significantly in what we see today. They should consequently figure in any thorough understanding, particularly if we envisage intervening in any way, as widely recognized, from the United Nations with its concerns for global governance to local bodies with interests in regional landscapes. Yet IK remains largely a 'known unknown' for many in the natural resources sector.

This book seeks to clarify what IK amounts to as seen in current cutting-edge research and to further understanding of its possible contribution to natural resources management. It is intended especially for those who may be unaware of its potential in addressing such current pressing issues as coping with ever more rapid change and ensuring global food supply with growing populations, and reversing environmental degradation and promoting sustainable practices. While the agricultural science community is already aware of IK – where, in my experience, many are frequently interested in, and willing to learn from, local farming arrangements – there is scope for an update and a need to inform those who are less aware of the approach. Indeed I accepted the invitation from CABI (Centre for Agriculture and Biosciences International) to edit this volume because it offers an opportunity to get the IK perspective even more widely known among agricultural scientists. CABI is a widely recognized and respected organization publishing extensively for over a century on technical aspects of natural resources management as part of its mission to improve 'people's lives worldwide by providing information and applying scientific expertise ... to find practical solutions to the most pressing problems in agriculture and the environment'.[1] It was a particularly welcome invitation in view of my recent somewhat pessimistic assessment of the state of IK research (Sillitoe, 2015).

In short, this volume seeks to advance understanding of IK in the context of management of natural resources: to promote it as a 'known known'. It addresses some key themes through case studies from bioculturally diverse regions of the world. The book links theory and practice in providing a state-of-play overview of the conceptual issues surrounding IK enquiries in the context of their contributions to sustainable agriculture and environmental conservation. In drawing together some of the various strands of biocultural diversity research into natural resources management, it also outlines a possible agenda to guide future work.

When I sat down to draft this preface, which involved organizing the book's contents page, I realized that this collection of chapters – contributed by some of the leading thinkers in the field of IK as it applies to the management of natural resources – represents the frequently overlooked, yet fundamental, complexity that characterizes such knowledge itself. The contents of the chapters connect with one another in several ways, similar to the complicated networks that characterize any local body of IK that feature varying concepts and categories. I shuffled the contents deck several times according to regions and various themes, and what I have come up with represents only one way in which the contents of this volume could be organized, which arranges the chapters according to four broad themes: change and dynamism; diffusion and extension; conservation and sustainability; and all-pervasive complexity.

In my opening chapter, I outline the content of the field of IK research, which goes under a plethora of sometimes disputed labels — such as local or traditional knowledge, indigenous environmental knowledge, local or traditional ecological knowledge and citizen science, among others. In contrasting indigenous with scientific knowledge, I draw parallels between them and point out how they may complement one another, while acknowledging the complexity of relations between them, notably with respect to variations between and within them. The project 'Understanding Predation' conducted by Scotland's Moorland Forum, which, at the time of writing, is looking at predation on the Scottish moorlands, serves to illustrate some of the issues surrounding collaboration between local land users and natural scientists, and efforts to accommodate their differing understandings and values.

The change and dynamism section starts with Victoria Reyes-García and colleagues, who consider how people are responding dynamically to contemporary rapid social and environmental change in the Congo and Amazon Basin regions. They discuss how the Baka and the Tsimane' peoples of the Congo and Amazon, respectively, have recently adopted and adapted agricultural practices. The Baka have incorporated subsistence farming into their traditional foraging livelihood regime and the Tsimane' have moved from subsistence-based to commercial agriculture, albeit local farming knowledge remains more widespread and evenly distributed than newly introduced farming knowledge. The authors argue that IK systems undergo constant change, featuring complex gain and loss of information, with reproduction and hybridization, innovation and erosion of specific components of the knowledge systems. They point out that attempts to use such local knowledge to advance sustainable agriculture need to consider that some associated practices may be exogenous and not time-honoured and tested.

In the next chapter, Roy Ellen stresses how IK systems are dynamic and not static, as often assumed. He uses his several decades of work with the Nuaulu people, who live on the Indonesian island of Seram, to illustrate how knowledge alters constantly in response to changing conditions and events. Mindful of the forest having ecological and cultural dimensions for people, he demonstrates how local forest knowledge features a 'process of continual engagement', which he illustrates in respect of galip nut trees, rattan climbing palms and firewood timber. He relates this engagement perspective to transformations in the islanders' material and social lives.

A concern to document and strengthen the intergenerational transmission of IK prompts Citlalli Binnqüist and Rosalinda Ledesma to consider the dynamism of local knowledge. They do so through a review of land use and tenure changes in Mexico's Veracruz region, and the introduction of various commercial crops to Nahua speakers. These illustrate how continuous innovation is necessary to ensure food production and land conservation in the face of economic and environmental change. They point out how tensions can occur, as seen between the milpa traditional subsistence farming regime (which is a resilient biodiverse agroecosystem featuring maize, beans and squash as main crops) and government environmental protection programmes (which promote planting of single tree species to advance reforestation and reduce soil erosion).

Recent advances in information technology have widened access to knowledge and have consequently, as Andrew Ainslie points out, increased rates of change. He considers the implications of the resulting rapidly evolving hybridized local/scientific knowledge in the context of the management of tick-borne diseases among cattle in the Eastern Cape Province of South Africa. Conflicts over knowledge and its use are increasing, with evermore questioning of 'expert' knowledge. What constitutes trusted knowhow and viable innovation is increasingly an issue, which relates to the process of certifying information as trustworthy, with certain knowledge increasing in strategic value. People are consequently more uncertain, having to make decisions informed by an unequal mix of local and scientific knowledge.

The diffusion and extension section opens with the development and spread of agricultural technology. Drawing on her long experience with Asian rice farmers, Florencia Palis discusses how IK may serve as a source of innovation. She also demonstrates how the incorporation of farmers' IK in the development of technology and extension work can further acceptance and adoption of new technologies. In the next chapter, Jeffery Bentley and colleagues focus on IK-informed extension work. They draw on a project in Mali to inform people about how the parasitic 'devil weed' *striga* reproduces and how to control it, using videos that feature local farmers explaining issues and their experiences. The authors follow up on changes that have occurred after people saw the video series, which involved farmers experimenting and modifying their cultivation practices.

In the third chapter of this section, Lars Otto Naess considers the part that local knowledge plays in people's adoption, or not, of development interventions. He focuses on vulnerability to climate change in the semi-arid Dodoma region of Tanzania, and what constrains and enables implementation of adaptation strategies. He relates how farmers' drought-coping tactics often rely on locally based knowledge and practices, not necessarily because they wish to continue with these as such, which they acknowledge are sometimes inadequate, nor because they fail to understand the benefits of outside interventions to increase community resilience. Rather, it is a question of asset limitations, labour shortages and lack of trust in outside schemes; regarding the latter, the actions of farmers signal resistance to external interference.

While the processes of diffusion and extension promote dynamism and change, concerns for conservation and sustainability may act in the opposite direction and encourage cautiousness and stasis. These opposed tendencies hint at the complexity of IK understandings, which mirror the contrariness of human behaviour generally. The third conservation and sustainability section starts with James Fairhead and colleagues considering indigenous soil knowledge, notably in respect to processes of soil enrichment, which they argue are more widespread in Africa and Asia than thought. They draw attention to two practices: the cultivation of abandoned settlement areas and the incorporation of anaerobic charred biomass together with other organic matter in the soil. They argue that these overlooked traditional agroecological soil management practices have the potential to contribute to sustainable agricultural development and strategies to tackle climate change.

In the next chapter, Doyle McKey and colleagues discuss raised-field agriculture in tropical wetlands, which some think affords a way to increase the productivity of an otherwise marginal environment without degrading it. Others consider this delusional, notably in the context of tropical America, where interest in such farming is found, even though it all but disappeared there 500 years ago. He points out that wetland raised-field cultivation occurs elsewhere in the tropics and reports on some of these present-day systems to evaluate conflicting judgements of it. They adopt a nuanced position that steers between an overly optimistic view of indigenous practices and an overly pessimistic one that sees them failing, indicating that these afford a viable way to farm wetlands while conserving their biodiversity, with unique advantages and disadvantages.

The following chapter addresses the increasingly acknowledged role of indigenous knowledge and practices in the conservation of agricultural biodiversity. In the context of Andean communities in southern Peru – custodians of considerable crop diversity – Chris Shepherd argues that negotiations between these local communities and conservation bodies need to give as much weight to the interests of farmers and the diversity of their livelihoods, as to the institutional goals and strategies that inform state and non-governmental development efforts, in order to maximize the role of IK in the conservation of agrobiodiversity. He proposes a 'cultural affirmation' and 'cultural integration' approach to assess on-farm conservation. The former seeks to strengthen and revive local subsistence traditions and practices beneficial to conservation, while the latter encourages farmers to adopt new technology and enter the market, albeit aware that these may be unfavourable to conservation.

The topic of crop biodiversity continues in the next chapter, where Stephen Brush addresses indigenous practices of crop selection, gene-pool maintenance and seed exchange. He focuses on the Andes and Mesoamerica, two regions of crop domestication (of potatoes and maize, respectively) where the wild ancestors of these crops occur together with the largest diversity of cultivars. He argues that understanding farmers' knowledge of plant diversity furthers insights into crop breeding and evolution. The cultural and nutritional salience of these plants gives clues to crop evolution, as do local knowledge of variations in crop species according to size, shape, colour, taste and so on, and cultivation practices that ensure conservation of a wide germplasm pool.

The complexity and variability section opens with Daniela Soleri and David Cleveland interrogating the assumptions that outsiders make in understanding local farmers' knowledge and practices, and their methodological implications. They investigate what farmers in the Oaxaca region of Mexico expect and achieve with maize seed selection, their perceptions of the risks that transgenic maize cultivation pose, and agreements and differences between them in identifying bean varieties. They argue that their assumptions and hypotheses did not match their empirically tested findings because the latter were too simplified to capture the complexities of farmers' knowledge and environmental relations. While outsiders may think that these complex variations suggest imprecise understandings and practices, they actually serve to protect crop genetic and phenotypic variety, which is central to the sustainability of local agriculture and furthermore contributes to the conservation of the world's crop diversity.

In the next chapter, Patricia Howard discusses the complex knowledge and skills that inform smallholder on-farm storage systems: an under researched topic. Contrary to popular negative assumptions, losses are low. The methods used are effective and sustainable, and appropriate to small farmers, representing long-term adaptations through interaction between local environmental, cultural and socio-economic circumstances. She outlines the complex management of these systems by women, largely as part of their domestic responsibilities, whose practices reflect knowledge of pests and diseases, and plant physical and chemical properties – knowledge that agrees with scientific knowledge. These systems promote food security and resilience, and contribute to the conservation of crop biodiversity, and we need measures to prevent erosion or loss of associated complex knowledge.

The irrigation of rice terraces on the island of Bali is a classic example of the complexity of indigenous resource management. Small farmer groups comprise local water-user associations called *subaks*, as Wayan Windia and colleagues describe, which meet in water temples to manage the irrigation of their terraced paddies. In their chapter, the authors discuss destabilizing changes in associated farming practices due to ill-informed outside interference, the main drivers of which are high-yielding varieties and the growth of tourism. The former, linked to the Indonesian government's drive for self-sufficiency in rice production, encouraged *subaks* to abandon traditional irrigation management. Subsequently, the government realized the tourist potential of the water temple managed system – underlined by its declaration as a UNESCO World Heritage Cultural Landscape – and reversed its policies. The *subaks* are currently in crisis, their survival threatened, caught up in contradictory ever-changing development plans that pressure them to adapt sophisticated traditional ways to fit new management policies that fail to grasp their function.

In the final chapter, Alder Keleman and colleagues address the political dimensions of IK, arguing that politics permeates all of our understandings, whatever the cultural context. They argue that designating any knowledge as indigenous is less a statement of content than a value-loaded relational statement. They use case studies from Ecuador, India, Indonesia and Ireland to explore how political power permeates agricultural practices and technology, notably how the powerful use labelling of knowledge as either indigenous or modern as a tactic to reinforce and legitimize their dominance, recognizing certain practices and innovations above others. They argue that to optimize the contributions of natural resource IK and associated management skills, it is necessary to acknowledge their political dimensions and set them within wider cultural context to promote the inclusion of people's values and aspirations in any development interventions.

Acknowledgement

I thank my wife Jackie for help with her careful copy editing of the volume and ensuring consistency throughout with the publisher's formatting guidelines. And Val Porter for her subsequent copy edit of the manuscript.

Note

[1] http://www.cabi.org/about-cabi/ I thank David Hemming, CABI commissioning editor, for inviting me to compile this volume, so giving us the opportunity to promote IK more widely in the natural sciences community.

Reference

Sillitoe, P. (2015) Indigenous knowledge. In: Stewart, P.J. and Strathern, A.J. (eds) *Research Companion to Anthropology*. Ashgate Publishing, Farnham, UK, pp. 343–368.

1 Indigenous Knowledge and Natural Resources Management: An Introduction Featuring Wildlife

Paul Sillitoe*

Since the 1930s there have been periodic bovine tuberculosis (bTB) outbreaks in British cattle herds, continuing into the 21st century. The search for vectors and their control has featured a long-running argument over the part that wildlife, notably badgers, play in transmitting the disease to cattle and how to prevent it. On one side, there are livestock farmers who think that badgers are significant in spreading the bTB bacterium (*Mycobacterium bovis*) to their herds. They advocate culling populations in areas adjacent to their pasture land, and organizations representing them such as the National Farmers' Union[1] have lobbied successive governments to waive wildlife protection legislation and allow the slaughter of badgers using a range of methods including shooting, trapping and gassing setts. On the other side, there are animal conservationists who think that badgers play a negligible role in spreading bTB to cattle, some even arguing the reverse. They maintain that culling is not only cruel, but also ineffective, and animal welfare bodies such as the Badger Trust[2] advocate vaccination if control of the infectious bacterium really is necessary in the badger population (Caplan, 2010, 2012).

The differences of opinion prompted governments to employ scientists to examine the evidence, authorizing the conduct of trials in some regions of the West Country, such as the 'Randomised Badger Culling Trials', to assess the role of badgers in spreading bTB to cattle and the effectiveness of culling in reducing infection rates (Ares and Hawkins, 2014). While the scientific evidence suggests that badgers may play a part in spreading bTB, it largely supports the protectionists' position, arguing that culling badgers is not an effective or cost-effective approach to controlling the disease. According to independent scientific experts, culling yields modest benefits that are short term without ongoing control programmes, which are more expensive than the financial returns gained from reduced herd infection rates. Furthermore, it can make matters worse on farms outside cull areas by disrupting animals' territories and movements, resulting in infected animals roaming more widely than previously. They argue that improving control of cattle movements and bTB testing could more effectively reduce herd infections.

The conclusions continue to fuel furious debate. 'Badger culling ... is a highly politicized arena, involving the national and local state, scientists, farmers and organizations such as farming unions, and those for animal protection and nature conservancy' (Caplan, 2012, p. 17). The farmers, who stand to lose tens of thousands of

* E-mail: paul.sillitoe@durham.ac.uk

pounds with herd infection, are annoyed by the outsiders' thwarting interference in their affairs, as their deep personal knowledge of animal management and extensive experience of the countryside convinces them that culling badgers on and around their farms reduces bTB infection of their herds. The scientists, on the other hand, with less at stake personally, seek to present the objective evidence of monitored trials dispassionately, albeit counteracted by strident activists with their sometimes disruptive demonstrations; both of these parties use the trial evidence to argue that culling is ineffective and even counterproductive in reducing the infection of herds with bovine tuberculosis. The evidence on either side of the argument is equivocal, particularly when seen from the other side. It is a stand-off: indigenous knowledge (IK) versus scientific knowledge (SK).

This book addresses such commonly encountered differences in the understanding of agricultural issues, focusing on IK. It seeks to further understand what IK amounts to, as shown by current cutting-edge research, and to showcase the part it plays in natural resources management, for those who may be unaware of the possibilities it offers in tackling, as pointed out in the Preface, such currently pressing issues as food security worldwide, promoting sustainable practices and conservation, and halting environmental degradation.

What is the Indigenous Knowledge Approach?

Although IK is increasingly acknowledged within natural resource research circles, it is perhaps advisable to start with a definition of the approach, which is not as straightforward as it sounds. The ongoing argument over appropriate terms for the field,[3] an indication of the flux within it, intimates the challenge, some arguing that 'indigenous knowledge' is inappropriate as it is difficult to define in a globalizing world and potentially divisive politically (Sillitoe, 2015, pp. 349–352). The semantics need not detain us: indigenous knowledge and IK are the term and acronym employed widely in development circles. Furthermore, people from a range of disciplines are contributing to the IK project – from anthropologists and human geographers to ecologists and environmental scientists, including agronomists and foresters – who, coming at it from a range of directions, give IK a diverse intellectual perspective and methodological heterogeneity. Nonetheless, whatever the differences, the fundamental premise behind all IK is unexceptionable, namely that an understanding and appreciation of local ideas and practices should inform any interventions in people's lives, as declared some years ago (Kloppenburg, 1991; DeWalt, 1994; Warren *et al.*, 1995). As a working definition, IK is any understanding rooted in local culture and includes all knowledge held more or less collectively by a community that informs interpretation of the world (Sillitoe, 2002, pp. 8–13). In this volume it concerns knowledge that relates to natural resource management. It is both mindful and tacit, often passed on through experience as the legacy of practical everyday life. It varies between communities; being culturally relative understanding learned from birth that informs how people interact with their environments. It comes from a range of sources and is a dynamic mix of past 'tradition' and present innovation with a view to the future. Although widely shared locally, its distribution is uneven, often according to gender, age, occupation and so on, maybe with political power implications.

The definition is redolent of anthropology, albeit focused on applied not academic problems, and in a sense IK research originated with the discipline. But as it relates to natural resources in development contexts, IK is more recent, appearing in association with some provocative work in the 1980s that marked a significant change of approach to development. This was from previous dominant top-down paradigms that were oblivious of IK issues – modernization with its transfer-of-technology model and dependency with its Marxist-inspired model of development – to bottom-up oriented participatory approaches (Chambers, 1997). These latter approaches attempt to bring the planning and implementation of interventions nearer to people, following growing discontent with expert-led approaches and expensive project failures. Participation features flexible methods that encourage local communities to analyse their own problems and communicate their ideas, promoting a better fit culturally and environmentally between research and technological interventions. It tackles some of development's most challenging problems today, albeit several problems attend participatory and, hence, IK approaches, centring on the facilitation of meaningful participation (Mosse, 2001). They vary widely in the scope they afford farmers to participate, from consultative (outsiders retain control), to collaborative (insiders cooperate as equal partners).

From an agricultural perspective, farming systems research, with affinities to participation, also contributed to IK's emergence (Collinson, 1985; Biggelaar, 1991). It promoted a holistic systems approach – encompassing agronomic, environmental, socio-economic, etc. components – given the complexity of natural resource management in different environments. It took research beyond the experimental station and on-farm to understand local practices and management constraints and advance more appropriate technological interventions. It encountered similar problems to participation, namely how to promote meaningful problem-centred farmer cooperation rather than expert-led scientifically driven analysis and intervention. Its systems concept was narrow and static, rarely extending beyond the farm boundary (whereas diverse farm-household livelihoods do), and overlooking their dynamic nature and scope for endogenous change. It also got bogged down in complex systems analysis, caught on the horns of the conundrum of how to identify and focus tightly on particular researchable constraints without losing the overall farming systems view. It is a paradox that IK addresses, being embedded by definition within the wider context. It also addresses the shortcomings of researching highly complex environmental–cultural systems using multidisciplinary teams that spend short periods of time on farms, which is crippling from an anthropological standpoint.

Indigenous and Scientific Knowledge

It is common, in the applied contexts where IK features, to contrast it, often unfavourably, with SK that informs many interventions. SK is characterized as global whereas IK is local (Sillitoe, 2007). The former is openly international and cosmopolitan in outlook while the latter relates closely to a particular cultural context. While SK has broad, universally generic, intellectual ambitions, IK has narrow, culturally specific, practical concerns. This contrasts with their approaches to understanding problems, where SK is reductionist, comprising the in-depth understanding of narrowly trained specialists, while IK is unitarist, comprising system-wide understanding of broadly informed citizens. One aspires to be objective and analytical, while the other is considered subjective and tacit. The scientist is formally taught in institutions that are keepers of knowledge, arranged in an orderly manner by discipline; the indigene is informally taught in the community where knowledge is organized in less systematic ways. Scientific method is more deductive with protocols agreed to test a consistent model of the world regularly through purposefully designed experiments (which are only predictable to varying extents, not always going to plan), while indigenous practice is more inductive with serendipitous assessment of a changeable world irregularly during everyday chance experiences (which are knowable to varying extents to others, as hotly disputed by postmodern thinkers).

In this comparison, SK is regularly characterized as dominant and IK as subordinate (Failing *et al.*, 2007). This judgement rests in considerable part on the scope that scientifically informed technology allows us to intervene in the world, as seen in such amazing achievements as organ transplant surgery, space exploration and electronic communications. In seeking to redress this judgement, the IK agenda is liable to misunderstanding. It is a common misapprehension, particularly among scientists and technocrats, that IK somehow implies denigration of these technological advancements, even advocating regression (Dickson, 1999; Anonymous, 1999; Ellen, 2004). A speaker at a conference in Bangladesh typified this view (Sillitoe, 2000), criticizing our interest in IK for promoting, it seemed to him, the undoing of the advances made in the scientific breeding of high-yielding varieties (HYVs) of rice and associated technology of fertilizers, biocides and so on, without which, he argued, the country would have been unable to feed its expanding population. An unspoken question was: what could IK research do to increase production similarly? It reflects the current confusion among many natural scientists, even those who are willing to countenance IK, who are often unclear what its contribution might amount to, how to access it and how to incorporate it effectively into their research.

It is necessary to clarify the possible role of IK in scientific research, which is one of this book's aims with respect to natural resources and environmental science. Those of us promoting IK research certainly do not intend to move communities backwards in any way. Indeed the opposite: for instance, with respect to HYVs in Bangladesh, collaboration with farmers during the rice breeding programme would arguably have helped avoid some of the problems that have subsequently emerged with their widespread cultivation, which include declining soil fertility and structure (exacerbated by reduced annual silt deposition with embankment construction diverting the monsoon floods) and increasing poverty among those unable to adopt the technology as too expensive (exacerbated by hierarchical social arrangements that support unequal politico-economic relations). Some trends in the scientific community are favourable; the award of the 2015 Nobel Prize in Physiology to the Chinese pharmacologist Youyou Tu for the discovery of the anti-malarial drug, artemisinin, which involved combing traditional Chinese medical texts, has prompted positive comments about the potential contribution of IK to science (Cesare, 2015). Furthermore, it is questionable to overly privilege SK in view of the increasingly evident costs of associated technology, such as climate change, land degradation and water pollution, when respect for IK could advance more sustainable ways, increasingly called for with such deleterious impacts becoming ever more worrisome.

Variations in Knowledge

The view of IK and SK as monolithic polar opposites distorts both. This stark discrimination misrepresents the distinctions and connections between them, even where used to argue for a review of the relationship to promote IK's equal participation. We do not have two tenuously connected knowledge traditions separated by a cultural–epistemological gulf, but rather a networked spectrum of relations. While at one end of the spectrum there are poor farmers with no formal education, who we may think represent 'pure' IK derived entirely from their own cultural tradition, and at the other end formally trained natural scientists, who may seek to accommodate aspects of local ideas and practices in their research, the majority of actors will fall between them with various intergradations of local insider and global outsider knowledge depending on community of origin and formal education. Many local people have some formal schooling and familiarity with science, which they will blend with their locally derived knowledge and cultural heritage. Many British farmers have gone through school to college and university, often to study agricultural subjects, environmental science and so on. And farmers worldwide receive scientifically informed extension advice via government agencies, non-governmental organizations and increasingly the mass media (Shepherd, 2005). In development contexts there are national scientists with extensive scientific backgrounds, some with higher degrees and occupying university posts, who as metropolitan native speakers are familiar with indigenous culture. Those from rural families may serve as a further pathway for SK to reach local communities, passing on some of their learning to relatives and friends. Foreign scientists working in local communities may do likewise, and those sympathetic to IK gain some understanding of local views in return. Both IK and SK are in a constant process of change, being continually influenced by new ideas. It is contemporary globalization in action, knowledge passing to and fro, blending with what is known locally to inform today's ideas and practice, such that natural resource management understandings are a difficult-to-disentangle mix of knowledge from various sources.

The variability is even more pervasive. The conflation of local knowledge traditions into an indigenous category and its contrast with global scientific knowledge overlooks differences within them. The knowledge held by people making up a local community is not all the same, being structured, as pointed out, according to gender, age, occupation, caste, class or whatever. And the knowledge of scientists varies between disciplines; the specialist knowledge of a soil chemist, for example, is different to that of a crop breeder and both differ markedly from that of a social scientist. Each has a unique perspective, with its own potential insights and blind spots. It is a complexity of relations, different stakeholders

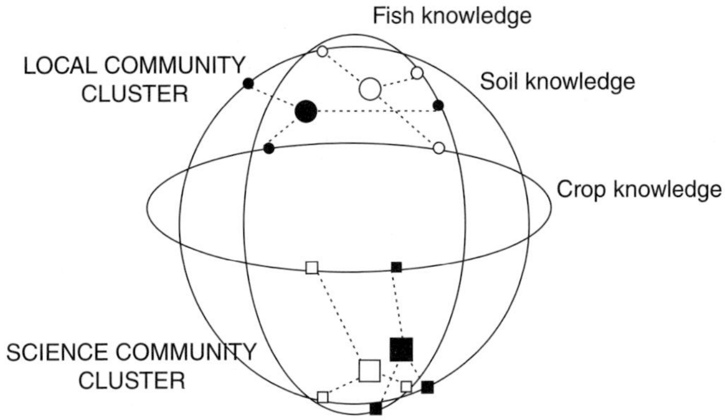

Fig. 1.1. The global knowledge meridians model (from Sillitoe, 2002, p. 119).

having a range of perspectives informed by their differing viewpoints that apprise their multiple objectives. A way to envisage this network of relations between different bodies of knowledge is as a series of meridians arranged around a globe (Fig. 1.1), each meridian representing a different knowledge domain; for example, the various fields in natural resources management, such as crops, soils, water resources, livestock and so on (Sillitoe, 2002). It accommodates variations in knowledge according to disciplines in the science cluster and life experience in the indigenous one. The meridians can represent any knowledge domain, from plants to animals and economics to politics, allowing the setting of enquiries as necessary within a holistic cultural context. The globe represents an interaction domain. It can plot the positions of individuals who interact within the domain, according to their knowledge of any range of meridian issues, extrapolating from these where they are located within the globe. While individuals' configurations vary, some will overlap more than others, comprising interest communities; for instance, plotting the global coordinates for farmers and scientists regarding natural resources knowledge in a region will result in two clustered points within the globe. The global model not only represents IK and SK as individually variable and not monolithic, but also subverts any hierarchical arrangement, the meridians arranged randomly about the globe, precluding any tendency to polarize clusters with dominant SK above subordinate IK, putting all on a par in a complex multidimensional network.

According to some critiques of development, it is dubious to conflate local knowledge traditions into an indigenous category and contrast it with global scientific knowledge, not only because it overlooks differences within them, but also because it overlooks similarities between them (Agrawal, 1995, 2009; Parkes, 2000; Sillitoe, 2007). They argue that these knowledge systems may be similar in rudiments and content, which is undoubtedly so because without substantial similarities and overlaps in the substantive contents of various knowledge systems, it is difficult to conceive of communication with one another. Also, there are some parallels between the methods that SK and IK use to explore reality and it is questionable to distinguish between them on methodological grounds; for example, that SK exclusively tests ideas through experimentation or is more objective, because, after all, local farmers are often keen experimenters and are among some of the world's most pragmatic people. Furthermore, SK is just as culturally located and value-laden as any other knowledge tradition, being rooted in European society where it largely took off, although contemporary globalization-driven hybridization is diminishing the influence of Western sociocultural heritage.

Accommodating Different Knowledge

Regardless of globalizing trends, and the differences within and similarities between IK and SK, it is defensible to distinguish between such local and global knowledge traditions. Many people do so, such as those in lesser developed countries who wish to share in the technological advances that SK underpins – allowing humans, as pointed out, unheard of ability to exploit resources – not only to increase their standard of living, but also sometimes to stave off starvation and sickness, particularly with the relentless population expansion. The dissemination of this technology is central to development, where awareness of IK can play a part in advancing appropriate interventions in accord with local ideas and practices. Well intentioned arguments that seek to redress the power imbalance between scientific and other perspectives are unhelpful in suggesting that it is illegitimate to distinguish between them. Indeed they are ironically supporting hegemonic relations by questioning people's cultural identity (Sillitoe, 2002), which they may deploy in their fights against cultural imperialism, including asserting a place for their knowledge. They are also liable to allegations of ethnocentrism, for implying that the 'they' of contemporary cross-cultural discourse is the same as 'us'.

People in different regions have unique cultural traditions and histories, which continue to inform significantly their understanding of being in the world. They concern different issues and priorities, reflect differing interests and experiences, formulated and expressed in differing idioms and styles, which outsiders may understand to varying extents. While individuals do not duplicate each other, they share a sufficient but indefinite amount of knowledge to make up a discrete cultural community sharing a common history, values, idioms and, likely, language. They are inculcated into these distinct cultural heritages, developed over generations albeit not in isolation but mutually influenced by other traditions that they have some connection and overlap with, while retaining their uniqueness, with the similarities between them correlating closely with geographical distance until recently. The rate of hybridization may have increased with the current boom in worldwide communications and associated acceleration in globalization processes eroding distinctions between different culturally specific knowledge systems (Dove, 2000; Shepherd, 2004; Thomas and Twyman, 2004), but these continue to inform different peoples' understanding of the world. So long as such communities exist with their differently framed cultural understandings, the struggle over the standing of different views – of which the IK versus SK debate is an aspect – is going to continue, being an aspect of contemporary global processes, extending to debates over such knotty issues as ideology, values and belief (Stiglitz, 2003; Rodrik, 2012). It follows that the IK and SK dichotomy is inescapable in some measure and to argue in effect that we should not distinguish between different knowledge traditions is unrealistic, however laudable the aim of overriding intellectual imperialism, and any privileging that occurs is not necessarily inevitable; it is dubious, as pointed out, to esteem overly scientific discourse as its technological costs become increasingly apparent.

The duck–rabbit image made famous in Wittgenstein's (1958) discussion of 'seeing' can be used to illustrate how the IK approach seeks to further the understanding of different views in both directions (Sillitoe, 2015, pp. 345–346) (Fig. 1.2).[4] It is mistaken, looking at the image, to ask: 'What view is "correct": is it a duck or a rabbit?' What you recognize may depend on what you are used to. If you are not accustomed to ducks, for instance, you will see a rabbit. If you see in turn a duck and then a rabbit, you make out the image's two different aspects. In the same way, the approach advocated here seeks to clarify the dual aspects of the IK and SK discrimination in natural resources management, both of which likewise focus on the same environmental issues 'out there'. The challenge of the duck–rabbit image – of striving to see both images when you can only see one or the other at any one time – conveys the ambiguity of IK research in attempting to get local and scientific understandings, which represent different perspectives on the same phenomena, to correspond

Fig. 1.2. The duck–rabbit (from Wittgenstein, 1958, p. 194e).

in some measure, or more likely, to complement one another. The inference is not that this approach advocates the translation of farming IK into agricultural SK, in all probability diminishing the former in the process, in addition to privileging the latter. Rather it attempts to connect them, as many farmers do who demonstrate the shortcomings of depicting IK as contrary to SK, intermingling both to produce many-sided hybrid knowledge. They may assimilate new information both coming from without and generated within to give a place-specific mix of local and global knowledge (Robertson, 1996).

Predation on Scotland's Moorlands

A recent review of predation on the Scottish moorlands illustrates the duck–rabbitness of differing IK–SK views of the same phenomena (Ainsworth *et al.*, 2016). The study of predation, organized by Scotland's Moorland Forum,[5] set out to compare and assess similarities and differences between scientific and local knowledge, and the scope for integration of different perspectives. It involved both natural and social scientists, the former engaging in an extensive zoological literature review of animal population trends, including an analysis of changes in Scottish wild bird populations using *Bird Atlas 2007–11* data (Balmer *et al.*, 2013), and the latter conducting a web-based questionnaire survey supplemented by a series of workshops and seminars (Fig. 1.3) to enquire into the issues with members of organizations concerned about bird predation in Scotland. The respondents were classified as either oriented to 'Local Knowledge' (e.g. land agents, gamekeepers, farmers and crofters) or 'Scientific Knowledge' (e.g. researchers, administrators, naturalists and green activists), according to what they identified as

Fig. 1.3. Scotland's Moorland Forum Workshop held at Scottish Natural Heritage Headquarters, Great Glen House, Inverness on 3 November 2015. (Photograph by kind permission of Simon Thorp, Forum Director.)

their primary source of predation information (either 'personal field management experience' or 'scientific peer-reviewed articles'), although predictably several relied on a combination of both sources, underscoring the point about hybridity.[6] The responses of the 'Local Knowledge' group were compared with those of the 'Scientific Knowledge' group and the data from the natural science review.

Some may query a discussion of moorland wildlife in an introduction to a volume on natural resource management that largely focuses on agriculture. While the culling of badgers to protect cattle herds may qualify, the protection of game birds seems to be of a different order, albeit the issues are arguably similar. They raise the question of what qualifies as agriculture. This takes on a particular salience in cross-cultural IK contexts, which can challenge conventional categories, opening up new ways to approach issues. The manipulation of the environment by hunter-gatherers, for instance, arguably makes them farmers; such as the actions of Australian Aborigines, who promote the growth of plants edible for both humans and hunted animals, often by firing vegetation and increasingly referred to as fire-stick farming (Gammage, 2011). The management of moorlands to encourage game birds is similar, involving intervention in predator/prey relations to protect game birds that supply food, albeit harvested largely as sport. It helps to think outside the box in this way because it encourages us to consider how other livelihood regimes may manage natural resources in ways not immediately apparent, which merit the same attention as more readily recognized conventional environmental management regimes. It also reduces the unequal relations between IK and SK to realize that tensions between these two ways of knowing feature in our culture too.

The Moorlands research found that there was broad agreement over population changes of the focal species[7] with predator numbers increasing and prey numbers decreasing, although the 'Local Knowledge' group thought that the predators had increased much more, and both groups differed over the *Bird Atlas* abundance data. The majority of 'Local Knowledge' respondents thought that predators, particularly crows and foxes,[8] had a 'medium to high negative impact' on prey species, including protected ones, followed by recreational disturbance (walkers, often with dogs; cyclists; birdwatchers, etc.), whereas 'Scientific Knowledge' respondents more often thought that habitat differences and interactions between a range of ecological and anthropogenic factors had the largest impact on predator–prey numbers (Ainsworth *et al.*, 2016, pp. 32–33, 231). Reasons suggested for the disagreements included differences in geographical and temporal perspectives, with 'Local Knowledge' holders considering restricted areas and events of immediate interest in the context of long-term experiences and intimate knowledge of the land, whereas 'Scientific Knowledge' holders focus on larger regions and processes over extended periods of data collection albeit with short-term field work and less familiarity with places. Also, predators and their impacts on prey are more immediately visible and locals may more readily perceive their numbers to have increased, whereas other environmental factors such as changes in habitat are less obvious and longer term in their effects on predator–prey populations. While the local concerns are direct and considered in straightforward cause-and-effect terms, the scientific ones are indirect and addressed in complex ecological feedback terms. The impact of predation is difficult to determine whatever your approach and experience, because a decline in prey populations is not necessarily due to increased predator numbers. Other drivers of demographic change include climate and weather, disease and parasites, and human activities involving land use and habitat change. These make assessment and resolution of different views problematic. They confound scientific studies on predation, making experiments difficult to devise and leading to ambiguous results. It is difficult to distinguish, for instance, between the effects of different predator species on the various prey populations that occupy an area. These sorts of issues may further account for the disagreements between survey respondents and the *Bird Atlas* data, concerning problems with species identification and bird counts (particularly of highly mobile or fluctuating populations).

While both respondent groups agreed that they wanted to ensure a sustainable long-term balance between healthy prey and predator populations, they disagreed on how to achieve it. The 'Local Knowledge' respondents predictably supported direct control of predator numbers to maintain prey populations, being of the opinion that predators posed the most immediate threat. They saw culling as a successful and swift management strategy. It has an immediate observable effect, which recommends it. They collectively had long-established experience of this management technique and its effectiveness, unlike other approaches, and this gave them confidence in it. Some studies support their faith in it; reducing predator pressure in the breeding season of prey species is particularly significant, though the effects are difficult to verify because if only one predator is targeted the numbers and behaviours of others may change, replacing it, especially if prey become more available, whereas if several predator species are culled it is impossible to know what the effect is of the reduction in each. An alternative strategy, particularly when breeding game birds, is to erect enclosures that exclude predators.

The 'Scientific Knowledge' respondents advocated landscape management to maintain healthy predator and prey populations by improving the quality of habitats for birds. They cited management of vegetation to increase plant and insect food availability and to provide nesting sites – from encouraging understory thickets and closed canopies in woodlands and controlled grazing and mowing of grasslands, to management of field margins and hedgerows on farmland, along with reduced pesticide and herbicide use. Research predictably supports the effectiveness of such habitat management for maintaining bird populations. Problems with it include the considerable time it takes for such ecological management to have a noticeable effect, and also it is necessary to cover considerable and preferably interconnected areas to be effective, which in turn implies cooperation between several parties, unlike predator control undertaken by a single gamekeeper. And teasing out the effects of particular measures in complex ecological systems can be a challenge. A wish to protect all species can complicate matters further.

Promoting Collaboration

The challenge is to facilitate cooperation and communication across the indigenous-to-scientific knowledge spectrum through the promotion of knowledge partnerships (Eversole, 2015) and beyond to planners, policymakers and politicians. It is unwise for scientists to underrate indigenous understandings, as it can breed defiant localism, even conflict, if those in power deny the validity of place-centred knowledge. While IK is more circumscribed than SK, it often matches and sometimes betters science-based understandings of, for instance, land use. It is increasingly recognized that indigenous peoples have their own effective 'science' and resource management systems (Sillitoe, 2007). There are many examples of the soundness of their knowledge and practices, and the need to respect them; some of them were previously thought 'primitive' and in need of modernization. It is widely acknowledged, for example, that local ways of managing natural resources are an integral part of any environment; notably in biodiversity management and conservation that may include culling, where the cessation of such practices may be as damaging as the loss of species (Posey, 1999; Knight, 2000; Anderson and Berglund, 2003). In reconciling IK with SK, we cannot assume that the one will be congruent with the other; rather we have to tease out parallels and contrasts, each potentially influencing the other. But some conflict is inherent in the process, because it is not just about furthering understanding, of advancing more rounded views, but of deploying knowledge to effect some action, and sometimes the values that underpin IK and SK are not readily reconcilable (Young *et al.*, 2010). The negotiations become far more complex but policies and interventions are more likely to be appropriate for more people, notably local actors, and so more sustainable (Harrison and Burgess, 2000; Taylor and Loe, 2012).

The Scottish Moorlands project illustrates the IK–SK tensions in advancing both rabbit and duck views. The management of predatory species is an emotive topic with, on the one hand, concerns about the negative impacts of predators on prey species and calls for their lethal control, and on the other, recognition of predation as an aspect of natural ecosystems and arguments that these benefit from conservation-minded human interference that aims to promote a sustainable balance. The subject is of policy interest to governments (the Moorlands review was presented to the Scottish Government) which typically rely on scientific evidence when making decisions. The assumption is that it is independent, unbiased and objective, undertaken by reputable academic bodies that use experimental methods and sound observation to collect data, statistically analysed for reliability, and scrutinized by peers before publication. Yet, whatever the quality of the evidence, other interested parties may reject decisions based on it as externally imposed by those ill-informed about local conditions who fail adequately to address relevant questions (Wynne, 1992). The reductionist approach of science tends to overlook wider context and may address issues that local resource managers think inappropriate, such as focusing on problems at the wrong spatial and temporal scale (Fig. 1.4). They value first-hand experience of dealing with predation, for instance, which they think gives them deep and reliable understanding. They mistrust science, thinking that the framing, reviewing and funding of research biases it, while scientists mistrust local views as subjective, lacking rigorous evidence, even featuring hearsay 'proof' (Failing *et al.*, 2007). Both are open to unintended bias informed by different values and understandings, of stewarding, or 'working with', versus managing, or 'working on', nature, which can polarize views.

Nonetheless, the majority of participants in the Moorlands review agreed that both viewpoints have their strengths and weaknesses and that cooperation might advance a better informed overall understanding of the role that predators play within ecosystems and the effectiveness of various management strategies to maintain healthy predator and prey populations. They identified a need to develop a new collaborative approach that includes locals and scientists from the outset in designing research, collecting data and interpreting results, to tackle perceived biases and generate new understandings acceptable to both sides. Building necessary trust is central through improved communication and networking, promoting the exchange

Fig. 1.4. Visit by Scotland's Moorland Forum to Invermark Estate (by kind permission of the Earl of Dalhousie) on 5 June 2015. (Photograph by kind permission of Anne Stoddart, Forum Administrator.)

of views and data, which aims to lead to 'co-production of knowledge'. The participants acknowledged the challenges of reconciling divergent views and objectives over issues such as what constitutes a suitable balance between predator and prey numbers, which relate to differing values, perceptions and experience of moorland environments; for example, some species, such as game birds (e.g. grouse, partridge and ptarmigan) are more important to some stakeholders than others. It is widely acknowledged that values influence understanding – such as the differing values signalled by affiliation to game management versus nature conservation organizations, which broadly correlate with relying on local and scientific knowledge, respectively – but accommodating the different views or beliefs engendered presents tricky problems and may even amount to trying to obviate paradox when parties hold opposed values, such as overcoming contradictory demands for managed versus wild environments. A particular challenge is to harmonize different views and experiences of scale, in respect of both time (short- versus long-term measures and outcomes) and space (interventions over small areas versus entire regions), which are as relative here on planet Earth as they are in the wider universe. The diversity of interests and priorities also varies between like-interested individuals, even those affiliated to the same conservation or shooting organization, which further complicates agreement.

Challenges of Integration

The IK approach presumes a methodology that mediates effectively between the contradictions that characterize the promotion of scientific research informed by an indigenous knowledge perspective. A range of eclectic approaches have been pioneered that favour techniques that directly involve local people (Sillitoe *et al.*, 2005), such as participatory mapping using all manner of media (mixing crayons with stones, beans and twigs) and drawing up calendars and activity diagrams, through to game playing and theatricals, and more conventional collaborative paper-and-pencil surveys that mix semi- to unstructured interviews with field observations. These are not culturally neutral but subject to external influence and may fail to access local knowledge with the subtlety expected by anthropology. Furthermore, deciding what to focus on is an individually informed judgement too, such that the drawer, game-player or whoever controls the representation may manipulate it according to their interests. Another approach is to encourage farmers' experimentation, though the connection between their problems and ideas and scientific research and possible technological interventions is not always clear, unorthodox farmer-led experiments being incompatible with scientifically designed trials and data analysis.

We not only have intercultural problems between scientists and locals, but also interdisciplinary problems among the scientists. This work often features a range of disciplines to promote different perspectives on complex problems and to facilitate action research (Sillitoe, 2004). The harnessing of an anthropological background to knowledge of a scientific–technical field that informs interventions, such as agriculture or environmental science, is useful in furthering interdisciplinary interfacing with local knowledge. The facilitation of such broad research is a major methodological challenge facing IK-through-SK research. It involves the resolution of longstanding strains between the natural and social sciences, conveying local knowledge to natural scientists such that they can appreciate its relevance. Both techno-scientific and socio-political issues feature, inextricably entwined, striving to reach a plausible consensus. It returns us to the issue of the domination of science and associated power issues, featuring on the one hand the association of technological interventions with natural scientists and on the other local community empowerment with social scientists. It is a battle of perspectives characterized as hard versus soft systems and so on.

At the local level, intercommunity problems further complicate aspirations of equitable negotiation central to such participatory research. Communities of interest are not homogeneous, as aforementioned. The distribution of knowledge and experience within a local community may be markedly uneven, which presents challenges in selecting participants, often of a political sort. The promotion of a locally informed perspective should extend to an awareness of local power structures. There is the possibility of self-selection by those with particular interests to promote these or the more influential or pushy dominating and directing enquiries and negotiations to their benefit. These are likely to be wealthier and more powerful persons in development contexts who seek to exclude the marginalized such as the poor and women. The Scottish predation project used a 'snowball sampling' strategy, for instance, drawing on the networks of the Moorland Forum organizations' members, which was subject to potential biases, not ensuring proportional representation from organizations according to membership size or extent of experience or differing interests.

The integration of local stakeholders' knowledge in the decision-making process can improve the quality of judgements (Huntington, 2000; Beierle and Konisky, 2001) and collectively agreed decisions that acknowledge local values and interests are more likely to be socially and politically acceptable to actors (Harrison and Burgess, 2000), and may lessen conflicts between parties (Young *et al.*, 2010). The challenge is to promote a rapprochement between different perspectives, playing off the advantages and disadvantages of different knowledge traditions, generating synergy to improve overall understanding of issues and problems, which at root comes down to reconciling differences in values and priorities. It is exciting work as the chapters in this book show. It is necessary to be open to the unexpected and new. Allowing the know-how and aspirations of local populations to inform development, for example, opens up the prospect of a redefinition of the meaning and aims of the very process. The IK agenda intimates such a shift, albeit there are concerns about the reaction of development agencies, which are likely to see it as subversive, reducing their control. It denotes the reduction of outsider hegemony, challenging the assumption that strangers have a right to impose interventions, through the promotion of what some call endogenous development (Haverkort and Reijntjes, 2006) that allows people to follow their own lines of enquiry and contribute meaningfully to the determination of 'development' objectives that concern them.

Acknowledgements

I thank the Earl of Lindsay, Chairman of Scotland's Moorland Forum, and Simon Thorp, the Forum's Director, for inviting me to participate in the Forum's Understanding Predation Project. Also Mark Wilson and Chris Wernham of the British Trust for Ornithology Scotland at Stirling University and Steve Redpath of the School of Biological Sciences at Aberdeen University for their helpful comments on a draft of this chapter.

Notes

[1] See, for instance, http://www.nfuonline.com/science-environment/bovine-tb/badger-cull-is-a-key-part-of-tackling-bovine-tb/.

[2] See, for instance, http://badger.org.uk/threats/bovine-tb.aspx.

[3] The alternative terms include local knowledge, rural people's knowledge, insider knowledge, indigenous technical knowledge, traditional environmental knowledge, peoples' science, local agricultural knowledge and folk knowledge.

[4] The source of the image, as Wittgenstein acknowledges, was Jastrow's (1900: 312) Victorian lithograph, used to argue that perception involves both eye (physical stimulus) and mind (mental activity).

[5] A partnership of 28 organizations that focuses on issues concerning the Scottish uplands with a view to informing and influencing policy, management and practices (http://www.moorlandforum.org.uk/).

[6] 'Local Knowledge' group = 211 respondents and 'Scientific Knowledge' group = 110 respondents; six of the nine workshops (involving 15 of the participating organizations) involved the former and three the latter, while the three 1-day seminars were joint.

[7] The focal prey species identified by the Moorland Forum were *Black Grouse Tetrao tetrix*, *Curlew Numenius arquata*, *Golden Plover Pluvialis apricaria*, *Grey Partridge Perdix perdix*, *Lapwing Vanellus vanellus* and *Oystercatcher Haematopus ostralegus*; predator species were *Common Buzzard Buteo buteo*, *Crow Corvus spp.*, *Northern Raven Corvus corax* and *Red Fox Vulpes vulpes*.

[8] The status of Ravens and Buzzards, like Badgers and Pine Martens also frequently mentioned by 'Local Knowledge' respondents as significant predators, is controversial as they are protected species under UK law.

References

Agrawal, A. (1995) Dismantling the divide between indigenous and scientific knowledge. *Development and Change* 26, 413–439.

Agrawal, A. (2009) Why 'indigenous' knowledge? *Journal of the Royal Society of New Zealand* 39, 157–158.

Ainsworth, G., Calladine, J., Martay, B., Park, K., Redpath, S. *et al.* (2016) *Understanding Predation: a Review Bringing Together Natural Science and Local Knowledge of Recent Wild Bird Population Changes and Their Drivers in Scotland*. Report for Scotland's Moorland Forum, Newtonrigg, Holywood, Dumfries.

Anderson, D.G. and Berglund, E. (eds) (2003) *Ethnographies of Conservation: Environmentalism and the Distribution of Privilege*. Berghahn Books, New York.

Anonymous (1999) Caution: traditional knowledge. Principles of merit need to be spelt out in distinguishing valuable knowledge from myth. *Nature* 401(6754), 623.

Ares, E. and Hawkins, O. (2014) *Badger Culling: TB Control Policy*. Library of the House of Commons, London. Available at: http://researchbriefings.parliament.uk/ResearchBriefing/Summary/SN05873 (accessed 5 June 2016).

Balmer, D., Gillings, S., Caffrey, B., Swann, B., Downie, I. and Fuller, R. (2013) *Bird Atlas 2007–11: The Breeding and Wintering Birds of Britain and Ireland*. British Trust for Ornithology, Thetford, UK.

Beierle, T.C. and Konisky, D.M. (2001) What are we gaining from stakeholder involvement? Observations from environmental planning in the Great Lakes. *Environment and Planning C* 19, 515–528.

Biggelaar, C. den (1991) Farming systems development: synthesizing indigenous and scientific knowledge systems. *Agriculture and Human Values* 8(1,2), 25–36.

Caplan, P. (2010) Death on the farm: culling badgers in North Pembrokeshire. *Anthropology Today* 26(2), 14–18.

Caplan, P. (2012) Cull or vaccinate? Badger politics in Wales. *Anthropology Today* 28(6), 17–21.

Cesare, C. (2015) From speculation to surprise, science Twittersphere reacts to Nobel prizewinners: online responses ranged from wonder to disappointment. *Nature* 526 (7573), 297.

Chambers, R. (1997) *Whose Reality Counts? Putting the First Last.* Intermediate Technology, London.

Collinson, M. (1985) Farming systems research: diagnosing the problems. In: Cernea, M.M., Coulter, J.K. and Russell, J.F.A. (eds) *Research–Extension–Farmer: a Two-Way Continuum for Agricultural Development*. The World Bank, Washington, DC, pp. 71–86.

DeWalt, B. (1994) Using indigenous knowledge to improve agriculture and natural resource management. *Human Organisation* 53(2), 123–131.

Dickson, D. (1999) ICSU seeks to classify 'Traditional Knowledge'. *Nature* 401(6754), 631.

Dove, M. (2000) The life cycle of indigenous knowledge, and the case of natural rubber production. In: Ellen, R., Parkes, P. and Bicker, A. (eds) *Indigenous Environmental Knowledge and Its Transformations.* Harwood Academic Publishers, Amsterdam, pp. 213–251.

Ellen, R.F. (2004) From ethno-science to science, or 'what the indigenous knowledge debate tells us about how scientists define their project'. *Journal of Cognition and Culture* 4(3,4), 409–450.

Eversole, R. (2015) *Knowledge Partnering for Community Development*. Routledge, Abingdon, UK.

Failing, L., Gregory, R. and Harstone, M. (2007) Integrating science and local knowledge in environmental risk management: a decision-focused approach. *Ecological Economics* 64, 47–60.

Gammage, B. (2011) *The Biggest Estate on Earth: How Aborigines Made Australia.* Allen and Unwin, Sydney.

Harrison, C. and Burgess, J. (2000) Valuing nature in context: the contribution of common-good approaches. *Biodiversity and Conservation* 9, 1115–1130.

Haverkort, B. and Reijntjes, C. (eds) (2006) *Moving Worldviews: Reshaping Sciences, Policies and Practices for Endogenous Sustainable Development*. Compas Worldviews and Sciences Series no. 4. ETC Foundation/COMPAS, Leusden, Netherlands.

Huntington, H.P. (2000) Using traditional ecological knowledge in science: methods and applications. *Ecological Applications* 10, 1270–1274.

Jastrow, J. (1900) *Fact and Fable in Psychology*. Houghton Mifflin, Boston, Massachusetts.

Kloppenburg, J. (1991) Social theory and the de/reconstruction of agricultural science: local knowledge for an alternative agriculture. *Rural Sociology* 56(4), 519–548.

Knight, J. (ed.) (2000) *Natural Enemies: People–Wildlife Conflicts in Anthropological Perspective*. Routledge, New York.

Mosse, D. (2001) 'People's knowledge', participation and patronage: operations and representations in rural development. In: Cooke, B. and Kothari, U. (eds) *Participation: The New Tyranny*. Zed Books, London, pp. 16–35.

Parkes, P. (2000) Enclaved knowledge: indigent and indignant representations of environmental management and development among the Kalasha of Pakistan. In: Ellen, R., Parkes, P. and Bicker, A. (eds) *Indigenous Environmental Knowledge and Its Transformations*. Harwood Academic, Amsterdam, pp. 253–291.
Posey, D.A. (ed.) (1999) *Cultural and Spiritual Values of Biodiversity: A Complementary Contribution to the Global Biodiversity Assessment*. Intermediate Technology Publications (for the UN Environment Programme), London.
Robertson, R. (1996) Glocalization: time–space and homogeneity–heterogeneity. In: Featherstone, M., Lash, S. and Robertson, R. (eds) *Global Modernities*. Sage, London, pp. 25–44.
Rodrik, D. (2012) *The Globalization Paradox*. Oxford University Press, Oxford, UK.
Shepherd, C.J. (2004) Agricultural hybridity and the 'pathology' of traditional ways. The translation of desire and need in postcolonial development. *Journal of Latin American Anthropology* 9(2), 235–266.
Shepherd, C.J. (2005) Agricultural development NGOs, anthropology, and the encounter with cultural knowledge. *Culture and Agriculture* 27(1), 35–44.
Sillitoe, P. (2000) Introduction: the state of indigenous knowledge in Bangladesh. In: Sillitoe, P. (ed.) *Indigenous Knowledge Development in Bangladesh: Present and Future*. Intermediate Technology Publications, London, and Dhaka University Press, Dhaka, pp. 3–20.
Sillitoe, P. (2002) Globalizing indigenous knowledge. In: Sillitoe, P., Bicker, A. and Pottier J. (eds) *'Participating in Development': Approaches to Indigenous Knowledge*. Routledge, London, pp. 108–138.
Sillitoe, P. (2004) Interdisciplinary experiences: working with indigenous knowledge in development. *Interdisciplinary Science Reviews* 29(1), 6–23.
Sillitoe, P. (2007) Local science vs global science: an overview. In: Sillitoe, P. (ed.) *Local Science vs Global Science: Approaches to Indigenous Knowledge in International Development*. Berghahn, Oxford, UK, pp. 1–22.
Sillitoe, P. (2015) Indigenous knowledge. In: Stewart, P.J. and Strathern, A.J. (eds) *Research Companion to Anthropology*. Ashgate Publishing, Farnham, UK, pp. 343–368.
Sillitoe, P., Dixon, P. and Barr, J. (2005) *Indigenous Knowledge Inquiries: A Methodologies Manual for Development*. Intermediate Technology Publications, London.
Stiglitz, J.E. (2003) *Globalization and Its Discontents*. W.W. Norton and Company, New York.
Taylor, B. and de Loe, R.C. (2012) Conceptualizations of local knowledge in collaborative environmental governance. *Geoforum* 43, 1207–1217.
Thomas, D. and Twyman, C. (2004) Good or bad rangeland? Hybrid knowledge, science and local understandings of vegetation dynamics in the Kalahari. *Land Degradation and Development* 15, 215–231.
Warren, D.M., Slikkerveer, L.J. and Brokensha, D. (eds) (1995) *The Cultural Dimensions of Development: Indigenous Knowledge Systems*. Intermediate Technology Publications, London.
Wittgenstein, L. (1958) *Philosophical Investigations*, 2nd edition (translated by G.E.M. Anscombe). Blackwell, Oxford, UK.
Wynne, B. (1992) Misunderstood misunderstanding: social identities and public uptake of science. *Public Understanding of Science* 1, 281–304.
Young, J.C., Marzano, M., White, R.M., McCracken, D.I., Redpath, S.M. *et al.* (2010) The emergence of biodiversity conflicts from biodiversity impacts: characteristics and management strategies. *Biodiversity and Conservation* 19, 3973–3990.

2 The Dynamic Nature of Indigenous Agricultural Knowledge. An Analysis of Change Among the Baka (Congo Basin) and the Tsimane' (Amazon)

Victoria Reyes García,* Isabel Díaz Reviriego, Romain Duda, Álvaro Fernández-Llamazares, Sandrine Gallois and Sascha Huditz

By definition, indigenous knowledge is an attribute of societies with historical and intergenerational continuity in resource management. Such a knowledge system constitutes an integrated corpus of knowledge, practices and beliefs that provide a holistic view of ecosystems (Toledo, 2002). Different works have documented the antiquity and ubiquity of indigenous wisdom and the many benefits it confers to people and societies (McDade *et al.*, 2007; Reyes-García *et al.*, 2016). Researchers have also argued that locally developed knowledge and skills about the environment are critical in explaining the logic of indigenous peoples' agricultural practices (Fox and Gershman, 2000; Long and Zhou, 2001; Hardwick *et al.*, 2004). Previous research has shown that traditional knowledge systems can contribute to more sustainable agriculture by, for example, increasing landscape biodiversity through the creation of a mosaic of different habitats (Wiersum, 2004), contributing to *in situ* conservation of crop varieties (Altieri and Merrick, 1987; Jarvis and Hodgkin, 1999), or helping to curb the clearance of tropical forest for subsistence agriculture (Pascual, 2005; Reyes-García *et al.*, 2011).

However, researchers have emphasized that indigenous knowledge systems should be considered neither static (Berkes *et al.*, 2000; Gómez-Baggethun and Reyes-García, 2013), nor isolated from other knowledge systems (Agrawal, 1995; Leonti, 2011). Rather, they should be understood as being in constant change, in a dynamic process that encompasses a complex combination of knowledge replication, loss, innovation, hybridization and transformation (Zent, 2013). Changes in indigenous knowledge systems can be triggered by multiple factors that range from individual learning and experimentation, to the adoption of new technologies, or to changes on the surrounding socio-ecological systems (Eyssartier *et al.*, 2011; Zent, 2013; McCarter and Gavin, 2014; Reyes-García *et al.*, 2014c).

Given that most indigenous societies today face rapid changes affecting their socio-environmental context, it becomes relevant to ask: how do indigenous knowledge systems respond to such changes? And how does change in indigenous agricultural knowledge relate to agricultural management? In this chapter, we address these questions by exploring the changes in the agricultural practices of two indigenous, small-scale, subsistence-based societies: the Tsimane' (Amazon Basin) and the Baka (Congo Basin). To date, the two societies have relatively little

* Corresponding author; e-mail: victoria.reyes@uab.cat

(albeit increasing and uneven) involvement in market economies (mostly through wage labour and the sale of forest products), school-based education, or modern healthcare systems (Reyes-García *et al.*, 2016). In addition, the two societies resemble one another in that they depend on the consumption of local natural resources through a combination of foraging and farming in an environment where such societies have historical continuity of resource use. The two societies differ in their historical process of adoption of agriculture; while the Tsimane' have practised horticulture since colonial times (Daillant, 2003), the Baka have only engaged in agriculture since the 1950s (Leclerc, 2012). Moreover, the two societies increasingly face new external pressures; the Tsimane' to engage in cash crop agriculture (Vadez *et al.*, 2008) and the Baka to abandon their dependence on wild resources and concentrate on subsistence agriculture (Leclerc, 2012; Oishi, 2012).

In such context, this work aims: (i) to describe the process of adoption of agricultural practices in the two groups; (ii) to examine the coexistence of agricultural knowledge related to local and newly introduced species and agricultural practices; and (iii) to test whether differences in knowledge of local and newly introduced agricultural practices relate to agricultural management.

Methodological Approach

Research context

Results presented here were collected within the framework of a large research project aiming to test the adaptive nature of culture (see Reyes-García *et al.*, 2016). Two researchers lived for 18 months in each of the two selected societies, each in a different village. Each researcher teamed up with local research assistants who helped in data collection and translations. We obtained 'free prior and informed consent' from each village and individual participating in the study, as well as agreement from the relevant political organization representing the indigenous groups. The research has received the approval of the ethics committee of the Universitat Autònoma de Barcelona (CEEAH-04102010).

Researchers devoted the first months in the field to becoming adapted to the local mores, building up trust with participants and collecting background ethnographic information. We also conducted semi-structured interviews with key informants on local livelihoods (i.e. techniques, division of labour, seasonality, and assets associated to subsistence activities) and on the content of indigenous knowledge (Davis and Wagner, 2003). The ethnographic information helped in the design of quantitative methods and to put our results into a broader context. The following 12 months were spent in quantitative data collection and visits to agricultural fields. To make our research locally specific, we adapted our protocols for each site (e.g. referring to local species and relevant management practices). However, to allow for comparability, the protocol's general structure and administration was identical across sites. The protocols used are described on the ICTA – Universitat Autònoma de Barcelona website.[1]

To capture variation, within each of the two societies we selected two villages at varying distances from the main market town where we worked with all male and female household heads who were willing to participate. The final sample with complete data for the analysis presented here includes 39 Tsimane' and 64 Baka households.

Agricultural knowledge

We measured the agricultural knowledge at the individual level. To capture changes in agricultural knowledge, we collected information related to two different knowledge systems: (i) knowledge and management practices of crop varieties present in the local agricultural system; and (ii) knowledge and management practices of newly introduced crop varieties. The approach, while downplaying the complexity of the agricultural knowledge system, allows comparison of information from two different knowledge systems and across societies.

To collect agricultural knowledge information, we used a knowledge test that included 30 questions on six different crop strains ($30 = 6 \times 5$), of which three were traditional and three were recently introduced varieties (for a similar approach see Reyes-García *et al.*, 2014a). We used

Table 2.1. Crop varieties used in knowledge tests.

	Tsimane'		Baka	
	Local	Newly introduced	Local	Newly introduced
Common variety	*Itsidye* (plantain)	*Dyurty'* (maize)	*Ndo* (plantain)	*Langa* (macabo)
Intermediate variety	*Cojo'* (plantain)	*Carolina* (rice)	*Boma* (cassava)	*Mebuta* (potato)
Rare variety	*Opuntye* (yucca)	*Tara' miqui' muntyi' tority'* (maize)	*Sapa* (yam)	*Woundo* (groundnut)

our ethnographic information to select one well known, one relatively known (i.e. intermediate) and one rare local variety in each site. We used the same criteria to select three newly introduced varieties (Table 2.1). For each item we requested informants: (i) to identify the variety by showing them the seed (or other propagation material such as bulbs or cuttings); (ii) to answer two questions regarding its management; and (iii) to answer two questions regarding its use. The test also included three general questions on local management practices (i.e. soil selection) and three general questions on newly introduced management practices (e.g. use of pesticides).

Agricultural management

To assess variations in agricultural management, we collected survey data regarding households' agricultural plots. We estimated the amount of area cleared for agricultural purposes by asking informants to report all the plots opened by the household during the last planting season and to estimate the surface of each plot and the type of forest cleared (i.e. fallow and old-growth forests).

We also asked respondents to estimate the number of person-days (i.e. the amount of work equivalent to one person working x days, or x people working for one day) invested in agriculture. We asked plot owners to recall the number of days and the number and age of people who participated in slashing, felling trees and cleaning debris in a plot in preparation for agriculture, as well as the number of days invested in weeding and harvesting. We computed adults as one full working day and children as a half-day. For owners of more than one plot, information was collected separately for each plot. We aggregated the information to generate a variable that captured the total number of person-days a household devotes to agricultural production. Baka informants, rather than reporting person-days, estimated the number of days in which they engaged in the different agricultural activities. So the variables should be interpreted differently for the two cases. We also asked respondents to report whether, in addition to the agricultural crops, they planted and or tended useful plants around their house compound.

Plot observations

We took direct measures in all household plots devoted to staple crops (i.e. plantain for the Tsimane' and manioc for the Baka). Specifically, we used a compass and a GPS to measure the total surface at the clearing of each plot. We then counted the number of plantain or manioc plants in each plot. As the Tsimane' grow up to ten plantain varieties within a plot, we counted the number of plantains per variety. The differentiation between varieties of manioc was not reported in the case of the Baka. We also noted the name of other crops in the field.

Data analysis

We used answers to the questions on traditional crops and management to generate a score of local agricultural knowledge (LAK) and answers to the questions on newly introduced crops and management techniques to generate a score of new agricultural knowledge (NAK). Specifically, we added a point to the respective score for each of the following: (i) the informant identified the propagation material by providing the folk name of the strain; (ii) the informant knew the management techniques of the strain; and (iii) the informant knew the use or preparation for that plant strain. Questions on management and use were originally designed as multiple-choice, but after unsuccessful attempts to apply such questions, we decided to collect data with open responses, which were later recoded into categories. To evaluate responses, we generated a measure of agreement with the group based on the number of times the informant's answer matched the modal response to a question (D'Andrade, 1987; Reyes-García *et al.*, 2016). Given that in the two studied societies agricultural fields are managed by both male and female household heads, we used the average of their knowledge scores in our calculations.

We performed a hierarchical cluster analysis to classify households according to their level of LAK and NAK. We used the Ward's method as agglomerative technique. Then we used Kruskal–Wallis and chi-square tests to examine potential differences between the groups in agricultural management practices obtained with the hierarchical cluster analysis. For the statistical analysis we used STATA 11.1 for Windows (Stata Corporation, Texas, USA).

Agriculture Among the Tsimane'

Our first case study society is the Tsimane', a small-scale indigenous society in the Bolivian Amazon. The Tsimane' number about 12,000 people living in about 100 villages of commonly about 20 households per village, concentrated along rivers and logging roads (Reyes-García *et al.*, 2014b). Up until the late 1930s, the Tsimane' maintained a traditional and self-sufficient lifestyle. However, their interactions with Bolivian society have steadily increased since the 1940s (Reyes-García *et al.*, 2014b).

Previously semi-nomadic, most of the Tsimane' are now settled in permanent villages, mainly established in commonly owned lands, with school facilities and increasingly connected to local towns. Today, the Tsimane' largely rely on swidden agriculture supplemented by hunting, fishing and gathering, though they are gradually entering the market economy (Godoy *et al.*, 2009) through wage labour in logging camps, cattle ranches and the homesteads of colonist farmers. Barter of thatch palm and sale of cash crops are also local important sources of income (Vadez *et al.*, 2008).

Tsimane' engagement with agriculture

Although Tsimane' have remained mostly dependent on forest resources until recently, references to swidden agriculture amongst the Tsimane' date back to the 16th century (Daillant, 2003). The long-term importance of agriculture for the Tsimane' is reflected in the large number of myths developed around domesticated plants such as plantain, manioc, or cotton (Huanca, 2008). While Tsimane' agricultural production was until recently mostly oriented to household consumption, over the past decades sales of crops have dramatically increased (Vadez and Fernández-Llamazares, 2015).

Our previous research in the area suggests that Tsimane' allocate about 22% of their daily productive time to agriculture, similar to the share of time allocated to hunting and gathering (21%), and with a larger involvement during the rainy season (27%) than during the dry season (10%) (Reyes-García *et al.*, 2009). According to the data collected for this study, Tsimane' households annually invest around 35 person-days in their agricultural plots (SD = 24): 14.5 person-days in plot preparation (i.e. slashing, felling, burning), 14 person-days harvesting and 6 person-days in plot maintenance (Table 2.2).

Nowadays, Tsimane' farming is extensive and includes cultivation in newly opened plots, fallow plots, and homegardens. The Tsimane' open new plots between May and August, burn them between September and October, and then plant them. Although sequence in planting varies, typically the first crop to be planted after burning is rice. After the rice harvest, fields may be partially replanted with maize, manioc or plantains. The margins of the fields are often planted with other crops, such as pineapple, peanuts, papaya, watermelon, squash, Amazonian yam bean and cotton. Households in our sample opened an average of 1.38 new plots, for a total of 0.60 ha (SD = 0.45), with an equally rough extension in old-growth (0.31 ha) and fallow (0.29 ha) forest. Crop fields are never cultivated for more than 2–3 years and then they are left aside for forest regeneration. Consequently, new

Table 2.2. Average (SD) Tsimane' agriculture-related characteristics, total and by level of new agricultural knowledge (NAK) and local agricultural knowledge (LAK).

	Households with high NAK (*n* = 17)	Households with high LAK (*n* = 23)	Pooled (*n* = 39)
Agricultural knowledge			
Household average level of LAK	9.17[(a)]	9.9	9.53
(ha)	(0.85)	(1.30)	(1.18)
Household average level of NA	7.11[(b)]	4.39	5.50
(ha)	(0.74)	(1.20)	(1.72)
Agricultural investments			
Total person-days	41.50	29.30	34.62
(ha)	(28.28)	(18.19)	(23.61)
Slash-and-burn/planting	14.09	13.84	13.95
(ha)	(6.23)	(7.65)	(6.98)
Weeding	9.59[(b)]	3.55	6.18
(ha)	(9.87)	(4.76)	(7.92)
Harvesting	17.82	11.91	14.49
(ha)	(17.09)	(10.62)	(13.93)
Households using chainsaw (%)	65	41	51
Households using sowing machine (%)	71	27	46
Households with homegarden (%)	76[(b)]	32	51
Av. no. crop species in homegarden	8.18[(b)]	2.50	4.97
(ha)	(4.92)	(4.95)	(5.65)
Newly opened plots			
No. of plots per household	1.47	1.32	1.38
(ha)	(0.72)	(0.57)	(0.63)
Total area cleared /household	0.79	0.46	0.60
(ha)	(0.50)	(0.35)	(0.45)
Old-growth	0.46	0.20	0.31
(ha)	(0.56)	(0.31)	(0.45)
Fallow	0.32	0.27	0.29
(ha)	(0.39)	(0.25)	(0.31)
Plots with staple crop (plantain)			
No. of plots per household	2.06[(a)]	1.38	1.66
(ha)	(0.94)	(0.64)	(0.83)
Total area with plantain/household	0.40[(a)]	0.27	0.32
(ha)	(0.30)	(0.38)	(0.35)
Old-growth	0.26	0.14	0.19
(ha)	(0.30)	(0.21)	(0.25)
Fallow	0.13	0.13	0.13
(ha)	(0.24)	(0.29)	(0.27)
Av. no. plantain varieties/plot	5.83	5.92	5.89
(ha)	(0.99)	(1.09)	(1.04)
Households practising multi-cropping	61	69	66
in plantain fields (%)	(0.50)	(0.47)	(0.48)
(ha)			
Av. no. other crop species. in plantain	1.93	2.20	2.09
fields	(1.14)	(1.16)	(1.15)
(ha)			

[(a)]$P < 0.05$; [(b)]$P < 0.01$

patches of old-growth or fallow forests are open every year.

Plots in transition to fallow are planted with a variety of crops, including plantains, bananas, peach-palm, arrow-cane and pineapple. Fallow plots are also used to plant fruit trees, which often have a long-term economic importance for Tsimane' families. Young fallows are managed more intensively than older ones, but the reuse of fallows is common and begins around 5 years after the land has been abandoned, depending on the crop and on village land availability (Huanca, 1999). A final important element in Tsimane' agriculture is homegardens (Díaz-Reviriego *et al.*, 2016). About half of the households in our sample had a homegarden, typically including fruit trees, cotton, medicinal plants and other useful species such as those that provide dyes, fibres, and fish poisons.

Tsimane' agricultural knowledge and management practices

Tsimane' local agricultural knowledge relates not only to management practices, but also to social institutions and spiritual beliefs. Thus, to engage in agriculture, the Tsimane' draw on their knowledge of soil types, topography, weather and the like to decide where and how much forest to clear (Reyes-García *et al.*, 2011); but before cutting trees down the Tsimane' also ask permission from the spirits who, according to their tradition, own them (Huanca, 2008). Accuracy in timing matters: if too late, brambles and brush will get too wet to burn well and leave too much debris when Tsimane' burn their fields. The potential weed burden also matters. Fallow forests are easier to cut but contain more weeds, whereas old-growth forests are harder to clear as they contain larger trees to fell, though they will eventually contain fewer weeds. To minimize peaks of work around weeding and harvesting time some Tsimane' opt to use old-growth forests. To reduce the risk of crop loss and smooth the work load, they stagger planting.

Agricultural practices, however, are constantly changing, a process that has increased in speed from the 1980s onwards, with the arrival of highlanders who introduced the use of new techniques (i.e. chainsaws) and external inputs (fertilizers, herbicides and pesticides) to boost agricultural production. For example, with the introduction of chainsaws, the Tsimane' are abandoning the use of axes to clear plots (51% of households in our sample used a chainsaw for clearing) and many (46% of the sampled households) have also adopted the use of a rice-sowing machine. The use of fertilizers and pesticides is still rare in Tsimane' fields, but, through their work as agricultural wage labourers in farm ranches in nearby areas, the Tsimane' are now cognizant of these newly introduced management techniques.

Overall, we found that average levels of local agricultural knowledge (LAK) are higher and show less variation than average levels of new agricultural knowledge (NAK) (Table 2.2). Nevertheless, Tsimane' households can be assigned to two different clusters regarding their agricultural knowledge. The largest cluster ($n = 23$) includes households with high levels of LAK (average = 9.93) and low levels of NAK (average = 4.39). We call this cluster 'households with high LAK'. The smallest cluster ($n = 17$) includes households holding slightly lower levels of LAK (average = 9.17) but higher levels of NAK (average = 7.17). We call this cluster 'households with high NAK.' A Kruskal–Wallis test suggests that there are statistically significant differences both in the average NAK and in the average LAK of households in the two groups ($P < 0.01$). Most households with high LAK (15 out of 23) are found in the most isolated village, whereas most households with high NAK (13 out of 17) are located in the village closer to town. In other words, while LAK seems to be ubiquitously distributed, NAK is relatively less common in both communities but growing in the village with higher market access.

Tsimane' traditional staple crops are manioc and plantain. In contrast to other Amazonian societies in which manioc is the main staple, the main use of manioc among the Tsimane' is to produce a fermented beer, called *shocdye*, a task undertaken by women, which plays a major role in Tsimane' culture and social life (Daillant, 2003; Zycherman, 2013). On the other hand, plantain is a fundamental ingredient in Tsimane' diet (Zycherman, 2013). We found that Tsimane' households had an average of 1.7 plots with plantain, with an average area of 0.32 ha/plot. While there are up to ten different plantain varieties in some fields (mean = 5.9 plantain varieties/plot), most plots contain a large number of plants of the two or three more common varieties and a lower number of the rarer varieties. As much as 66% of the households with plantain plots practise some form of multi-cropping, with about 2.1 other crops in plantain fields.

Rice is the most important newly introduced crop for the Tsimane' and currently plays a vital role for subsistence and as a cash crop. Rice was first introduced in the area by Jesuits in the 17th century (Thomas, 2012), but Tsimane' only started to use it widely from the 1950s onwards, when missionaries brought improved rice varieties. In this sense, it is interesting to notice that the Tsimane' do not have traditional beliefs related to rice (Huanca, 2008). They argue that their rapid adoption of rice relates to the fact that this crop can be easily stored and sold on the local markets, which provides them with financial security.

As suggested by the existence of the two clusters of households with different types of knowledge, Tsimane' households have integrated new agricultural technologies in different ways. Interestingly, our data show that households in the two clusters display differences not only in levels of agricultural knowledge, but also in agricultural management. Thus, households with high NAK invest more time in agriculture (41.5 versus 29.3 person-days), and specifically in weeding their fields (9.6 versus 3.6 person-days, $P < 0.01$), than households with high LAK. They also clear larger surfaces for new plots (0.79 ha versus 0.46 ha) and have larger plots with plantain (0.40 ha versus 0.27 ha). A higher percentage of households with high NAK also use modern technologies (chainsaws and sowing machines) and have homegardens (76% versus 32%) where they grow a larger number of species (8.18 versus 2.50). Interestingly, households with high LAK cultivate more diversity, in terms of number of both plantain varieties and other crops. Households with high LAK are also more likely to practise some form of multicropping in plantain fields (69% versus 61%).

Agriculture Among the Baka

The Baka, who are Oubanguian speakers, are one of the former hunter-gatherer groups indigenous to the tropical rain forests of the Congo Basin. Spread in four different countries, most Baka live nowadays in south-eastern Cameroon, where their population is estimated to be between 30,000 and 40,000 (Leclerc, 2012), and where our study was carried out. Traditionally Baka lived in semi-nomadic groups and depended mainly on wild resources. They were, however, closely associated with local sedentary swidden agriculturalists, always Bantu-speaking people, with whom they maintained a relationship of mutual interdependence: Baka provided bushmeat and other wild resources to Bantu in exchange for agricultural products and commercial goods such as iron tools, salt and clothes (Althabe, 1965; Bahuchet, 1993). Beyond the economic relationship, such alliances were constructed on political partnerships of reciprocal supports through pseudo-kinship bonds and ritual exchanges (Joiris, 2003; Rupp, 2003). From the 1950s, the Baka progressively settled along logging roads and initiated the process of adoption of agriculture – phenomena that, while complex and apparently generated by different causes, are often concomitant (Leclerc, 2012).

Nowadays, the Baka in our study area maintain a high level of mobility, constantly alternating between villages and forest camps. They are associated with the Nzime, a Bantu-speaking group. Most Baka continue to combine hunting-gathering with work for the Nzime and cultivation of manioc and plantains, their major staple crops, for subsistence (Robillard and Bahuchet, 2013). The Baka are also increasingly involved in wild-products trade in an economic system that is increasingly monetarized (Kitanishi, 2006).

Baka engagement with agriculture

As other hunter-gatherer populations of the Congo Basin, the Baka have traditionally presented a paradigmatic example of systematic cohabitation with local agriculturalists. This ancient proximity with farmers is mirrored by a high complexity and accuracy of lexical domains of terms used to refer, for example, to plantain and bananas, the first crops adopted by the Baka (Leclerc, 2012). It is worth noticing that, despite the absence of past agricultural practices among the Baka, researchers have reported the paracultivation of wild yams (Dounias, 2001). Thus, without cultivating them, the Baka seem to have actively managed wild yams to maintain productivity, by (for example) extracting the tuber without killing the plant. Interestingly, although

Baka myths and tales do not include any sign of past agricultural involvement, wild yams are embedded in Baka cosmology, where this species holds a central place between humans, elephants and forest spirits (Joiris, 1993).

The first forms of agricultural adoption among the Baka are associated with a restructuring of their relationships with the neighbours due to the massive involvement of the latter in French colonial economy (rubber and cacao production), which created both the farmers' demand for a labour force and the Baka's need to obtain agricultural products no longer provided by their neighbours (Leclerc, 2012). Within this context, some Baka groups adopted agriculture in the 1950s (Althabe, 1965), while others engaged in farming only in the late 1970s (Leclerc, 2012). At about the same time, between the 1960s and 1970s, in our study area there was a clustering of different Baka groups which settled along roads (Leclerc, 2012). With this establishment, the Baka started to spend more time close to their neighbours and increasingly to work in their agricultural fields.

Nowadays, most Baka adults engage in agriculture either in their own plots, in Nzime plots, or in both. For example, Gallois *et al.* (2016) stated that only 5% of the women and 10% of the men in their sample reported never having performed subsistence agriculture; likewise only 7% of women and 18% of men had never worked in Nzime fields.

Consequently, when discussing Baka engagement in agriculture, it is important to differentiate between work in their own plots and work in Nzime plots, activities that seem to be equally important in terms of time distribution (Gallois *et al.*, 2016). Overall, women seem to be more involved in agriculture (24% of those in the reports working in Nzime fields and 21% in their own plots) than men (10% and 17%, respectively); although men invest more time during the opening of new plots, women perform agricultural work year round, including opening the plots, weeding and harvesting. Differentiating between these two forms of engagement in agriculture is important for two reasons. Firstly, as the peak of agricultural work occurs simultaneously in Baka and Nzime fields, the Baka constantly face the need to decide between investing in their own plots (with future reward) or working for the Nzime (which provides immediate payment, often in the form of baskets of agricultural products, mostly manioc and plantain). Secondly, while the Baka need to mobilize their agricultural knowledge to take management decisions (e.g. when, what, or where to plant) when working in their own plots, this is not the case when they work in Nzime fields, where decisions are likely to be taken by the plot owner. In a recent development, the government and local NGOs have tried to introduce new agricultural techniques among the Baka, including the cultivation of cacao, a main cash crop in the region (Oishi, 2012) and groundnut, a traditional crop grown by the Nzime. However, such programmes have not achieved large success (Leclerc, 2012).

Because the Baka have only recently adopted agriculture and because of their close relation with the Nzime, Baka agricultural management techniques closely resemble those of their neighbours, albeit with low levels of external inputs. Thus to open their plots, the Baka, like the Nzime, rely on swidden agriculture, in which men cut the big trees with axes while women and adolescents cut brush with machetes. None of the households in the sample used a chainsaw. The Baka open their plots in January, by the end of the dry season, and plant them in March, with the arrival of the first rains. Once the plots are planted, the investment in agricultural work decreases and the work mostly consists of weeding. During June–August, the minor dry season, a minority of Baka open new and smaller plots. The major rainy season typically arrives in September, marking the second main period of high mobility among the Baka, who might spend several weeks in their forest camps. During this season, the agricultural work mostly consists of weeding.

The Baka mostly grow manioc in their plots. Groundnuts and maize are common in Nzime fields but rare in Baka plots. According to our field observations, Baka plots are small and do not contain high agricultural diversity. Households in our sample opened an average of 1.12 new plots per year, for a total area of 0.17 ha (SD = 0.12). Plots were mostly opened in fallow (0.15 ha) rather than in old-growth forest (0.02 ha) (Table 2.3). Almost the totality of the plots contained manioc: the average manioc plots comprised 0.16 ha versus the 0.17 ha of the total.

Table 2.3. Average (SD) Baka agriculture-related characteristics, total and by level of new agricultural knowledge (NAK) and local agricultural knowledge (LAK).

	Households with high NAK (n = 29)	Households with high LAK (n = 35)	Pooled (n = 64)
Agricultural knowledge			
Household av. level of LAK	12.20[(a)]	12.81	12.54
(ha)	(1.55)	(1.39)	(1.48)
Household av. level of NAK	7.75[(b)]	5.34	6.42
(ha)	(1.04)	(0.87)	(1.54)
Agricultural investment			
No. days when household head conducted agricultural work	208.4	192.6	199.5
(ha)	(101.6)	(85.81)	(92.42)
Slash-and-burn/planting	35.48	48.39	42.71
(ha)	(24.39)	(44.80)	(37.45)
Weeding	20.41	20.79	20.62
(ha)	(10.18)	(7.42)	(8.66)
Harvesting	152.5	123.4	136.2
(ha)	(85.10)	(72.59)	(78.87)
Newly opened plots			
No. plots per household	1.09	1.14	1.12
(ha)	(1.29)	(0.36)	(0.33)
Total area cleared	0.18	0.16	0.17
(ha)	(0.15)	(0.09)	(0.12)
Old-growth	0.005	0.04	0.025
(ha)	(0.02)	(0.09)	(0.07)
Fallow	0.18	0.13	0.15
(ha)	(0.15)	(0.10)	(0.13)
Households with garden (%)	23	29	26
(ha)	(0.43)	(0.46)	(0.44)
Av. no. crops in garden	0.27	0.54	0.42
(ha)	(0.55)	(1.07)	(0.88)
Observed manioc plots			
No. plots per household	1.087	1.038	1.061
(ha)	(0.29)	(0.20)	(0.24)
Area with manioc	0.164	0.158	0.161
(ha)	(0.14)	(0.08)	(0.11)
Old-growth	0.005	0.03	0.02
(ha)	(0.02)	(0.07)	(0.05)
Fallow	0.159	0.124	0.140
(ha)	(0.144)	(0.107)	(0.126)
Households practising multi-cropping (%)	100	100	100

[(a)]$P < 0.05$; [(b)]$P < 0.01$

Although all Baka households in the sample practised some form of multi-cropping, fields only had an average of 3.8 crops in addition to manioc – mostly plantain, cocoyam, sweet potatoes and domesticated yams. Finally, it is interesting to notice that, overall, Baka households reported not having a homegarden, with only 26% of the households reporting that they intentionally grew some crops (e.g. chilli) around their houses.

Baka agricultural knowledge

Even beyond their territory, the Baka are largely known for their detailed and intimate knowledge of the forest environment. For example, for a long time the Baka have been in demand as valuable guides for foreigners who want to access wild resources (ivory, game, ebony), or for their use of medicinal plants. At the same time, the Baka are generally considered poor farmers, although interestingly enough some of their forest expertise does seem to be valued in farming. For example, farming neighbours value Baka's abilities to (supposedly) control the rain and the wind, or to ensure good luck in farming.

We found differences in how agricultural knowledge is distributed across Baka households; these differences, however, mostly related to NAK. So, as for the Tsimane', the cluster analysis divided Baka households in two groups: one (n = 35) with high LAK (average = 12.81) and low NAK (average = 5.34), and another (n = 29) also with high LAK (average = 12.20) but higher NAK than the other cluster (average = 7.75). A Kruskal–Wallis test suggests that there are statistically significant differences in the average NAK of households in the two groups ($P < 0.001$), but not in the average LAK. Households in both clusters were equally distributed in the two villages. We have used the same labels as before to name these groups.

In contrast to the Tsimane', however, we could not observe any statistically significant differences in agricultural management practices related to the two groups of households. Thus, irrespective of their levels of LAK and NAK, Baka households seem to have the same level of time investment, low use of external inputs (e.g. chainsaw), surface cleared, or crop diversity in their plots (Table 2.3).

We argue that the lack of differentiation in Baka agricultural management in relation to their agricultural knowledge relates both to the recent adoption of agriculture among the Baka and to the place agriculture takes in Baka livelihood. On the one side, agricultural knowledge is a recent domain of knowledge for the Baka. Such knowledge has been mostly adopted from farmers' neighbours and, in recent times, from NGOs working on agriculture. In this sense, Baka agricultural knowledge combines agricultural knowledge from different sources and hybridizes it with their own cultural conceptualization of the environment. But Baka agricultural knowledge probably lacks the coherence of other, long tested, indigenous agricultural systems that might ultimately reflect differences in agricultural management. On the other side, agriculture is not the main Baka livelihood activity, as most Baka continue to abandon their fields for weeks or months and move to their forest camps. Yasuoka (2012) argued that, in harmony with Baka understanding of the world, this behaviour just reflects that indeed Baka adoption of agriculture is only another strategy to increase 'diversification of resources usage'. In this line, the fact that some households differ in their level of NAK might also reflect the diversity of livelihood strategies among the Baka, with the households with higher level of NAK corresponding to those who engaged in projects encouraging the adoption of cash crops.

Conclusion

As other domains of local knowledge (Gomez-Baggethun and Reyes-García, 2013), agricultural knowledge is fundamentally dynamic, which implies that not all indigenous agricultural practices currently observed have the same temporal depth. Changes in indigenous agricultural knowledge systems directly relate to, and should be interpreted in, the context of other socio-ecological transformations that indigenous societies face. Furthermore, the analysis of changes in indigenous agricultural knowledge systems should be considered not only in relation to the agricultural knowledge system itself, but also in relation to agricultural management and even to the mere place of agriculture in indigenous subsistence systems. Thus, for the Tsimane', changes in agricultural knowledge seems to relate to a shift from a logic of subsistence to a logic of commercial agriculture, and this is reflected in changes in agricultural management techniques. For the Baka, the shift seems to be the acceptance that subsistence agriculture can be a complementary activity in their economic system. The dynamic nature of agricultural knowledge should be taken into account not only when studying indigenous groups relatively new to agriculture, such as the Baka, but also when studying those who, such as the Tsimane', have long practised agriculture. Projects aiming to use indigenous systems of knowledge to promote sustainable agriculture should be aware that not all of these systems are time-tested through long-term practices, since many of them might include exogenous forms of agricultural knowledge new to these societies.

In this sense, the second important finding of our work relates to the interactions of local and newly introduced agricultural knowledge. For the two case studies, we found higher and more homogeneously distributed levels of local, rather than new, agricultural knowledge. While local knowledge seems to continue to provide the knowledge base for agricultural practices, some households are exploring newly introduced agricultural knowledge. Such dynamism should not necessarily be seen in a logic of the substitution of one type of knowledge by another. Rather, the integration of different knowledge systems could offer a chance to find innovative answers to old and new problems (Davidson-Hunt *et al.*, 2013) and thus help to deal with socio-ecological change (Olsson *et al.*, 2004). At the same time, we should not neglect that changes are not equally distributed within the group. While here we have not explored the factors that drive such differences (e.g. sociodemographic characteristics of households), we do show that such differences are important as they might relate to actual agricultural management practices. Exploring such differences could help researchers to understand changes in indigenous knowledge systems and their potential effects on sustainable agriculture.

Acknowledgements

The research leading to these results has received funding from the European Research Council under the European Union's Seventh Framework Programme (FP7/2007–2013)/ERC grant agreement no. FP7-261971-LEK.

We extend our deepest gratitude to the Baka and the Tsimane' people and villages for their friendship, hospitality and collaboration. We thank CIFOR and CBIDSI for logistical help. Reyes-García thanks the Dryland Cereals Research Group at ICRISTAT-Patancheru for providing office facilities.

Note

[1] The general protocols used by researchers across sites can be accessed at http://icta.uab.cat/etnoecologia/lek

References

Agrawal, A. (1995) Dismantling the divide between indigenous and western knowledge. *Development and Change* 26, 413–439.

Althabe, G. (1965) Changements sociaux chez les Pygmées baka de l'Est-Cameroun. *Cahiers d'Etudes Africaines* 5, 561–592.

Altieri, M. and Merrick, L.C. (1987) *In situ* conservation of crop genetic resources through maintenance of traditional farming systems 1203. *Economic Botany* 41, 86–98.

Bahuchet, S. 1993. L'invention des Pygmées. *Cahier* 33, 153–181.

Berkes, F., Colding, J. and Folke, C. (2000) Rediscovery of traditional ecological knowledge as adaptive management. *Ecological Applications* 10, 1251–1262.

D'Andrade, R.G. (1987) Modal responses and cultural expertise. *American Behavioral Scientist* 31, 194–202.

Daillant, I. (2003) *Sens Dessus Dessous. Organization sociale et spatiale des Chimane d'Amazonie boliviane*. Société d'Ethnologie, Nanterre, France.

Davidson-Hunt, I.J., Idrobo, C.J., Pengelly, R. and Sylvester, O. (2013) 'Kaysheeyuhahnookeymuhkuhg', knowledges working together: pikangikum teachings for producing and sharing environmental knowledge in Northwestern Ontario. *Ecology and Society* 18(4), 44.

Davis, A. and Wagner, J.R. (2003) Who knows? On the importance of identifying 'Experts' when researching local ecological knowledge. *Human Ecology* 31, 463–489.

Díaz-Reviriego, I., González-Segura, L., Fernández-Llamazares, A., Howard, P.L., Molina, J.L., *et al.* (2016) Social organization influences the exchange and species richness of medicinal plants in Amazonian homegardens. *Ecology and Society* 21(1), 1.

Dounias, E. (2001) The management of wild yam tubers by the Baka Pygmies in Southern Cameroon. *African Study Monographs* 26, 135–156.

Eyssartier, C., Ladio, A.H. and Lozada, M. (2011) Horticultural and gathering practices complement each other: a case study in a rural population of Northwestern Patagonia. *Ecology of Food and Nutrition* 50, 429–451.

Fox, J. and Gershman, J. (2000) The world bank and social capital: lessons from ten rural development projects in the Philippines and Mexico. *Policy Sciences* 33, 399–419.

Gallois, S., Duda, R. and Reyes-García, V. (2016) 'Like Father, Like Son'? Baka children's local ecological knowledge learning in a context of cultural change. In: Reyes-García, V. and Pyhala, A. (eds) *Hunter Gatherers in a Changing World*. Springer, Dordrecht, Netherlands, pp. 195–211.

Godoy, R., Reyes-García, V., Gravlee, C.C., Huanca, T., Leonard, W.R., *et al.* (2009) Moving beyond a snapshot to understand changes in the well-being of native Amazonians panel evidence (2002–2006) from Bolivia. *Current Anthropology* 50, 560–570.

Gomez-Baggethun, E. and Reyes-García, V. (2013) Reinterpreting change in traditional ecological knowledge. *Human Ecology* 41, 643–647.

Hardwick, K., Healey, J.R., Elliott, S. and Blakesley, D. (2004) Research needs for restoring seasonal tropical forests in Thailand: accelerated natural regeneration. *New Forests* 27, 285–302.

Huanca, T. (1999) *Tsimane' Indigenous Knowledge. Swidden Fallow Management and Conservation.* PhD dissertation, University of Florida, Gainesville, Florida

Huanca, T. (2008) *Tsimane' Oral Tradition, Landscape, and Identity in Tropical Forest,* La Paz, Imprenta Wagui.

Jarvis, D.I. and Hodgkin, T. (1999) Wild relatives and crop cultivars: detecting natural introgression and farmer selection of new genetic combinations in agroecosystems 1254. *Molecular Ecology* 8, s159–s173.

Joiris, D.V. (1993) The mask that is hungry for yams: ethno-ecology of *Dioscorea mangenotiana* among the Baka, Cameroon. In: Hladik, C.M., Hladik, A., Linares, O.F., Pagezy, H., Semple, A., *et al.* (eds) *Tropical Forests, People and Food: Biocultural Interactions and Applications to Development.* Parthenon Unesco, Paris, pp. 633–641.

Joiris, D.V. (2003) The framework of central African hunter-gatherers and neighbouring societies. *African Study Monographs Supplementary issue* 28, 57–79.

Kitanishi, K. (2006) The impact of cash and commoditization on the Baka hunter-gatherer society in southeastern Cameroon. *African Study Monographs* 33, 121–142.

Leclerc, C. (2012) *L'adoption de l'agriculture chez les pygmées Baka du Cameroun; dynamique sociale et continuité structurale.* FMSH édition, Paris.

Leonti, M. (2011) The future is written: impact of scripts on the cognition, selection, knowledge and transmission of medicinal plant use and its implications for ethnobotany and ethnopharmacology. *Journal of Ethnopharmacology* 134, 542–555.

Long, C.L. and Zhou, Y.L. (2001) Indigenous community forest management of Jinuo people's swidden agroecosystems in southwest China. *Biodiversity and Conservation* 10, 753–767.

McCarter, J. and Gavin, M.C. (2014) Local perceptions of changes in traditional ecological knowledge: a case study from Malekula Island, Vanuatu. *Ambio* 43, 288–296.

McDade, T.W., Reyes-García, V., Blackinton, P., Tanner, S., Huanca, T. and Leonard, W.R. (2007) Ethnobotanical knowledge is associated with indices of child health in the Bolivian Amazon. *Proceedings of the National Academy of Sciences of the United States of America* 104, 6134–6139.

Oishi, T. (2012) Cash crop cultivation and interethnic relations of the Baka hunter-gatherers in Southeastern Cameroon. *African Study Monographs Supplementary issue* 43, 115–136.

Olsson, P., Folke, C. and Berkes, F. (2004) Adaptive comanagement for building resilience in social-ecological systems. *Environmental Management* 34, 75–90.

Pascual, U. (2005) Land use intensification potential in slash-and-burn farming through improvements in technical efficiency. *Ecological Economics* 52, 497–511.

Reyes-García, V., Godoy, R.A., Vadez, V., Ruiz-Mallen, I., Huanca, T., *et al.* (2009) The pay-offs to sociability. *Human Nature – an Interdisciplinary Biosocial Perspective* 20, 431–446.

Reyes-García, V., Pascual, U., Vadez, V., Huanca, T. and Team, T.B.S. (2011) The role of ethnobotanical skills and agricultural labor in forest clearance: evidence from the Bolivian Amazon. *Ambio* 40, 310–321.

Reyes-García, V., Aceituno-Mata, L., Calvet-Mir, L., Garnatje, T., Gomez-Baggethun, E. *et al.* (2014a) Resilience of traditional knowledge systems: the case of agricultural knowledge in home gardens of the Iberian Peninsula. *Global Environmental Change – Human and Policy Dimensions* 24, 223–231.

Reyes-García, V., Paneque-Galvez, J., Bottazzi, P., Luz, A.C., Gueze, M. *et al.* (2014b) Indigenous land reconfiguration and fragmented institutions: a historical political ecology of Tsimane' lands (Bolivian Amazon). *Journal of Rural Studies* 34, 282–291.

Reyes-García, V., Paneque-Galvez, J., Luz, A.C., Gueze, M., Macia, M.J., *et al.* (2014c) Cultural change and traditional ecological knowledge: an empirical analysis from the Tsimane' in the Bolivian Amazon. *Human Organization* 73, 162–173.

Reyes-García, V., Gueze, M., Diaz-Reviriego, I., Duda, R., Fernández-Llamazares *et al.* (2016) The adaptive nature of culture. A cross-cultural analysis of the returns of local environmental knowledge in three indigenous societies. *Current Anthropology* 57(6), 711–784.

Robillard, M. and Bahuchet, S. (2013) Les Pygmées et les autres: terminologie, catégorisation et politique. *Journal des africanistes* 82, 15–51.

Rupp, S. (2003) Interethnic relations in Southeastern Cameroon: challenging the 'hunter-gatherer'–'farmer' dichotomy. *African Study Monograph Supplementary issue* 28, 37–56.

Thomas, E. (2012) The impact of traditional lifestyle, provenance and contact history on plant use knowledge and management: a cross-cultural comparison of two small-scale societies from the Bolivian Amazon. *Human Ecology* 40(3), 355–368.

Toledo, V. (2002) Ethnoecology: a conceptual framework for the study of indigenous knowledge of nature. In: Stepp, J.R., Wynham, R.S. and Zarger, R.K. (eds) *Ethnobiology and Biocultural Diversity*. University of Georgia Press, Athens, Georgia, pp. 511–522.

Vadez, V. and Fernández-Llamazares, A. (2015) De la agricultura de subsistencia a la comercialización. In: Reyes-García, V. and Huanca, T. (eds) *Cambio Global, Cambio Local. La Sociedad Tsimane' Ante la Globalización.* Icaria, Barcelona, Spain, pp. 147–175.

Vadez, V., Reyes-García, V., Huanca, T. and Leonard, W.R. (2008) Cash cropping, farm technologies, and deforestation: what are the connections? a model with empirical data from the Bolivian Amazon. *Human Organization* 67, 384–396.

Wiersum, K.F. (2004) Forest gardens as an 'intermediate' land-use system in the nature-culture continuum: characteristics and future potential. *Agroforestry Systems* 61, 123–134.

Yasuoka, H. (2012) Fledging agriculturalists? Rethinking the adoption of cultivation by the Baka hunter-gatherers. *African Study Monographs Supplementary issue* 43, 85–114.

Zent, S. (2013) Processual perspectives on traditional environmental knowledge: continuity, erosion, transformation, innovation. In: Ellen, R. Lycett, S.J. and Johns, S.E. (eds) *Understanding Cultural Transmission in Anthropology: A Critical Synthesis*. Berghan, New York, pp. 213–265.

Zycherman, A. (2013) The changing value of food: localizing modernity among the Tsimané Indians of lowland Bolivia. PhD thesis, Columbia University, New York.

3 Contingency and Adaptation over Five Decades in Nuaulu Forest-Based Plant Knowledge

Roy Ellen*

Older studies of local knowledge in traditional societies often describe it as if it were a fixed ahistorical quantum. But we now know enough about the systems in which such knowledge is embedded to realize that, on the contrary, it presents a dynamic changing landscape, constantly altering in response to new circumstances and events (Ellen and Harris, 2000); less a 'thing' than a process of continual engagement. This chapter examines several ways in which the forest knowledge of the Nuaulu (a people of central Seram in the eastern Indonesian province of the Moluccas) has shifted over the past five decades, while in other ways has remained remarkably stable. This is illustrated with reference to three very different examples (the genus *Canarium* harvested for its proteinaceous nuts; the category 'firewood'; and rattan as a construction material) but in each case the context of how forest as a more encompassing cultural and ecological entity is also being transformed in people's material and social lives.

Documenting Forest Knowledge

That part of indigenous knowledge scholarship that relates to understanding the biological world has its origins in the accumulation of information on local uses of species by those working for colonial governments (Ellen and Harris, 2000). The underlying motives were generally to increase the database of imperial science, provide a service to industry and commerce and find solutions to the practical problems of what we would now call 'development'. In the middle of the 20th century, this body of work was given greater theoretical meaning and intellectual rigour through the linguistic 'turn' in ethnobiology. This provided a framework for the study of folk classification, which subsequently contributed in significant ways to explaining cultural cognition of the natural world. It was these concerns that first encouraged me to study ethnobiology in the 1970s. However, from the 1980s onwards there was an increasing global alliance between students of ethnobiology, the environmental movement and the politics of indigeneity. This shifted the focus to how we might better ensure what was increasingly being described as 'sustainable development', recognizing ecological knowledge as a cultural and economic resource for local peoples themselves. Within the ethnobiological community (and perhaps for some outside it) there was a strongly optimistic assumption – though as often implicit as spelled-out – that

* E-mail: rfe@kent.sac.uk

simply through codification of local knowledge we might somehow create an ex situ reservoir or 'memory bank' that could be tapped, used and transferred depending on the project at hand. It is fair to say that, though not entirely misplaced, this optimism has failed to deliver the outcomes anticipated: the IK initiative was 'flagging instead of burgeoning' (Sillitoe, 2015, p. 343).

In my own teaching in anthropology I have tended to focus on the contribution of ethnobiological knowledge to our understanding of human cognition, on its importance for the creation and maintenance of biocultural diversity, its role in parataxonomy and as an adjunct to conservation. I have stressed that understanding local people requires attending to their perceptions and knowledge of the world (especially how they make decisions with respect to natural resource management), and that this is vital in avoiding basic mistakes in the context of particular projects and general development policy. This kind of ignorance and its consequences are well demonstrated by Paul Richards (1985) in his study of *Zonocerus* infestation of cassava in Ghana, but there are many similar examples. I have argued that there is a strong ethical case for documenting the significance of such knowledge in order to demonstrate its often complex, sophisticated and insightful character, both for the benefit of outside agencies but as importantly for those whose knowledge it is. Local people are increasingly motivated to document their patrimony, both in the abstract as 'lists' (Tsing, 2005) and to use it for political ends (Miller, 2016). It has also become a legally and morally contested domain. For these reasons we should always regret the loss of such knowledge, attempt to preserve it in some form before it disappears, but not to compartmentalize it. It is not always wise to be too proactive in making recommendations based on a partial understanding of cultural knowledge and even less to transfer locally embedded knowledge that has worked successfully in one place to somewhere else entirely. The most important applied contribution that work on ethnobiological knowledge systems can make to solving the practical problems of development is in reinforcing the message that it is essential to build in a core capacity for local independent adaptation.

Embedded Knowledge

It has taken me 50 years to acquire a rather incomplete inventory of Nuaulu knowledge of the forest. Most of the relevant data are organized in my notes and databases in a way that makes some sense in the context of a realist ethnography of knowledge and according to the methodologies and epistemologies of biological and social science. However, on the whole these have not been organized in ways that traditional Nuaulu would recognize, until recently. In order to simplify and analyse in a format familiar to modern anthropological and scientific scholarship I have had to dis-embed it from its complex social and landscape context and reorganize it in a form conducive to scholarly and scientific analysis. This is not to say that Nuaulu do not themselves organize knowledge in ways that are consistent with naturalist ontologies, but that they also tend to experience and represent nature in other ways as well, in ways that are equally meaningful to them.

Just as there is a strong motivation to translate the detail of peoples' uses of biological taxa into the organizational conventions of global science, so there has been a tendency to apply that same model to peoples' knowledge of forest: to describe an ethnoecological inventory of fixed ecotypes that mirrors the way in which inventories of plants and animals are conceived. I have discussed some of the difficulties of such an approach elsewhere (Ellen, 2010). Since all peoples, and especially peoples of the tropical forests, have a multi-dimensional knowledge of ecology this is not an unreasonable assumption and has its uses; but the very process of description can sometimes eliminate what is special about such knowledge: that it is local, embedded in an organizational matrix that goes beyond biology, and which provides what Atran and Medin (2008) have called (rather understatedly) 'cultural support' for cognition. For example, mythic narratives are not simply colourful stories that impair access to folk-scientific knowledge, they are inseparable from accounts of landscape; the symbolic features of old swidden plots and settlement sites and notions of ownership are part of the very fabric of thinking about forest ecology, not just a superimposed layer. What a Nuaulu 'sees' in the forest is not just plants and animals and fungi, creeks, rocks, waterfalls, lakes, hills and valleys, but paths trodden by ancestors, trees planted by grandfathers, boundaries between clans, areas that are sacred and prohibited. All is simultaneously physical and mental, biological and social. Neither is this embeddedness static or unaffected by change. It is constantly changing. Nuaulu read into the forest in a very literal way their recent (as well as mythic) history, through an understanding of tree maturation and forest succession, swidden cycles, settlement histories, the way old paths change their course due to flooding, land slippage and tree fall, all such events and processes providing a framework through which to measure time (cf. Condominas, 1954). A real part of Nuaulu historical knowledge is an understanding that much of the forest of central Seram, from Mount Binaia in the east to Waraka in the west, from several kilometres inland from the south coast to the watershed in the Manusela national park, is forest intimately and irrevocably belonging to Nuaulu clans, evident from the distribution of old settlement sites throughout the interior, but from which Nuaulu had moved in the late 19th century at the behest of the coastal rajas and Dutch colonial government. These areas continued to be part of the Nuaulu extractive area and patrimony after they vacated them, reflected in stories and acknowledged by adjacent political groups.

The instability caused by the Second World War and the ensuing separatist Republik Maluku Selatan insurrection meant that there was little change in the forests of central Seram and development for a period of 30 years. The Dutch had introduced rubber and copra production, while clove and nutmeg cultivation had been the traditional exchange crops since the 17th century. There had also been a trade in forest products. By 1970, and probably since they came to the coast, Nuaulu have had an ambiguous relationship particularly with the domain of Sepa, but much occluded by ongoing internal disagreements between individual Nuaulu clans about authenticity of practice and political leadership.

By 1970, the official state legal position was that, depending on what law was invoked (the Forestry Law or the Agrarian Act), Nuaulu theoretically had access rights, or were denied them. In practice this did not seem to matter, as the rights of Nuaulu over most of this area were recognized in the customary law of other local peoples, not the least being the rajas of the larger local coastal settlements of Sepa, Tamilouw, Amahai and Makariki. This was the position in 1970, but by the early 1980s important changes were taking place, as the policies of the Indonesian New Order government were rolled out. Logging began to expand, roads were built and transmigration zones established, all encouraging in-migration, both spontaneous and planned. These all had consequences for Nuaulu. They could, for example, access parts of their forest more easily, but the better transport and incoming population led to contestation and conflict over resources. The combination of these trends together with the introduction of regular schooling and a massive growth in population led to new ways of thinking about the forest and their own situation with respect to it, internalizing government and non-governmental organizations' rhetoric about traditional peoples and their 'indigenous knowledge'. At this time, some clans moved and established new settlements in the lower Nua valley, close to sago and other resources.

The fall of the New Order government in 1998, and a new emphasis on *otonomi daerah* (local autonomy), provided the impetus for Nuaulu to re-open the debate around their position in the local government system, and in 2012 they finally achieved an independent *desa* (local government unit) of their of own. This was Nuanea, in 2013 occupying 125 km^2 along the Nua valley, and under its own government-recognized *raja*. This made it the third largest administrative unit in the sub-district of South Seram, following Tamilouw at 596 km^2 and Sepa at 223 km^2, out of a total of 1149 km^2 in the sub-district as a whole (Badan Pusat Statistik, 2014). The political settlement was far from perfect. Many clans did not recognize the political authority of the Nuanea *raja* and continued to occupy settlements under the jurisdiction of the *raja* of Sepa, but the important change was the re-conceptualization of forest by Nuaulu as a cartographically authenticated zone, part of which was subservient to no other polity. Ecological knowledge had been officially reconnected with official political sovereignty, but in the context of an economy in which land could be sold and maps used to represent it in the abstract.

Nuaulu have never been in a position in which their biological knowledge as such has been commoditized, for example as herbal remedies, and there has never been concern regarding theft of this kind of knowhow, but they are well aware that cultural knowledge in general is valuable. Since all knowledge is embedded in other kinds of property that are contested, it is inevitably also political; and since both knowledge and resources are about things that underpin production, circulate and are consumed, they are intrinsically part of a dynamic economy of supply and demand. The rest of this chapter looks at three case studies that illustrate just how contingent the changing circumstances of local forest knowledge can be: (i) harvesting *Canarium* nuts; (ii) the domain of firewood; and (iii) rattan. The first is a classic example of intricate biocultural interaction, showing how oscillating patterns of extraction of different species of the same genus have been influenced by the politically driven patterns of movement already described. The second case – the category 'firewood' – raises the vexed issue of what ethnobotanists mean when they measure 'usefulness', and how this is not always best examined in terms of standard models typical of the ethnobiological literature, which tend to promote the individual species as the unit of analysis. The third example examines a more familiar scenario, namely the impact of market volatility, commodity price and the decline in subsistence demand for a particular non-timber forest product: rattan.

Changing Places: Shifting Preferences in Harvesting *Canarium* Species

In his grand *Herbarium Amboinense* (1741–1750), Rumphius (2011, pp. v2, 215–251) documented the diversity and complexity of the genus *Canarium* (in Malay, *kenari*), several species of which he noted were used by the inhabitants of the central Moluccas for oil, nuts and resin. Amongst these species, *Canarium indicum* in particular was reportedly further subdivided in folk classification, reflecting its overarching importance in the lives of 17th century Ambonese. Rumphius also mentioned a further species, now known as *Canarium hirsutum* (the *Kenari Seran*), which was less widely distributed and used and which had larger edible, hairy fruits.

For Nuaulu in the early 21st century this account is well reflected in their own folk classifications and patterns of use. The dioecious *C. indicum* is known as *iane*, but subdivided into at least two types of female fruiting tree and a male tree. Its significance in gift-giving and ritual feasting in the form of a flat bread (*maea*), made from grinding the nuts with sago flour, is exemplified in the numerous synonyms for each of the kinds of fruiting tree, associated with taboos placed by particular clans, houses or descent lines on uttering one name or the other. In addition, Nuaulu place *Canarium maluense* (which they call *ananate*) and *C. hirsutum* (which they call *kamine*) within the same folk grouping, often describing them all as *iane* in the wider sense. During the 1970s, *C. indicum* was the most common and widely concentrated species of the genus around current settlements and around abandoned settlements in the mountains. By comparison, *C. maluense* was (and still is) known as a tree of the upper reaches of rivers flowing south into the Banda Sea, while *C. hirsutum* was associated, in 1970 at least, with more remote and largely inaccessible areas further inland.

In 1970, of all folk varieties, it was *C. indicum* that was regularly being harvested, but 45 years later, as the economy shifted towards a greater focus on plantation crops, competing demands for labour time exerted pressure on harvesting and encouraged less time-consuming alternatives. Climbing large trees to harvest nuts had always been hazardous, resulting in many injuries, and this made harvesting even less attractive. By 2015 *maea* made from sago and groundnut (*Arachis hypogaea*) was considered acceptable for some lesser ritual and social obligations, both in the new settlement of Nuanea and when provisioning birth rituals in the neighbouring non-Nuaulu south coast village of Yalahatan by affinally related clans from Rouhua. However, the movement of many Nuaulu clans from the south coast to the new settlements in and around Nuanea positioned them in a different ecological zone, with access to forest of a different composition, and which had previously (for over 100 years) been considered 'remote'. In this area there were fewer *C. indicum*, but a larger number of *C. hirsutum*, which were now considered perfectly acceptable (and indeed preferable) for ritual purposes. Thus, following settlement movement, a species that had for a long time delivered a major feasting food of ritual significance was being effectively replaced by a species that hitherto had been considered too remotely located to be used on a regular basis, but which had known and appreciated properties.

In the case of *Canarium*, although people are here relying on a generic knowledge of the properties of closely related taxa (in this case a polymorphic scientific genus), the particular species extracted can alter depending on local ecology, settlement circumstances and economic pressures. Moreover, the historical shift in interest between different species of the genus (perhaps oscillating backwards and forwards over the centuries) and the consequent intermittent selective pressure this places on each of the species give credence to the view that sometimes domestication must be understood not as a simple mono-specific linear process focused on particular species, but as a pattern of human–plant interaction over time with several useful species of the same genus. Recent research on *Canarium* suggests that desirable fruit characteristics of cultivated and wild-harvested edible species have evolved multiple times within a polyphyletic genus, with characteristics useful for edible nuts dispersed throughout the phylogeny (Ellen, 2015). The utility of particular species in a genus can therefore vary depending on their accessibility across different ecologies, as forest composition varies, as human populations move around, and in terms of what changing infrastructures and political circumstances permit.

Measuring Utility: Firewood

One of the difficulties in measuring utility is to know what a 'use' is. In ethnobotany and economic botany it has become standard to employ a tight empirical definition of what might constitute a 'use' (e.g. Cook, 1995), with established protocols to assess the number of uses a plant has, such as 'multiple use curves' (Balick, 1994). Such protocols permit enumeration and other kinds of measurement, while meeting the requirements of science and politically driven accountancy. But this is not how ordinary people living in tropical forest view the utility of their plant world, who seldom think in terms of fixed finite species uses. Rather, all plants have uses, but particular plants can have multiple but often ad hoc and indefinite uses, not all of which may be the best application of their properties. What is useful often depends on what is available. For example, in studies of medicinal plants there is a widely held but generally unstated assumption not only that the local knowledge corpus is something like a fixed list of species with particular applications, but also that somehow many species and the treatments with which they are associated are found over a wide area, such that we might speak of 'regional pharmacopoeias'. However, in a recent comparative study of data drawn from eight ethnographic locations in Borneo and Seram (including the Nuaulu), Raj Puri and I (Ellen and Puri, 2016) found, counter-intuitively, that medicinals used for particular treatments were extremely locally specific. While genera used medicinally overlapped widely, there was little overlap in terms of species used. Even within local cultural groups defined by language boundaries, there was variation in species harvested and used for particular purposes.

Something similar applies to patterns of firewood use. Table 3.1 lists some named categories of woods used by Nuaulu as fuel, but as we shall see this is a very partial and misleading representation of how Nuaulu think about and select firewood, or of how much they use individual species. Nuaulu call firewood *menie* (literally, 'that which is dry'), and we need to understand that *menie* is not just trees (*ai*) but other burnable vegetable material as well, including charcoal (*okone*), kindling and firelighters. Moreover, most species of woody plant provide Nuaulu with potential fuel; therefore a large number of species could be listed as being used for firewood, and 'fuel' entered in the relevant 'use' field in a database. But whether and when Nuaulu use particular species depends on the options and opportunities available as much as the particular purpose to which it will be put. If you asked a Nuaulu to list the total number of woody plants that could potentially be used as fuel you would likely get a figure of around 340, that is almost all trees and some bamboos and palms. If you ask what fuel woods had actually been used in the past 12 months you would likely arrive at a figure considerably less than this, and in my intermittent records of species

Table 3.1. Selected Nuaulu firewood species.

Nuaulu name	Family[(a)]	Scientific name[(a)]	Firewood status[(b)]	Notes
ai ane	Rosaceae	*Prunus arborea*	2	
ai numa	Lauraceae	(no ident.)	2	
ai osi	Meliaceae	*Aglaia parviflora*	3	
ai polo	Moraceae	*Antiaris toxicaria*	3	
ai nona	Alangiaceae	*Alangium griffithii*	1	
ai suto	Verbenaceae	*Premna integrifolia*	3	Prohibited for clan Peinisa
akahante	no ident.		2	
anametene	Ebenaceae	*Diospyros amboinensis*	1	
ansaha	Rubiaceae	(no ident.)	2	
asahune	(no ident.)		2	
asatua	(no ident.)		2	
asihata	Euphorbiaceae (no ident.)		2	Burned green because of its oil content
hana (ai unate)	Rhizophoraceae	*Bruguiera gymnonorrhiza*	1	Gives off fierce heat
hatae	Sapindaceae	*Allophylus* sp.	1	Used for lime-burning
hisa onate	Moraceae	*Broussonetia papyrifera*	2	
iane	Burseraceae	*Canarium indicum*	2	
ipina onate	(no ident.)		2	
ito	Apocynaceae	*Alstonia scholaris*	4	
nahani nani	(no ident.)		4	
neune	Casuarinaceae	*Casuarina equisitifolia*	1	
pasane	Verbenaceae	*Vitex cofassus*	1	Particularly for charcoal
suenie	Poaceae	*Schizostachyum brachycladum* *S. lima*	3	Most frequently used fuel
tunene (warata)	Annonaceae	*Annona muricata*	2	
tunene	Bombacaceae	*Durio zibethinus*	2	

[(a)](no ident.) = no identification
[(b)]1, excellent; 2, good; 3, average; 4, poor.

actually observed on particular occasions, the number was 24 – that is the number recorded in Table 3.1. Regardless of whether the number of species used as fuel is thereby underestimated or overestimated, any form of enumeration is likely to be distorting and misleading.

Thus, although the list in Table 3.1 samples the diversity of species used, there are other materials it does not refer to; nor does it make much sense as a basis for the division of a semantic domain in the terms sometimes expected from studies of folk classification, comprising instead a number of convergent and cross-cutting evaluative criteria, such as: hardness (hard: high density to soft: low density); burning quality (quick–slow); drying quality (fast–slow); heat quality (cold–warm–hot); harvesting characteristics (ease of cutting, breaking and splitting); and size (small–large).

Fast-growing trees, for example, provide low-density wood and instant heat when burned, but release a lot of smoke. The advantage of these fuels for Nuaulu is that they are rarely in short supply and readily to hand. However, for slow-burning hot fires with less smoke for cooking certain foods and for keeping fires in overnight more dense woods are appropriate, derived from slow-growing species, for example *Casuarina equisetifolia*, which provides an excellent fierce heat and a long-lasting charcoal; and *Erythrina variegata*, which keeps the fire in, as it does not burst into flames, but also burns completely to ashes. *Ficus altissima* and *Avicennia marina* are also regarded as being effective in keeping a fire going; both burn slowly and well in a dry wind. Consequently, it is not that the cultural domain Nuaulu call 'firewood' does not have a distinctive internal organization reflecting a broad and relevant knowledge base; it is just that its complex polythetic character cannot easily be represented as a quasi-taxonomy of 'special purpose' use categories of the kind that we have become familiar with from, say, studies of medicinal plants, dye plants or culinary herbs, and any other of the standard codifications of 'indigenous knowledge'.

As a general rule, what species are used by Nuaulu as fuel and what physical form the wood takes depend less on the purpose of the fire or on the seeking out of particular species for their qualities. Most woody species used as fuel are not explicitly tagged as such: most are useable in some circumstances, depending on their condition, and Nuaulu apply generic judgements of acceptability based on the condition of the wood (including the diagnostic criteria listed above) and what is readily available. The fuels of preference, therefore, are often the most humble (bamboo, palm leaves – what Nuaulu call 'rubbish fuel', *menie ai yehue*). But even where materials are available they have to be harvested, and these costs can be high. Ease of harvesting is therefore a factor in preferring particular species. Most Nuaulu firewood is collected by women and moved around in large back baskets from nearby forest and the edges of swidden land. Piles of firewood are a common sight along the edge of paths, where they are protected with prominent scare charms (*wate*). Until the 1980s the only tool used for cutting firewood was the 47 cm or 68 cm *parang*. Most households would have access to a felling axe but these would seldom be used when harvesting firewood. However, with the arrival of the petrol-driven chainsaw in the 1980s, trees could be considered for firewood that would have previously been avoided, for example the coconut palm, sawn into short lengths and then split.

For Nuaulu, bamboo is the single most important fuel and that most commonly seen on fires. But bamboo as a fuel is often neglected in inventories of economic uses. For example, Dransfield and Widjaja (1995) in their survey of the economic botany of south-east Asian bamboo did not mention it at all, although in the companion volume on 'auxiliary' plants, Hanum and van der Maesen (1997, pp. 28–30) did mention some of its convenient properties as fuel. Bamboo has several advantages, not least being its abundance. It is also light to carry and easy to harvest, cut and split. Its abundance, of course, is linked to it being fast growing, readily colonizing patches where land has been cleared for gardens, or where light intrudes elsewhere through the forest canopy and on the edge of plantations. Thus, while it is not actively planted, it is managed indirectly, through the creation of post-swidden fallow and subsequent cutting back when harvesting for fuel and other uses. Being fast growing, it is easily replenished and thus sustainable. Thin-walled species such as *Schizostachyum brachycladum* and *S. lima* grow the fastest and are particularly abundant around villages, and along paths, and these tend to be the most often selected for fuel. Bamboo provides a good fuel, as it reduces pressure on other biomass fuels, the rotation cycle of bamboo usually being shorter than for other fuel species.

Beyond this, factors determining species are heavily circumstantial. For example, fires lit when travelling to distant sago swamp are made on floating rafts of sago palm bark, and given that these areas are almost mono-specifically composed of sago palm (*Metroxylon sagu*), most fuel for these fires is derived from parts of the palm itself. However, since sufficient dry tinder may not be reliably available in swampland, some dry bamboo for this purpose will actually be taken to the location where the palms are to be cut, along with lengths of green bamboo in which to cook. A similar example is the disproportionate use of dry coconut husks and shells as fuel when drying copra (both *Cocos nucifera*).

Outside of this core of default firewood used in village fires, where species is less important than the application of generic knowledge, and the special circumstances of the kind just illustrated, there are instances of species-specific related knowledge. For example, some species are flagged precisely because they should be avoided. It is important to know not only what good firewood to prefer, but also what bad firewood to avoid, either due to poor physical burning and heating qualities (e.g. *ito*: *Alstonia scholaris*), or because there is a taboo placed upon it, as in the case of *ai suto* (*Premna integrifolia*) used commonly as fuel, but prohibited for the clan Peinisa. On the other hand, there are species that are singled out as excellent fuels for particular purposes, especially as charcoal (*okone*). Different species of woody plant collected as *menie* can be transformed into distinctively different types of charcoal. For example, the stony endocarp (shell) of the coconut makes excellent compact charcoal, as does mangrove wood (in this case *hana*: *Bruguiera gymnonorrhiza*). Nuaulu have few applications for high-temperature fires, as they do not work metal, but there is a market demand for suitable dense woods that Nuaulu occasionally exploit. However, they do have other specialist applications for fire for which different fuels are appropriate. Smoking is important for preserving meat and to a lesser extent for driving away insects and other animals. Coconut husks are excellent for this purpose. Nuaulu also require a special kind of fire (*usa nosa*) for producing mineral lime (*kotu nosa*) used in betel chewing. This involves burning mollusc shells, coral or soft limestone. The fuel of preference for this is coconut shells (on top), and a pyramid of stacked split bamboo raised off the ground on several lengths of sago leafstalk to allow for good ventilation. Bamboo fires, though not of this structural complexity, are preferred for fixing the vegetable dyes used to prepare *Pandanus* leaf strips when making ritual objects, for hardening arrow points, for roasting plantain or yams, and for heating stones used when cooking *maea* (see above). Nuaulu do not make pottery, but in Moluccan villages that do, coconut husks too are used to provide a steady heat, while a quick blazing heat is obtained by using dry fronds and leafstalks of both coconut and sago palms, including old pieces of thatch made from these materials, which provide instant low-temperature bonfires suitable for potting using local clays.

Dense fuel woods are the fuels of choice in those areas of the Moluccas where fuel of all kinds (including driftwood) are in short supply. This is typical of small coral and volcanic islands that are resource deficient, but which are important centres of trade or fish production. For example, at the easternmost tip of Seram there are specialist mainland villages engaged in extracting mangrove wood for inter-island trade to small tree-deficient islands such as Geser. Although there are significant costs to consumers in relying on such dense woods, the trade would not be viable were it to be restricted to those fast-burning fuels often preferred elsewhere. The quantities involved would have to be huge, increasing costs for producers, traders and consumers alike. Not only are these denser wood fuels important for basic subsistence in such places, but they also support the local boat-building industry, which has need of fires for wood-bending.

We finally turn to adjunct materials employed to light fires. This is no trivial matter, and the regularity with which different cultural groups distinguish kindling (in Nuaulu, *men ikine*: 'small firewood') from firewood *sensu stricto* is significant. In a Nuaulu village context fire is generally always alight somewhere, and dead fires are usually re-lit by transferring from one fireplace to another, from one domestic space to another. The most usual way of doing this is to employ a firebrand (*osi*) made from a bunch of dried coconut leaves, and in 1970–1971 it was common to see young children sent off to bring fire from the house of, say, a sibling or a cross-cousin. Fire, therefore, followed lines of kinship and affinity. Outside the village, however, fire making had to start from scratch, as in sago swamp fires described above. In 1970 this was still mainly achieved through the use of flint strike-a-lights (*kinonote*; see Ellen and Glover, 1975, pp. 52–53) and a piece of iron sparking against a small wad of *panua*, the fine dry soft earthy-coloured lichenous tinder scraped from the bark of coconut, sago and aren palms[1]. Once the *panua* is smouldering, ignited matter can be transferred to a *sakete*[2] made from *nahue*, the fibre found in clumps around the base of the petiole of an aren palm (*naha niane*: *Arenga saccharifera*, '*gamatu*' in Ambonese Malay) and thereafter transferred to a pile of wood shavings and kindling (noun: *masahini*; verb: *asanasi menie*), followed by the progressive addition of small lengths of wood obtained by splitting (verb: *akusina menie*) and finally larger cut pieces (verb: *sene menie*).

In the 1970s *panua* and *sakete* were part of the ubiquitous equipment carried around by every adult male in their betel-chewing pouch. By 2015 these technologies were obsolete, indeed virtually extinct, replaced first by matches and then petrol-fuelled cigarette lighters. Only in the domain of ritual and within sacred clan houses were these forms of fire lighting prohibited, and where the traditional forms survived. It is also in sacred houses that another form of firelighter, the resin torch (*kamane*), is found, made from various kinds of *dammar*. Resins from certain *dammar* species are prohibited for use in sacred houses, for example *kama onie* (*Shorea selanica*) and *kama wae* (*Agathis dammara*), though they may be used for illumination and to transmit fire outside the village. Amongst the most common resins permissible and used within sacred houses is *kama kamine* from the *kenari* species *C. hirsutum*, which we have already met. Overall, the qualities sought in materials used as fire-lighters are that they are dry, will ignite easily and remain alight through smouldering.

Ecology, the Market and Social Change: a Rattan Case Study

The changing shape of Nuaulu environmental knowledge, as well as my capacity to document it, has been influenced by ecological and economic change, both in terms of cycles and linear trends.

In 1970 it was common to come across the large eroded and colourless shells of the giant African land snail (*Achatina fulica*, which Nuaulu call (*nunu*) *keon*, from Malay *keong*) in gardens and along forest paths. These shells were being used at the time, along with other freshwater and terrestrial shells, to make mineral lime by burning (see above). The lime served, and continues to serve, as an ingredient when chewing the betel quid as a mild stimulant (usually the crushed fruit of *Areca catechu* and the catkin-like inflorescence of *Piper betle*). However, *Achatina* shells were not otherwise obviously being used and I did not enquire further. I saw no live specimens in 1970, nor on subsequent field trips during the 1970s. By 1990 the species was everywhere, both live specimens and shells, on plantation land especially, where it was impossible to avoid treading on them. This prompted me to enquire further, and it was clear that the snails were be being harvested for their edible flesh, but more importantly were also the cause of significant crop damage in garden land. What this pattern reflected, therefore, was a boom-and-bust cycle, a population explosion typically accompanying initial introduction followed by a crash. The utility of the species, what Nuaulu knew about it, and what I was able to find out, were all being affected by its reproductive dynamics.

The same is true for the market economy. During the early 1970s there was little outside demand for wild honey from *Apis cerana* bees, which was used mainly by Nuaulu as a medicinal product. In 1996 there was a great demand, due mainly to the growth of the administrative centre at Masohi on Elpaputih Bay. However, the obvious demand quickly encouraged many others to take advantage of market conditions, and soon supply was outstripping demand. This in turn led to disappointment amongst harvesters, who lost interest in supplying the market. Like the virtues of *Achatina* shells, this knowledge would not have come my way had it not been for the serendipity of my field visits coinciding with the right phases in cycles of growth in demand. Though harvesting honey employed existing Nuaulu knowledge of nests, bee behaviour and extraction techniques, at the same time it stimulated and augmented knowledge that had become moribund.

A related point can be illustrated with respect to rattan. The range of species available to Nuaulu on the island of Seram is limited compared with the diversity found in the two large islands on either side, Borneo and New Guinea. For example, in a recent study Schreer (2016) reported Ngaju Dayak in Indonesian Kalimantan as recognizing and using 32 named folk types of rattan corresponding to six scientific genera. In terms of species per genus these were: *Calamus* (17), *Korthalsia* (six), *Daemonorops* (six), *Plectocomiopsis* (two) and *Ceratolobus* (one). By comparison, Nuaulu distinguish nine named folk-categories, indicating six species of *Calamus* and four species of *Daemonorops*. This difference arises partly because Seram lies in Wallacea, a biogeographical zone of transition between the tectonic plates of Sunda and Sahul which has served as a barrier to species movement, and which is depauperate for many groups of terrestrial macro-organisms (Ellen and Puri, 2016). It also reflects the fact that Schreer's study site lies in the centre of an area well known for rattan production and export. Nuaulu rattan knowledge and production was in 1970 also extremely important, a vital resource for house building, basketry and for the manufacture of a wide range of other objects. Rattan was also in demand by outsiders, including the Indonesian government, who would recruit Nuaulu to collect rattan for the construction of bunkrooms (*asrama*) for police and the military. There were also some small-scale local enterprises in Ambon importing rattan from Seram to manufacture cane furniture. Although rattan exports have increased over the last few decades (along with other non-timber forest products) in Indonesia as a whole, there was little evidence that this had impacted on the central Moluccas. By 2015, government demand for rattan in south Seram had dried up. Figures for export are absent from consulted records for 1983–1993, and for 2013 only 12 t are noted as having moved through the port at Amahai (Badan Pusat Statistik, 2014), despite the holding of an INIDO-INBAR Rattan Furniture and Crafts Skills Training Workshop in Ambon during 2006, and government attempts to encourage post-harvesting processing by banning export of the raw product. Moreover, by 2015 most ordinary Nuaulu houses no longer relied on rattan for securing posts, beams and rafters; they used nails and sawn timber instead, while small baskets and other containers formerly made of rattan and bamboo, were increasingly being replaced by plastic. Finding, harvesting and preparing rattan for use is a time-consuming and labour-intensive occupation that requires technical skill in the harvesting and post-harvest processing, as well as in identifying and locating the correct species. Once such skills are lost they are difficult to replace. But despite all the factors that might be expected to lead to a decline in rattan knowledge, in 2015 it was still thriving, competence being widely distributed and its transmission apparently unimpaired. The reason for this was ritual rather than economic. Every clan has at least two sacred houses, and major rituals require the use of baskets made in part from rattan. As the population had tripled since 1970, without substantial religious conversion (Ellen, 2012), the need for rattan to support ritual objectives had actually increased. The reasons for such robustness in ritual is an issue that I have addressed elsewhere, but to provide some idea of its impact on rattan production, on one occasion alone in April 2015 over 40 lengths of rattan (6 m long) were collected over a 3-day period to rebuild the Matoke-pina sacred house in the village of Rouhua. This case exemplifies a principle described as cross-domain resilience (Ellen, 2009): if a resource is used across a range of divergent purposes (say, construction, food and healing) it is less likely to be vulnerable to knowledge loss than where the resource is supporting a single purpose.

Conclusion

There are several threads connecting these various cases, each of which seem to advocate caution in regard to the established enthusiasm for documenting local or indigenous environmental knowledge.

Firstly, all of the examples described engage with the theme that knowledge changes, but what is also clear is that what changes about the knowledge is not always predictable. Change may take place as part of established ecological or economic cycles, or it can be apparently linear and innovatory. In either case, while some fairly predictable factors may influence its decline or modification (schooling, the market, religious conversion), in other cases circumstances conspire to resist change. Thus, while new forms of charcoal and paraffin-fuelled stoves are replacing conventional wood fires, and fires are lit with cigarette lighters, the domain of firewood knowledge is likely to remain resilient for the foreseeable future. On the other hand, while changes in local subsistence and market conditions diminish reliance on traditional materials such as rattan, a constant theme throughout these cases is the way in which robust ritual practice is maintaining knowledge in sometimes quite unexpected ways. While resin is no longer a source of general house illumination, it is still essential in sacred houses; and while traditional skills (e.g. rattan-stripping) are less and less relevant for house-building, ritual needs are keeping these alive. It is still forbidden to use any form of cooking or heating device in ritual houses other than the wood fire. What all this speaks to is the importance of recognizing that knowledge is never free-floating and always embedded in wider contexts, whether these be the political consequence of the Indonesian policy of *otonomi daerah* that has permitted Nuaulu to re-connect with forest last regularly traversed 150 years previously, or the maintenance of temporarily protected areas of forest in order to deliver the supplies required for major feasts and ceremonies (*sin wesie*), but which serendipitously conserve species and patch diversity in lowland forest (Ellen, 2010). We must always be circumspect regarding claims for generic patterns of change.

Secondly, all change involves knowledge hybridization, a concept that is now seriously in danger of losing its value through overuse.

The kinds of hybridity heralded in an earlier generation of local knowledge studies were significant mainly in pointing to how macro-scale historically separated systems of knowledge conjoined through asymmetric social relations could combine in productive ways, as at the interface between trader and producer, or between say the incorporation of the outcomes of field trials in agricultural research centres with local farmer knowledge. But often the hybridized outcomes concealed a more complex and longer history of mutual knowledge interaction. And even within local systems, hybridization is a continuous feature of biocultural knowledge creativity, as lessons learned with respect to one group of organisms are transferred to another, or knowledge is shared between social groups and individuals of various degrees of remoteness.

Finally, what we construct as local knowledge is in part a consequence of the opportunities that arise to observe and record it, partly a consequence of what we are seeking, and partly a consequence of the models we introduce to make sense of it. For me, *Canarium*, firewood and rattan each presented themselves in different fieldwork and temporal contexts, and analysis has required dispensing with one aspect or another of conventional approaches to studying local knowledge. Fifty years after the much maligned paper by Metzger and Williams (1966) on Tzeltal firewood categories, we still find that firewood is a neglected and under-researched plant resource category, and in some sense not a proper 'use' at all. Metzger and Williams, of course, were mainly concerned with the development of protocols for ethnographic research in the ethnoscience/ethnosemantic tradition, and with showing how the lexical and semantic content of a particular cultural domain had a distinctive structured character. They were not particularly concerned with understanding how poor farmers make decisions about the selection of fuel, or with seeking to demonstrate the relevance of a particular area of indigenous knowledge for applied anthropology.

Notes

[1] British Museum Specimen 1972 As.1.3.1.

[2] British Museum Specimen 1972 As.1.30.

References

Atran, S. and Medin, D. (2008) *The Native Mind and the Cultural Construction of Nature*. The MIT Press, Cambridge, Massachusetts.

Balick, M.J. (1994) Ethnobotany, drug development and biodiversity conservation – exploring the linkages. In: Chadwick, G.D. and Marsh, J. (eds) *Ethnobotany and the Search for New Drugs*. Wiley, Chichester, UK, pp. 4–18.

Badan Pusat Statistik (2014) *Kecamatan Amahai Dalam Angka 2014*. Badan Pusat Statistik (Central Bureau of Statistics), Amahai, Kabupaten Maluku Tengah.

Cook, F.E.M. (1995) Economic Botany Data Collection Standard. Royal Botanic Gardens, Kew, UK.

Condominas, G. (1954) *Nous Avons Mangé la Forêt de la Pierre-Génie Gôo (Hii saa Brii Mau-Yaang Gôo)*. Mercure de France, Paris.

Dransfield, S. and Widjaja, E.A. (eds) (1995) *Plant Resources of Southeast Asia, Volume 7: Bamboos*. Prosea, Bogor, Indonesia.

Ellen, R. (2009) A modular approach to understanding the transmission of technical knowledge: Nuaulu basket-making from Seram, Eastern Indonesia. *Journal of Material Culture* 14, 243–277.

Ellen, R. (2010) Why aren't the Nuaulu like the Matsigenka? Knowledge and categorization of forest diversity on Seram, Eastern Indonesia. In: Johnson, L.M. and Hunn, E.S. (eds) *Landscape Ethnoecology: Concepts of Biotic and Physical Space*. Berghahn, Oxford, pp. 116–140.

Ellen, R. (2012) *Nuaulu Religious Practices: the Frequency and Reproduction of Rituals in a Moluccan Society*. KITLV Press, Leiden, Netherlands.

Ellen, R. (2015) Landscapes of exchange and the domestication of Kenari. Unpublished paper presented at the Clusius Symposium on 'Other ways of gardening', University of Leiden, 29 May 2015.

Ellen, R. and Glover, I.C. (1975) Ethnographic and archaeological aspects of a flaked stone collection from Seram, Eastern Indonesia. *Asian Perspectives* 18(1), 51–61.

Ellen, R. and Harris, H. (2000) Introduction. In: Ellen, R., Parkes, P. and Bicker, A. (eds) *Indigenous Environmental Knowledge and its Transformations: Critical Anthropological Perspectives*. Harwood, Amsterdam, pp. 1–33.

Ellen, R. and Puri, R. (2016) Conceptualising 'core' medicinal floras: a comparative and methodological study of phytomedical resources in related Indonesian populations. *Special issue of Conservation and Society* 14(4), 345–348.

Hanum, I.F. and van der Maesen. L.J.G. (1997) *Plant Resources of Southeast Asia, Volume 11: Auxiliary Plants*. Prosea, Bogor, Indonesia.

Metzger, D. and Williams, G.E. (1966) Some procedures and results in the study of native categories: Tzeltal firewood. *American Anthropologist* 68, 389–407.

Miller, T. (2016) 'Living lists': coming to know indigenous Canela encounters with plants through ethnobotanical classification. In: Miller, T. et al. (ed.) *Botanical Ontologies. Special Section, Journal of Ethnobiology* 36(1), 105–124.

Richards, P. (1985) *Indigenous Agricultural Revolution: Ecology and Food Production in West Africa*. Unwin Hyman, London.

Rumphius, G.E. (2011 [1741–1750]) *The Ambonese Herbal, Volumes 1–6* [translated by E.M. Beckmann]. Yale University Press, New Haven, Connecticut, and London.

Schreer, V. (2016) Learning knowledge about rattan (*Calamoideae, Arecaceae*) and its uses amongst Ngaju Dayak in Indonesian Borneo. In: Miller, T. (ed.) *Botanical Ontologies. Special Section, Journal of Ethnobiology* 36(1), 125–146.

Sillitoe, P. (2015) Indigenous knowledge. In: Stewart, P.J. and Strathern, A.J. (eds) *Research Companion to Anthropology*. Ashgate Publishing, Farnham, UK, pp. 343–368.

Tsing, A. (2005) Friction: *An Ethnography of Global Connection*. Princeton University Press, Princeton, New Jersey.

4 'Keeping our *Milpa*': Maize Production and Management of Trees by Nahuas of the Sierra de Zongolica, Mexico

Citlalli López Binnqüist*, Rosalinda Hidalgo Ledesma and Fortunata Panzo Panzo

This chapter aims to contribute to the discussion about the importance of recognizing knowledge systems as part of diversified agroforestry landscapes. Our focus is *milpa*, a maize-centred polyculture system of pre-Hispanic origin which has historically constituted an essential source of food and has persisted until the present. Our area of study is the Sierra de Zongolica (SZ), an understudied, mountainous, indigenous Nahua region within the state of Veracruz, Mexico.

Milpa is a traditional food production system that is in constant adaptation. In SZ, timber trees have been increasingly incorporated, a change that has reconfigured the regional landscape and is modifying the forms of agroecological production. The first part of this chapter describes the area's history of land use and the *milpa* production system as the central part of the productive landscape and life of the indigenous Nahua families of SZ. Next, the main factors that are promoting the introduction of timber trees into *milpa* and the associated adaptations of land management are described. The conclusion elaborates on the various impacts of the introduction of trees and current trends in SZ agroecological practices.

Most of the information in this chapter was gathered during our teaching-facilitating projects in SZ. Our teaching has used material from research conducted by outside academics, as well as the theses of Nahua students from Universidad Veracruzana Intercultural (UVI), which has a campus in SZ. Since 2009, we have collaborated with forestry technicians and students from UVI to strengthen the intergenerational transmission of indigenous knowledge (including local history, and natural resource management as applied to agroforestry, and to the production of handicrafts). We have worked as much as possible in the Nahuatl language. We use diverse forms of peer-to-peer learning including storytelling, map making, student-produced video documentaries and exchanges with indigenous communities outside of SZ.

Agroecological Knowledge in Context

Agroecological knowledge generated by indigenous communities is part of what Berkes *et al.* (2000) called traditional ecological knowledge (TEK). Continuous observation of the environment and ecological cycles, combined with trial-and-error processes, leads to an accumulation of practices that are transmitted generationally, often becoming critical for group survival and often ending up transforming landscapes and ecosystems (Perfecto *et al.*, 2009).

* Corresponding author; e-mail: cilopez@uv.mx

At present TEK and livelihood strategies in peasant societies cannot be isolated from the international and transnational processes of globalization (Bacon *et al.*, 2012). New productive activities can usually be traced to these external processes which incentivize the production of goods for global markets – products that in many cases were first introduced by migrants (Trujillo, 2008; Bacon *et al.*, 2012). As noted by Alexiades (2009), the most powerful market forces tend to increase homogeneity in knowledge systems and in lifestyles.

Rural landscapes come to reflect current agreements, negotiations, tensions and conflicts at different levels: among the members of each family, and within and between communities, regions and nations. Therefore, as mentioned by Hecht *et al.* (2012), rural households manage social and landscape ecologies at an increasingly complex and global scale.

In Central America and Mexico, one of the characteristics of agricultural landscapes, even when fragmented and small, is that they can maintain a multiplicity of land uses and often are populated with trees in various ways (hedgerows, demarcation of land, orchards, agroforestry systems, etc.) (Hecht *et al.*, 2012). Relatively small areas can develop considerable floral taxonomic and structural diversity (Toledo *et al.*, 2003), either by intensification in handling certain species or combinations of species, or by the introduction of new commercial species. The case presented in this chapter is part of an ongoing pattern in Latin America in which forest and tree management dynamics are linked to social, political and economic changes occurring at different scales, prompting new agroecological practices and economic activities at the regional level.

We work in an indigenous area noted for its biocultural richness. Mexico ranks fifth in the world amongst 'megadiverse' nations with between 10% and 12% of the world's endemic species, and is equally notable for its ethnic diversity with 68 indigenous official national languages (Boege, 2008; Sarukhán *et al.*, 2009). Within Mexico, the areas of greatest biodiversity fall mostly in areas with a majority of indigenous inhabitants (Sarukhán *et al.*, 2009). In addition, across Mexico the best predictor of maize race diversity is ethnicity diversity (Ureta *et al.*, 2013). Therefore indigenous communities play a central role in the management and preservation of cultural and biological diversity in their territories.

Location and Climatic–Altitudinal Zones of the Sierra De Zongolica

SZ constitutes the southern end of the Sierra Madre Oriental mountain range (Fig. 4.1) and has been inhabited by the Nahua since pre-Hispanic times. SZ comprises 14 municipalities in a 957 km^2 area, with 181,485 inhabitants (INEGI, 2010). It is the region with the highest concentration of Nahuas and has the highest percentage of bilingual speakers in Mexico (Alatorre Frenk *et al.*, 2015). Geologically, SZ is a karst area with steep slopes, deep valleys, closed depressions without apparent external runoff, and large numbers of caves and springs. Its rugged terrain overlooks the Gulf of Mexico with altitudes ranging from 500 m to 2500 m above sea level. SZ has one of the highest rainfall values among indigenous regions in Mexico (4000 mm annually) (Boege, 2008) and it is the headwaters of the culturally and industrially important Papaloapan River.

The region can be subdivided into three altitudinal and climatic zones (Ortiz Espejel, 1991): (i) the 'cold zone' over 1800 m above sea level, where the vegetation is dominated by pines and oaks; (ii) the 'temperate zone' between 800 m and 1800 m above sea level that includes most of the region's cloud forest; and (iii) the 'warm zone' under 800 m above sea level, where most of the vegetation is middle- to low-elevation tropical evergreen forest. This differentiation in altitude and climates has channelled distinct histories of land uses and land tenure. In each zone, the Nahua have developed different techniques for the management and use of the diverse wild species, and the cultivars of domesticated species, suited to each microclimatic region.

Settlement History and Livelihood Strategies

It is not known with certainty when the first humans settled in SZ, but there is evidence of

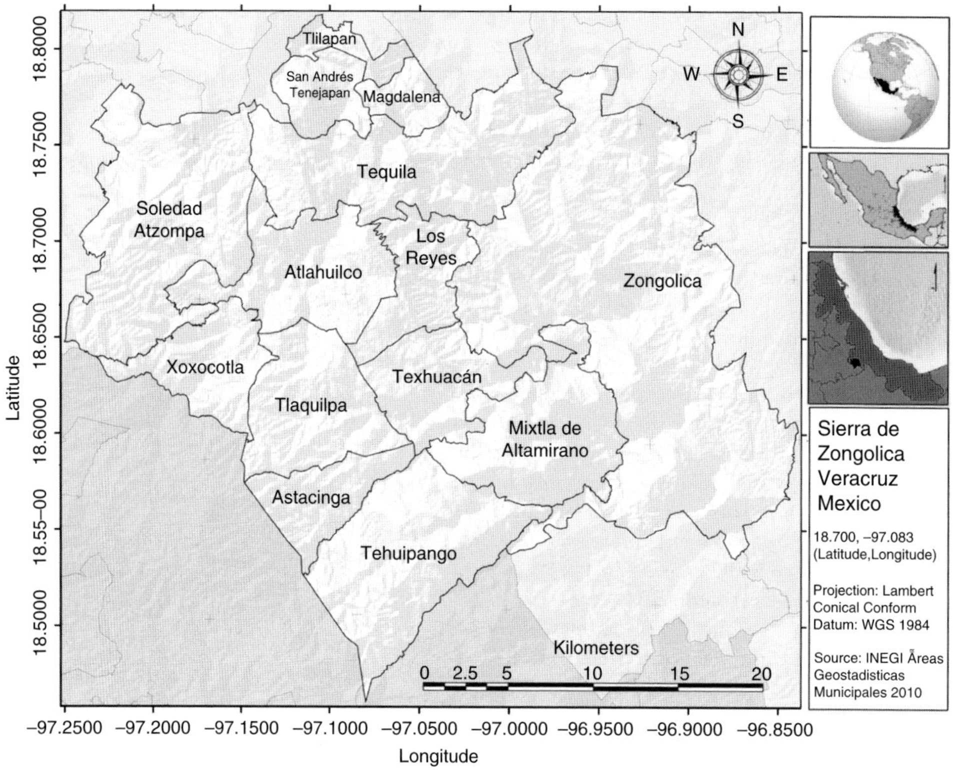

Fig. 4.1. Sierra de Zongolica location. Municipalities located in the cold zone: Tehuipango, Astacinga, Tlaquilpa, Xoxocotla, Soledad Atzompa, Atlahuilco. Municipalities located in the cold and temperate zones: Mixtla de Altamirano Texhucan,Tequila, Los Reyes. Municipalities located in the temperate and warm zones: Zongolica, San Andres Tenejapa,Tlilapan, Magdalena.

northern groups moving into the region and trading with the Nahua inhabitants of the nearby Tehuacán Valley around the time that these Nahuas were domesticating maize, between 6000 and 10000 BP (Morales Carbajal, 2014). During the Spanish Colony, the first areas in SZ that were occupied and exploited by the Spanish were in the warm zone. The cold zone was not suitable for commercial and productive purposes, was isolated by rugged terrain and was difficult for transportation. In the warm zone, Spaniards first introduced sugar cane and cattle. Later, during the early 18th century, tobacco was grown in large haciendas using indigenous slave labour. The low areas of SZ became the main tobacco supply for the Spanish Crown until the production fell when the market became saturated, and when commercial relations were severed with independence from Spain (Early, 1982; Rodríguez López, 2003). (Mexican independence was recognized in 1821, and slavery formally ended in 1829.)

The Mexican Revolution and the Agrarian Law of 1915 marked the beginning of the redistribution of land back to the Nahua. Peasants who had worked for the owners of the large haciendas in the warm zone organized into communities (*ejidos*) that continued to work the same land – which they now owned communally. Other peasants moved into and claimed previously uninhabited areas. In the cold zone, land was distributed in the form of private property to Nahuas who had a legitimate claim of affiliation with the area (Aguirre Beltrán, 1995; Rodríguez López, 2003). Later in the 20th century there was no more land to distribute. With plots transmitted from parents to children there has been increasing fragmentation and smaller average landholdings over time (Rodríguez López, 2003). Today, in the warm zone municipality of

Zongolica, 14 *ejidos* represent 88% of the total area, while private land tenure dominates in the cold zone.

Social and economic conditions changed dramatically in the 1980s when new neoliberal policies led to the restructuring of regional economies. The prices of traditional agricultural exports (coffee, sugar, cotton, banana) fell, and government programmes began to promote new agricultural products for export (vegetables and flowers). The peasants in several regions of Mexico, including SZ, were forced to diversify their occupations, seeking jobs in urban areas and via international migration. Many began grazing small livestock, and manufacturing and distributing handicrafts and forest products such as charcoal within the region and rustic wood furniture at the national level.

Throughout history, the main form of livelihood of families in SZ has been subsistence agriculture. *Milpa* and backyard gardens play an important role in food production for both humans and domestic animals even when the maize production is insufficient. The practice of *milpa* has persisted despite the introduction of commercial crops and despite more recent lifestyle changes associated with urbanization processes. *Milpa* continues to be at the centre of the productive and cultural life of almost all SZ households, especially in the cold zone (Rodríguez López, 2003, 2016; Martínez Canales, 2013).

Milpa – Food Crop Biodiversity and Farming Practices

Milpa (from Nahuatl 'maize field') is a Mesoamerican agroecosystem where maize, beans and squash are the principal crops. In addition to food, *milpa* provides fodder for animals, medicinal products, dye ingredients, flowers/ornamentals and wood for agricultural tools. *Milpa* integrates specialized knowledge of the requirements of the diverse managed plants, sophisticated soil management and the development of cultivation techniques that are effective at all levels – from species to ecosystem. In SZ all of the above facets come into play because of the high environmental heterogeneity and the diverse socio-economic conditions.

Milpa constitutes a biologically diverse habitat, with native Mesoamerican species (some widespread, others local/endemic) and introduced species from around the world (some introduced recently, others brought during the Spanish conquest). In the *milpa*, domesticated plants grow as well as incipient and semi-domesticated plants, such as the traditional Mesoamerican 'edible weeds' known as *quelites*. Working in two different SZ cold zone municipalities, Pérez Pacheco (1992) and Navarro Pérez and Avendaño Reyes (2002) found that residents were making use of 154 plant species in Astacinga and 197 plant species in Tlaquilpa. Larios *et al.* (2013) found 281 species in Nahua home gardens in Tehuacán Valley 50 km away.

Dzib Aguilar (1994) identified eight races, one sub-race and a variant of maize in SZ and found that their distribution varies with the type of cultivation (if tilled, or slash-and-burn), the type of terrain, the type of slope, the zone and the cultivation period. Boege (2008) noted that SZ is one of the indigenous areas where local maize races have diminished. However, even though government programmes have introduced standard maize seeds, the peasants in some localities are, in effect, resisting this standardization. As reported by Álvarez Morales (2013), peasants keep their seeds, continue exchanging seeds amongst themselves, and in some cases are experimenting with hybrids of government and local cultivars.

As for beans planted in the mountains, Araujo and Baez (1994) identified four species in three genera (with six varieties/subspecies): *Phaseolus vulgaris* L. (wild), *P. vulgaris* var. *vulgaris*, *P. coccineus* L. subsp. *coccineus*, *P. coccineus* L. subsp. *darwinianus* Hdez. X. and Miranda C., *Cajanus cajan* (L.) Millsp. and *Vigna umbellata*. Among other species growing as part of the *milpa* are squash (*Cucurbita moschata* Duch ex Poir, *C. mixta* Pang., *C. ficifolia* Bouché), fava bean (*Vicia faba* L.), pea (*Pisum sativum* L.), barley (*Hordeum vulgare* L.), oats (*Avena sativa* L.), potato (*Solanum tuberosum* L.) and manzano hot pepper (*Capsicum pubescens* R. and P.). In the warm and temperate zones, there are also coffee plants (*Coffea arabica* L., *C. canephora* Pierre ex A. Foehner) and banana (*Musa* spp.). The Nahua consume wild and domesticated avocados (*Persea* spp.) and SZ is an important area for preserving avocado germplasm diversity (Aguilar Gallegos *et al.*, 2007).

The two main maize cultivation techniques are the swidden-fallow system (also known as slash-and-burn fallow) and the tilling system. The latter varies with regard to the tools used: a plough or a hoe. The hoe is more common since a plough can only work in level areas and less rocky soils. Plot size and location often affect how maize is cultivated. For example, burning is not applied when a plot is very small or near vulnerable crops such as fruit trees or coffee. Owners of 1 ha or less of land plant every year without letting the land lay fallow, as they have no other land to turn to. In contrast, those who have more plots or a larger area utilize only one part of their land while letting the other parts lay fallow.

The swidden-fallow system is the most common cultivation system in the rocky soils in the cold zone. The cycle begins with the clearing of the fallow land (*acahuales*) with a machete. Useful felled timber is salvaged from the woodpiles before burning and some tree saplings are left alive and pruned. These thin tree stalks act as supports for such crops as beans and squash and kick-start growth when the field is returned to a fallow condition. On sloping land, a burning technique called *contra fuego* is performed where flame fronts coming from the upper and lower parts meet and die out in the middle (Álvarez Morales, 2013). A digger is used to make the holes to deposit the maize seeds.

Rotation (used here to mean cycling between cultivation and fallow periods) compensates for the many areas with poor soils in the mountains (by returning nutrients, especially nitrogen). Diversification (cultivating many plant species) compensates for the limitations of the nutritional properties of the principal crop, maize (by including plants with a complementary balance of nutrients, especially amino acids). By practising both rotation and diversification, the Nahua of the SZ have 'fed' their soils and fed their families for thousands of years.

However, fallow periods have been decreasing (Dzib Aguilar, 1994), often to as little as 2–4 years (Rodríguez López and Álvarez Santiago, 1992). This can be ascribed to several factors: (i) with each passing generation, fathers have been dividing private plots amongst their sons, and the average plot size has been getting steadily smaller, with many plots only being capable of sufficient production if constantly cultivated; and (ii) greater demand has disincentivized letting land lie fallow, particularly now that chemical fertilizers can substitute for rotation.

Milpa – Socio-cultural Dimension

Milpa is not just an important source of food but, as expressed by Rodríguez López (2016, pp. 74–75): 'Cultivating a piece of land refers to more than its objective representation in terms of survival and consumption. It also refers to its symbolic and ritual dimension. While the land is life-giving through the maize, sacred food is offered to the same ground as a sign of thanks in the different stages of the agricultural cycle.' The author continued by saying that even if the land and the *milpa* have lost their original function as the main source of resources for survival, for the Nahua, possession of the land has a deep meaning that is expressed in cultivating, harvesting and consuming the bounty of the *milpa*.

According to Báez-Jorge (2000), the pre-Hispanic cosmovision persists in the rituals dedicated to land and these rituals are preserved by the permanence of the practice of cultivating maize in the *milpa*. Land and maize constitute a unity in ritual terms and are part of a set of operational and symbolic meanings. Present-day Nahuas maintain a worldview where the mountains, forests, springs, animals, plants and all living forms are safeguarded by owners who work in duality: Tlalokan Tata (the male deity) and Tlalokan Nana (the female deity). The *milpa*, and the land where it grows, is at the centre of this Nahua cosmovision.

The relationship of human beings with everything else, a relationship that is governed by the two deities, is protected through requesting permission, apologizing and expressing gratitude during the *xochitlalis*. These rituals consist of offering flowers on to the ground and are carried out for agricultural purposes, such as asking for forgiveness for hurting the earth and asking for good harvests. These rituals weave and reinforce the social fabric and affirm the unity between the supernatural and earthly.

At the heart of the *milpa* and the Nahua worldview is the concept of reciprocity which is the hub that connects the relationships between the individual, society and the environment.

Reciprocity is a way to keep family ties and to maintain networks within and between communities. It is observed on the symbolic level with the organization of the *xochitlalis* and other rituals, and materially in the daily life of the Nahua. For example, as mentioned by Rodríguez López (2016), milpa enables families to participate in community festivals – preparing food and offering ceremonial meals.

Even in an unprofitable *milpa* harvest, salvaged seeds, fruit, cereals, flowers and medicinal resources can be shared with relatives and neighbours. This exchange may be more highly valued than money, because some of the products are not commercially available, or are products that happen to be key to survival at that time, and because practicing *milpa* and engaging in such exchanges serves to reassure the Nahua that they are maintaining a close and collaborative relationship with their neighbours.

Historical Background of Forest and Tree Management

One of the more noticeable changes in recent decades in the SZ landscape is a result of the increasing interest of peasants in growing trees on their land. According to our observations and interviews in the field, this trend is mainly due to: (i) the interest of the inhabitants of SZ to regain control over their forest resources that had been exploited clandestinely by external intermediaries; (ii) the work done by regional organizations to improve the lives of the inhabitants of SZ by fostering economic alternatives, and by incorporating natural resource management; and (iii) national government programmes promoting timber tree plantations.

Commercial logging began in SZ in the late 19th century with the arrival of external logging businesses who supplied lumber and fuelwood to the nearby Cordoba–Orizaba industrial corridor. All over Mexico commercial logging became increasingly exploitative and in 1952 the federal Ministry of Agriculture and Forestry Development declared an indefinite ban throughout the country, which, in the state of Veracruz, lasted until 1978. During the ban, 90% of the forest areas in the state were declared closed, but one of the few areas where it was not enforced was SZ. This was a terrible time for the region. Loggers often used violence to get their way, frequently opening new roads and removing trees without permission. The inhabitants were paid little or nothing. These loggers followed a practice that removed the largest, most valuable trees in the fastest, cheapest and most destructive way. Tree removal tore up the soil and the remaining trees in a way that decreased forest regeneration and increased erosion. The long-term effects were not only the degradation of standing forests and their ecosystems, but also outright deforestation because the torn-up soil incentivized conversion to non-forest uses (Hidalgo Ledesma, 2016).

Starting in the early 1980s, there were important improvements in the prospects for forestry. There was the start of a national logging policy with new regulations and standards aimed at increasing production, with more active involvement of forest owners (Jardel, 1986). In the 1990s, in order to stop the advance of deforestation and to secure and expand both environmental services and products from forests, the largest part of governmental support was channelled to the establishment of tree plantations. In this millennium, policies have begun to prioritize sustainable development with funds earmarked for reforestation and/or preservation, especially in areas at risk, and of economic importance, such as mountain areas and their watershed headwaters (Merino Pérez, 2004). Reforestation in SZ picked up pace after regional forestry organizations began to promote the benefits, and after federal support for reforestation and timber tree plantation programmes began to be channelled through municipal government institutions.

After three decades, the resulting extensive reforestation is one of the more noticeable changes in the region's landscape.

> It changed the hillsides, before we had to look at bare rock and mudslides, and now it looks more beautiful here in Tlaquilpa, it looks greener. We already started timber and others have begun to plant trees. Springs that were dry now have water again and now we even have trout.
>
> (Alejandro Salas Romero, Tlaquilpa Municipality, 2012)

Within Mexico, Veracruz is the state that has received the most support for the introduction of forest plantations, accounting for 20% of the total planted area nationwide between 2000 and 2013 (CONAFOR, 2012) and SZ is considered a priority region within the state. In 1994, 599 ha were reforested in SZ; by 2010 this area had increased to 3556 ha (INEGI, 1994, 2010). These figures do not include the many hectares of trees planted by Nahua families without government support.

In the cold zone, most of the planted trees in commercial plantations and reforestation areas are a single species of pine known locally as *ocote* (*Pinus patula*). In the temperate and warm zones, the two main species are mahogany (*Swietenia macrophylla*) and cedar (*Cedrela odorata*). In these zones, apart from the species obtained through the national forestry programmes, peasants are experimenting with the introduction of native timber species, especially white *xochikuawitl* (*Cordia alliodora*) and black *xochikuawitl* (*Cordia megalantha*), as shade trees in coffee plantations (Elizondo Salas, 2015).

Changes and Continuity in the Use and Management of *Milpa* and Trees

Trees are a prominent part of the landscape in SZ, naturally growing or managed under regimes of different intensity in a wide range of land uses; Fig. 4.2 attempts to represent the wide management practices related to trees. At the left are the traditional land uses where diverse tree species are growing as successional phases of *milpa*/*acahuales* or as part of remnants of forest patches. At the right are the most recent strategies for introducing timber trees, generally a reduced number of species. The circle in the middle represents combinations of management practices (traditional and recent ones), where different uses and goals combine or conflict.

The size and number of plots dedicated to timber trees is decided by each family, according to the number and size of parcels they own. Labour is one of the most important factors and this depends on the family composition and the mosaic of activities carried out by the various members. The household's food requirements, and the production costs of the maize and other crops, as well as the need among some families to reserve spaces for grazing sheep and other domestic animals, are also taken into account.

Some timber tree plantations cover a large area, and sometimes belong to a single owner. Others are owned by consortia of peasants who have organized sets of parcels to access government support for introducing timber trees. However, as most of the inhabitants have plots measuring an average of 1–2 ha, trees are more often individually planted in the spaces that are available to the household: home gardens, fallow fields, maize fields and edges. In these spaces, the trees become an integral part of agricultural and domestic uses. During their growth they supply firewood. Thin trunks are used as poles in huts and as props during building construction.

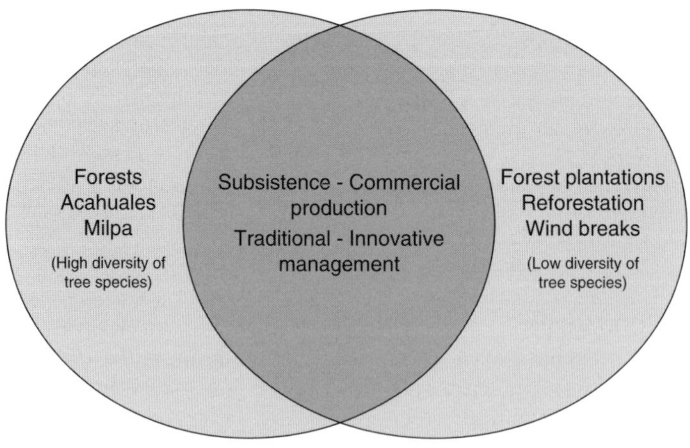

Fig. 4.2. Forest and tree management in the Sierra de Zongolica.

Along property boundaries they function as fences, wind barriers, etc.

In more recent years in the cold zone, pines (*ocote*) may be combined with maize. Usually the two are interspersed and the maize is planted among the pine trees only during the first 5 years of tree growth (while the trees are short enough that they do not shade out the maize). The latest strategy consists of planting pine rows with more separation in such a way that after the first 5 years of maize production, maize is replaced with pasture, with the trees preventing freezing of the grass during winter, enabling year-round grazing by the sheep.

> I lend this field to my neighbors to plant maize and at the same time I plant ocote. They clear the maize field at the same time I clear the ocotes, so everything grows faster and I save labor. I can plant maize and ocote for 5 years for as long as the young trees don't overshadow the maize. Also the ocote is very jealous; when it has grown to the point where it has covered the ground with needles nothing else can grow. For me it is good because I can tend maize and ocote together as little brothers. If I lend out a part of my land where I have planted timber to another peasant who plants maize in between my trees he acts as a guard keeping an eye on my trees.
>
> (Edgar Xocua Antonio, Tequila Municipality, 2014)

In the warm zone, peasants are growing tropical timber and other tree species as shade trees within coffee plantations:

> In the past, in order to plant *milpa* we cut down trees without exploiting the felled trees. Then we started to think about how we could work to get more out of the earth, so we organized and diversified production. So for five years we have planted camedor palm trees, timber trees, banana and coffee. The timber trees take the longest to make a profit. Coffee and camedor palm are medium term. The short term are the bananas, which can take a year and we can eat and sell as we harvest them.
>
> (Jesús García Choncoa, Zongolica Municipality, 2013)

Conclusion

This chapter has explored the importance of *milpa* as part of a traditional knowledge system, its adaptation and innovation and its most recent dynamics related to the introduction of timber trees. Today Nahua diversified forestry and agricultural production systems integrate a wide range of traditional and recent management practices to grow tree species at different management intensities (Fig. 4.2).

From a landscape history perspective, diversified forestry and agroecological systems have evolved as peasants responded to local needs and external social and economic pressures. From previous research and local narratives, we know, for example, that present shaded coffee plantations are located in areas that were devoted to tobacco production in the 19th century (Dzib Aguilar, 1994). Other coffee plantations are located in parcels opened more recently to cultivation by slash-and-burn, especially the ones on the slopes. Rodríguez López and Álvarez Santiago (1992) and Dzib Aguilar (1994) observed during the 1990s that patches of natural forest used for charcoal production were often transformed into *milpa* as the next step in their management. Other parcels are reverting to forest cover after being abandoned due to migration (Cordova Plaza, 2012). In other cases, parcels owned by migrants are being planted with timber trees (Martínez Canales, 2013; Elizondo Salas, 2015). As explained by some peasants from the cold zone, some of the first areas where timber trees were introduced about 20 years ago are now used once again for *milpa*, while other areas have a long history of continuous *milpa* cultivation. This sequential utilization of useful plant species, including trees, corroborates the practical significance of local knowledge as a mode of adaptive management (Berkes *et al.*, 2000). Peasants are exploiting both spatial and temporal opportunities, taking advantage of the large number of plant species and varieties that will grow in a fragmented landscape with different microclimates and soil types, and at different stages of plant succession (Ellis, 1998). In this way the management of *milpa*, in the cold zone of SZ, has constituted the pivot point for agroecological changes at the landscape level.

Forests and trees have always been a fundamental part of subsistence in SZ and are perceived as an important part of Nahua life and identity. Starting in the 1980s, new perceptions of their importance have emerged. Given the historical dispossession of their resources, cultivating trees became a strategy to regain and maintain control over Nahua territory. In addition, peasants are more aware that timber trees can be a commodity. Most of the Nahua have no certainty that they will eventually be able to sell the timber but they consider that these trees are a valuable asset for their children to inherit.

The interest in introducing trees is increasing and is being carried out by Nahua peasants with or without government support. We have observed significant differences between government programmes' goals and local objectives and expectations. Government programmes and extension services measure success in terms of the number of trees planted. On the other hand, Nahua peasants introduce trees to maintain vegetation cover, to control soil erosion and conserve water sources while obtaining different uses from the trees such as firewood and building materials. The downside of the introduction of timber trees can be lowering of biodiversity and the displacement of *milpa*, especially in parcels where migrant owners leave SZ for long periods and devote the land entirely to timber tree plantations.

Preserving the practice of *milpa* in the current context depends on many factors in a complex local and global environment and will be increasingly influenced by decisions regarding the removal, maintenance or introduction of vegetation cover. Micro and macro dynamics are taking place forcing or promoting changes regarding tree management, including government programmes that support the establishment of environmental services, reforestation and conservation areas, especially at the origin of catchment basins of high ecological and economic importance.

A constant in SZ and throughout the Mexican countryside is the multiplicity of activities carried out by rural families for their daily survival. Through mutual support and networking, households create a broad portfolio of activities to generate income and produce various goods and services. In today's globalized economy, peasants no longer depend on their land alone but countless informal or extra-rural activities ranging from migration to integration into long-term employment or temporary commercial activities in their own communities (Hecht *et al.*, 2012).

International migration has very different effects in different households. In some cases, cultivation of *milpa* continues uninterrupted in the hands of women, or labourers paid with remittances. But our experience suggests that migration is a threat to the persistence of the knowledge and practices associated with *milpa*, which depends on intergenerational transmission. The departure of migrant parents usually occurs when children are young (Martínez Canales, 2010; Cordova Plaza, 2012) and they grow up with little or no experience accompanying their parents in their agriculture–forestry activities. At the same time, current school programmes, which are designed at the national level, are decontextualized from the students' local social and natural environments. Within this context, we continue working with Nahua peasants, especially youth, as their enthusiasm to learn their language and agroecological traditional knowledge are key to preserving their territory and biocultural inheritance.

Acknowledgements

The work presented in this chapter was carried out as part of the project 'Community-based forest management and conservation in Central Mexico; building links, networks and capacities' financially supported by the Overbrook Foundation from 2010 to 2015 and conducted in collaboration with teachers-researchers and students from Universidad Veracruzana Intercultural and People and Plants International (www.peopleandplants.org). We thank Mark Stowe for help with polishing our English.

References

Aguilar Gallegos, N., Barrientos Priego, A., Núñez Colin, C. and Nieto Ángel, R. (2007) Distribution and potential areas for collecting of *Persea americana* Mill. Germplasm in Mexico. *Proceedings of the 6th World Avocado Congress*. Vina del Mar, Chile, pp. 12–16.

Aguirre Beltrán, J. (1995) *Cuatro nobles titulados en contienda por la tierra*, Centro de Investigaciones y Estudios Superiores en Antropología Social, México DF, Mexico.

Alatorre Frenk, G., García Campos, H., Hidalgo Ledesma, R., López Binnqüist, C., Negreros-Castillo, P. *et al*. (2015) *Diálogos sobre cuencas, bosques y agua en Veracruz: una propuesta metodológica intercultural*. Universidad Veracruzana, Xalapa, Veracruz, Mexico.

Alexiades, M.N. (2009) The cultural and economic globalization of traditional environmental knowledge systems. In: Heckler, S. (ed.) *Landscape, Process and Power: Re-evaluating Traditional Environmental Knowledge*. Berghahn Books, New York, pp. 68–98.

Álvarez Morales, L. (2013) La milpa, conocimiento, saber y pensamiento indígena en San Andrés Tenejapan. Bachelor thesis, Universidad Veracruzana Intercultural, Veracruz, Mexico.

Araujo Rodríguez, C. and Baez Vázquez, J.E. (1994) *Colecta y caracterización preliminar de cultivares de frijol en la sierra de Zongolica, Ver*. Sexta Reunión Científica del Sector Agropecuario y Forestal del Estado de Veracruz, Veracruz, Mexico.

Bacon, C.M., Getz, C., Kraus, S., Montenegro, M. and Holland, K. (2012) The social dimensions of sustainability and change in diversified farming systems. *Ecology and Society* 17(41). doi: 10.5751/ES-05226-170441.

Báez-Jorge, F. (2000) *Los Oficios de las Diosas (Dialéctica de la religiosidad popular en los grupos indios de México)*, 2nd edn. Universidad Veracruzana, Veracruz, Mexico.

Berkes, F., Colding, J. and Folke, C. (2000) Rediscovery of traditional ecological knowledge as adaptive management. *Ecological Applications* 10, 1251–1262. doi: 10.1890/1051-0761.

Boege, E. (2008) *El patrimonio biocultural de los pueblos indígenas de México: hacia la conservación in situ de la biodiversidad y agrodiversidad en los territorios indígenas*. Instituto Nacional de Antropología e Historia, Comision Nacional para el Desarrollo de los pueblos indigenas, México DF, Mexico.

CONAFOR (2012) *Inventario Nacional Forestal y de Suelos. Informe de Resultados 2004–2009*. Comisión Nacional Forestal, México DF, Mexico.

Córdova Plaza, R. (2012) 'Sin el bosque no queda más que irse': migración internacional entre nahuas de Atlahuilco, Veracruz. *Migraciones Internacionales* 6, 209–241.

Dzib Aguilar, L.A. (1994) La tecnología de cultivo en la sierra de Zongolica, Ver. *Revista de Geografia Agrícola* 20, 35–54.

Early, D.K. (1982) *Café: dependencia y efectos: comunidades nahuas de Zongolica, Ver., en el mercado de Nueva York*. Instituto Nacional Indigenista, México DF, Mexico.

Elizondo Salas, A.C. (2015) Conocimiento campesino de árboles maderables en cafetales diversificados de la zona templada de la Sierra de Zongolica. MSc thesis, Universidad Veracruzana, Xalapa, Veracruz, Mexico.

Ellis, F. (1998) Household strategies and rural livelihood diversification. *The Journal of Development Studies* 35, 1–38. doi: 10.1080/00220389808422553.

Hecht, S., Kandel, S., Morales, A. and Cuéllar, N. (2012) Introduction. Migration and natural resources: comparative case studies in politics, localities and ruralities. In: Hecht, S., Kandel, S. and Morales, A. (eds) *Migration, Livelihoods and Natural Resources*. PRISMA Foundation, International Development Research Centre, Ford Foundation, San Salvador, El Salvador, pp. 15–32.

Hidalgo Ledesma, R. (2016) El campesinado Zongoliqueño en la disputa territorial, saberes territoriales y manejo campesino del monte. MSc thesis, Universidad Autónoma Metropolitana, México DF, Mexico.

INEGI (1994) *Anuario estadístico del Estado de Veracruz*. Instituto Nacional de Estadística, Geografía e Informática, Gobierno del Estado de Veracruz, Mexico.

INEGI (2010) *Anuario estadístico del Estado de Veracruz*. Instituto Nacional de Estadística, Geografía e Informática, Gobierno del Estado de Veracruz, Mexico.

Jardel, E.J. (1986) Efecto de la explotacion forestal en la estructura y regeneracion del bosque de coniferas de la vertiente oriental del Cofre de Perote, Veracruz, Mexico. *Biotica* 11, 247–269.

Larios, C., Casas, A., Vallejo, M., Moreno-Calles, A.I. and Blancas, J. (2013) Plant management and biodiversity conservation in Náhuatl homegardens of the Tehuacán Valley, Mexico. *Journal of Ethnobiology and Ethnomedicine* 9, 1. doi: 10.1186/1746-4269-9-74.

Martínez Canales, L.A. (2010) La dinámica comunitaria vista desde la migración en la sierra nahua de Zongolica, Veracruz: análisis y perspectivas de estudio. *Migraciones Internacionales* 5, 7–37.

Martínez Canales, L.A. (2013) Cultura y economía para la sobrevivencia: procesos y relatos desde el etnoterritorio nahua de Tehuipango, en la sierra de Zongolica, Veracruz. *Anales de Antropología* 47, 73–108.

Merino Pérez, L. (2004) *Conservación o deterioro. El impacto de las políticas públicas en las instituciones comunitarias y en las prácticas de uso de los recursos forestales*,. Secretaria de Medio Ambiente y Recursos Naturales, Instituto Nacional de Ecología, Consejo Civil Mexicano para la Silvicultura Sostenible AC, México DF, Mexico.

Morales Carbajal, C. (2014) Las voces de la montaña: diálogos a través del ritual entre los nahuas de Zongolica, Veracruz. Doctoral thesis, Universidad Nacional Autónoma de México, México DF, Mexico.

Navarro Pérez, L.d.C. and Avendaño Reyes, S. (2002) Flora útil del municipio de Astacinga, Veracruz, México. *Polibotánica*, 67–84.

Ortiz Espejel, B. (1991) Los paisajes naturales de la sierra de Zongolica. In: Boege, E. (ed.) *Cultura, naturaleza y sociedad en la sierra de Zongolica: Una contribución para el aprovechamiento sostenido de los recursos naturales*. Instituto Nacional Indigenista, Centro de Investigaciones y Estudios Superiores en Antropología Social, México DF, Mexico, pp. 9–22.

Pérez Pacheco, A. (1992) Paisaje y aprovechamiento del suelo y la vegetación en Tlaquilpa, Veracruz. MSc thesis, Colegio de Posgraduados de Chapingo, Montecillos, Estado de México, Mexico.

Perfecto, I., Vandermeer, J. and Wright, A. (2009) *Nature's Matrix: Linking Agriculture, Conservation and Food Sovereignty*. Earthscan, London.

Rodríguez López, M.T. (2003) *Ritual, identidad y procesos étnicos en la sierra de Zongolica, Veracruz*. Centro de Investigaciones y Estudios Superiores en Antropología Social, México DF, Mexico.

Rodríguez López, M.T. (2016) Prácticas territoriales y dinámicas migratorias en la sierra de Zongolica, Veracruz. *ULÚA Revista de Historia, Sociedad y Cultura* 25, 69–91.

Rodríguez López, M.T. and Álvarez Santiago, H. (1992) Estrategias productivas entre los nahuas de Zongolica. *La Palabra y el Hombre* 84, 127–144.

Sarukhán, J., Koleff, P., Carabias, J., Soberón, J., Dirzo, R. *et al.* (2009) *Capital natural de México, Síntesis: Conocimiento actual, evaluación y perspectivas de sustentabilidad*. Comisión Nacional para el Conocimiento y Uso de la Biodiversidad, México DF, Mexico.

Toledo, V.M., Ortiz-Espejel, B., Cortés, L., Moguel, P. and Ordoñez, M.D.J. (2003) The multiple use of tropical forests by indigenous peoples in Mexico: a case of adaptive management. *Conservation Ecology* 7, 9. Available at: http://www.ecologyandsociety.org/vol7/iss3/art9/ (accessed 20 September 2016).

Trujillo, L. (2008) Coffee production strategies in a changing rural landscape: a case study in Central Veracruz, Mexico. In: Bacon, C., Méndez, V.E., Gliessman, S.R., Goodman, D. and Fox, J.A. (eds) *Confronting the Coffee Crisis: Fair Trade, Sustainable Livelihoods and Ecosystems in Mexico and Central America*. MIT Press, Cambridge, Massachusetts, pp. 69–98.

Ureta, C., González-Salazar, C., González, E.J., Álvarez-Buylla, E.R. and Martínez-Meyer, E. (2013) Environmental and social factors account for Mexican maize richness and distribution: a data mining approach. *Agriculture, Ecosystems and Environment* 179, 2534. doi: 10.1016/j.agee.2013.06.017.

5 The Contested Space that Local Knowledge Occupies: Understanding the Veterinary Knowledges and Practices of Livestock Farmers in the Eastern Cape Province, South Africa

Andrew Ainslie*

When concepts such as 'the knowledge economy' and 'intellectual property' roll so easily off the tongues of such diverse social actors as politicians, bureaucrats, scientists, the legal fraternity and the advocates and spokespersons for 'indigenous' rights around the world, it is clear that the production and 'management' of knowledge has moved centre stage. As innovations in information technology continue their rapid evolution, there are now arguably more instances of contestations over knowledge management, a situation which points to the ever greater strategic value of certain knowledges.

But a significant product of the 'knowledge economy' is the uncertainty that arises because of the surfeit of (often conflicting) information available via the World Wide Web, radio, television and print media (cf. Beck, 1999). Access to such information allows us all to fundamentally question scientific orthodoxies and accepted 'expert' knowledge systems and to weigh up evidence for credibility, salience and legitimacy (Cash *et al.*, 2003). As a result, what is suddenly more important than ever is the need for criteria by which to evaluate, i.e. make value judgements about information that comes to us 'packaged' as knowledge (Sillitoe, 2010).

The concern in this chapter is that ever increasing rates of technological innovation may distract attention from the contested social and political-economic contexts in which knowledge is produced, reformulated and disseminated (Shepherd, 2010). It is precisely these contexts which shape or deliver the criteria for the evaluation of what might be called the certification of knowledge. The case study presented here demonstrates that both the political-economic context and complex social relationships between bearers/users of different knowledges are important to consider in this regard. This chapter examines how livestock keepers come to hold different registers of basic and specialist local knowledge in areas of their agricultural practice. Attention is focused on the tensions that exist between local knowledge and scientific ideas whilst recognizing that the knowledge that herders have and use is a hybrid of these two idealized forms. It is suggested that such tensions persist in many, perhaps all, farming contexts worldwide.

The chapter examines the local knowledge system implicated in the production and husbanding of livestock, and specifically in relation to the management of tick-borne diseases (TBDs) of cattle. The focus falls on the largely rural Ngqushwa

* E-mail: a.m.ainslie@reading.ac.uk

Municipality (formerly 'Peddie District'), which lies on the eastern seaboard of South Africa and forms part of the Eastern Cape Province (Ngqushwa IDP, 2002; Ainslie, 2005, 2013); see Fig. 5.1. This part of the country is recognized as having a virulent TBD challenge with respect to livestock. To shed light on these issues, the chapter explores some salient components of rural farmers' existing veterinary knowledge and indeed their assumed 'ignorance', which is the flipside of knowledge.

To begin with, it is especially puzzling why, after nearly a century of a very regular, government-sanctioned and (in part at least) scientifically endorsed dipping programme for all cattle, most rural African farmers still fail to make an explicit causal connection between the heavy loads of ticks on their cattle and the prevalence of TBDs like redwater, heartwater and gallsickness. Rather, ticks are seen as a serious irritation in that they attack the cattle, sucking their blood and

Fig. 5.1. Map of the Eastern Cape Province of South Africa.

robbing them of strength. Although this lack of association between ticks and TBDs seems highly unlikely after decade upon decade of an official dipping programme, it is telling that the cattle farmers interviewed did not see redwater or heartwater as a problem, with many saying they had not encountered it in their own herds. This is surprising, since this area is part of the coastline that is characterized by virulent heartwater and redwater which kills cattle in significant numbers every year.

This, it turns out, is not simply a translational issue that could be resolved through dialogue relating to causes and symptoms between practitioners using different or hybrid knowledge systems. At its heart lies a fundamental disconnect about what ticks can and cannot do to a bovine animal. In this chapter, the considerable uncertainty around knowledge relating to animal disease must be understood in relation to: (i) the political context/legacy of South Africa, which over the past century has not only shaped the (communal) livestock health management regimes and limited the agency of cattle farmers, but also, as a system of knowledge, been predicated on weak regimes of trust on the part of farmers in relation to authoritative knowledge pronouncements; (ii) substantive (though inadequately documented and shared) shifts in the scientific understanding of tick-borne livestock disease – see below (Tice *et al.*, 1998; Brown and Gilfoyle, 2007); and finally (iii) the dynamic, intrinsically social character of herders' specific (natural and supernatural) challenges and thus practices for managing stock diseases, ticks and TBDs.

What are Regarded as the Essential Characteristics of Local Knowledge?

Anthropologists have long problematized research on local (or 'indigenous', 'customary', 'traditional', etc.) knowledge (Marchand, 2010, p. S3). More critical work has focused on the hybridity or 'admixtures' of knowledge and cultural practice (Agrawal, 1995, 2002). Indeed, in constructing his model for the analysis of knowledge, Barth (2002) avoided the increasingly sterile dichotomy of scientific versus indigenous knowledge. Attention is now regularly paid to the various ways in which both scientific knowledge and folk knowledge come to be integrated in people's existing epistemologies and knowledge pathways (Brown *et al.*, 2013). Sillitoe (1998) made the important point that much of this mixing and learning is non-verbal: people 'transfer much knowledge between generations by tradition learnt and communicated through practical experience and are not familiar with trying to express everything they know in words… Knowledge is passed on by informed experience and practical demonstration; more often shown than articulated, it is as much skill as concept' (Sillitoe, 1998, p. 229).

What is clear is that if knowledge is to serve people in the ever changing circumstances of their daily lives, it must be constantly reconfigured, updated and refreshed, whether through tacit practice or more conscious, formal learning. But what is it about the provenance and method/media of transmission of new knowledge, improvisations and technologies that allow some types/modes to be assimilated into people's existing knowledge 'systems' and their daily repertoires while others are rejected? In his influential paper, Barth (2002) insightfully argued that knowledge consists of three aspects.

Firstly, it exists as a 'corpus of substantive assertions and ideas about aspects of the world that rests partly on valid inference, notably through believing what the people we trust tell us they know' (Barth, 2002, p. 3). He pointed out that in analysing knowledge systems, we tend to focus on 'generalisation, consistency, and a logical coherence' when in reality, knowledge is quite rarely so systematic (Barth, 2002, p. 7). Indeed, other research has shown that we all get by with a surprisingly large degree of what social psychologists and economists refer to as 'cognitive dissonance' (see for instance, Akerlof and Dickens, 1982).

Secondly, knowledge 'must be instantiated and communicated in one or several media as a series of partial representations in words, symbols, pointing gestures, actions' (Barth, 2002, p. 3).

Thirdly and crucially, knowledge is 'distributed, communicated, employed and transmitted within a series of instituted social relations… [o]f trust and identification, and instituted authority positions of power and disempowerment' (Barth, 2002, p. 3).

Barth further observed that people within a single community may participate in different ways and modes in multiple social knowledges. In addition, he noted that 'branches' of knowledge are not evenly distributed across groups of people (cf. Sillitoe, 1998, p. 232). Warren (1998, p. 244) expanded helpfully on this notion of multiple social knowledges, by distinguishing between three types of knowledge for any domain: '*basic core knowledge* is possessed by virtually all members of a community and provides the basis for communication on a given topic; *shared knowledge* that expands on the core knowledge and allows persons occupying related occupational niches (like herders) to communicate using a more nuanced vocabulary and conceptual apparatus and *specialised knowledge* exists within an occupational niche that most others in the community do not require' (italics added). This is a useful differentiation of knowledge types to which I shall return below in relation to the practices of cattle herders in rural South Africa.

For Barth, while the processes that underpin these distributions of knowledge are of considerable interest, it is the agency of 'the knowers – the people who hold, learn, produce, and apply knowledge in their various activities and lives' – that are central in his analysis (Barth, 2002, p. 3). Critically, it is in the social contexts in which people interpret and act in the world that the all-important criteria of validity for the knowledge (i.e. the knowledge about knowledge) that people hold are generated (cf. Sillitoe, 2010). In some social settings, they have latitude for improvisation and innovation; in others, they have virtually none.

Sillitoe (2010) usefully extended in several ways Barth's analysis of how knowledge is transmitted through time and beyond a specific social grouping. In particular, whereas Barth alludes to the issue of trust in the acts of validating inferential knowledge, it is this aspect that is central to Sillitoe's argument. He showed that the Wola people of Papua New Guinea, like other peoples (such as the indigenous Australians and Quecha-speakers in the Andes of Latin America), have an abiding concern – and one that is embedded in their complex language – for validating and assessing the trustworthiness of any communication. Given that they live in a stateless polity, the Wola have no recourse to an authority (such as that provided by state-endorsed education systems), which can 'standardise what is known and can adjudicate when persons disagree and settle who is right' (Sillitoe, 2010, p. 22; cf. Harris, 1997, pp. 113–114). Foreign as it may seem to our sensibilities, a social context and a lived reality in which there is no recognition of experts and their claims to superior knowledge is thus the context for a constant interrogation of people's knowledge statements. This forms an integral and meaningful part of Wola everyday speech. Is there a parallel in the context of the rural Eastern Cape?

The Role of Ignorance – the Flipside of Local Knowledge

In its crudest sense, 'ignorance' (Kirsch and Dilley, 2015) of formal, rational, cause-and-effect science on the part of the rural Xhosa-speaking cattle herders in the Eastern Cape, as elsewhere in rural South Africa, rests on the cruel legacy of colonialism and apartheid's racist, impoverished 'Bantu Education' system (Nasson and Samuel, 1990). Historically, few African people managed to secure more than a primary education. Those who did were exposed to the didacticism of the few 'Mission' schools that dotted the rural countryside. It is only in the past 30 years that a significant minority have had the opportunity to attend universities and other tertiary institutions, and even then the general state of rural primary and secondary schooling has remained abysmal: as recently as 2005, the mean number of years of formal schooling for adults in Ngqushwa Municipality was a mere 6 years. Functional literacy stood at a low 55% across the board and a paltry 13% of rural residents had 12 years of secondary schooling or more.

So 'ignorance' (understood as being poorly uneducated in the formal sense) exists here, as it does in many of the impoverished areas of South Africa. The migration to cities and towns of those fortunate enough to complete post-secondary education, leaving the bucolic rural life behind, further reinforces the rural educational deficit.

An important consequence of this is that those who have progressed through the formal higher education system – for our purposes in the agricultural sciences – have consistently encountered a denigration of their cultural experiences and their local practical knowledge of agriculture in favour of a textbook-based 'scientific' model (in, for instance, plant and animal diseases and their control) of learning and knowing. They have been force-fed via textbooks and study materials an alienating emphasis on economic efficiency, using the individual profit-oriented cattle farmer as the model for what a 'modern, commercial farmer' should resemble. This approach has undermined their confidence in their 'core' local knowledge and, with it, their agency and shared practical knowledge of bovine management and disease control in a village context. They have repeatedly been told that their prior local knowledge was a form of ignorance that lacked scientific certification and should be wholly abandoned.

In addition, their new learning has driven a wedge between the knowledge systems and practices of less educated men and the young men with tertiary qualifications in agricultural sciences. Many of these educated young men continue to keep cattle at their rural homes and a sizeable number of them are employed in technical and extension positions in the Department of Agriculture's Animal Production and Animal Health teams (Modisane, 2009; Beinart and Brown, 2013; Brown *et al.*, 2013).

There is a third layer of 'ignorance' in the form of a significant de-skilling (Stone, 2007) over time, which rests on other parameters: the long years that most rural men spend working as migrants, often down the mines or in industry, in distant townships and cities, has a telling impact on their lives and agency. Many of the older men have spent decades away from their home village, returning only in December or sporadically when their employment contracts expires or some family crisis require their presence in the rural village. Some have left small herds of cattle in the care of wives, young sons or elderly kinsmen, learning of the fortunes of their cattle through letters from home or verbal messages passed along by other migrants to the city or mining compound.

Many of them have never lost their boyhood connection to, and basic knowledge of, rural environments that they acquired as herders of cattle. In particular, they have valuable local knowledge about where to drive their herds of cattle in times of drought or when the rangeland is freshly burnt and grazing is low (Bennett *et al.*, 2010, 2013). They have also inherited, from their fathers and grandfathers, long-standing relationships with kinsmen in other parts of the district with whom they can exchange animals when the consecutive years of low rainfall threaten their animals. But in their years of absence from the village and in their retirement, their store of dynamic, adapting local knowledge, in this case specifically their ability to deal with bovine diseases, has tended to ossify over time.

A fourth perspective on the way ignorance 'works' as a key part of local knowledge that differs from a scientific causality relates to people's ways of inferential knowing and acting that specifically invoke the supernatural (Dilley, 2010; Ainslie, 2014). In this category, beliefs in both the positive supernatural power of ancestor intercession and, to a lesser extent, clan totems, and the negative supernatural power of witchcraft, albeit both strongly intermeshed with Christian beliefs in this area, inform and shape many Xhosa-speaking people's daily practice and understandings of causality. This is the case because cattle feature centrally in Xhosa ritual practice, including slaughtering for the ancestors, and as a symbol of the homestead. Arriving at explanations of both common livestock diseases and extraordinary events, for instance why a particular cow is struck by lightning, frequently involves invoking the supernatural. Moreover, the process of rationalizing between the possible causes to arrive at a solitary explanation for a particular event (such as the unexplained illness or death of a prized heifer) is seldom final or unequivocal, certainly not in relation to the observable causes. Rather, it remains open to a protracted social dialogue and a process of assessing the efficacy of the various ritual actions to counter the misfortune.

Is Local Knowledge about Tick-Borne Diseases a 'Specialized' Knowledge?

It is immediately clear that, in the villages of Ngqushwa Municipality, animal health knowledge is not a 'shared' knowledge that all or even most livestock owners possess in equal measure

(Masika *et al.*, 1997). While all these farmers both 'inject' (with biomedical veterinary products) and '*seza*' (treat with 'traditional' Xhosa remedies using decoctions of indigenous plants) their animals as part of a disease preventive regime (Dold and Cocks, 2001; Moyo and Masika, 2009), they are often reluctant to share knowledge about specific medicines and remedies with each other. Local cattle farmer, Dumisa, mentioned his uncle Bilman who he said was a 'cruel' man. Bilman had good medicine: he knew how to cure cattle using *amayeza yesixhosa* (Xhosa medicine). Dumisa had tried to soften him up, by doing favours for him and hoping Bilman would help him out with his cattle. But Bilman did not like to share his knowledge with anyone. He treated his own cattle with a mixture that he made himself and his cows were always very fat and each cow produced a calf every year. Dumisa explained, 'When you ask him – even as a close relative – to give you some of the same medicine, he either refuses or does so reluctantly and you are never sure if the stuff he gives you is the full-strength medicine or just a watered down version.' He went on to say that when he used this medicine on his herd, it did not have the same effect as it did on Bilman's own herd.

However, with his elderly female neighbour, Ngenelwa, Dumisa enjoyed more success: inside the entrance to Dumisa's cattle byre he kept a bottle of *mayeza* known as *magonsana*. It was made with the root of a plant that Ngenelwa knew where to find in the forest. He asked her for it and she went to find some and gave it to him. The problem he had was that one of his animals was stabbed by another animal in the *kraal* (cattle enclosure) after being pinned to the inside of the *kraal*. Dumisa noticed that when it urinated, the urine was red, i.e. was mixed with blood. He was worried that it was also looking weak. He grated the *magonsana* root and then boiled it in water and decanted it into a 2 l bottle. Dumisa planned to administer it to the affected cow over 2 days in two 750 ml doses to effectively remedy its problem. According to Dumisa, this mixture would clean the animal's blood and its internal system and within a few days it should stop urinating blood and be sturdy again.

Matshoba (a community animal health worker) revealed that he knew certain *amayeza yesixhosa* remedies that his father had taught him, but these were his secret because if he gave them to others for their sick animals and the animal died, they would 'come back' to him with questions. He had shared some remedies with his wife's sister in a nearby village when her goats needed medicine and it worked fine. He had also helped some men in his own village, but he did not like doing it. Also, he felt that you should not mix 'a needle' (purchased medicine) with 'a bottle' (*amayeza*), but his explanation of his own practice contradicted this. He used Xhosa *mayeza* for the tick-borne disease known as 'redwater', but when a cow was already sick, a needle with Hi-tet was more effective, because it worked faster. Asked to explain this, he said that *amayeza* were better for prevention than for curing very sick animals. When someone stabbed his young ox at night, it was bleeding internally (*isivubeko* – an internal wound), which he deduced because it was excreting blood. He used a needle and injected it with Hi-TET (Bayer) and also made a Xhosa medicine called *umhlavuthwa* 'to clean it out inside'.

One of Xolani's animals had an *intambo* (a leather strop tied around its neck). He said it got rid of the *intsumpa* (warts) on the head, and also around the vagina and ribs of a cow. He could not explain how it worked, but it had been advised to him long ago and it definitely worked. He said that his neighbour's ox had had huge *intsumpa* all over its dewlap. Xolani had advised him to use *intambo*, but the man also bought a salt solution to rub on the *intsumpa*. The ox was clean now, but it was not clear which thing did the trick, in this case. Either one or both working together could have worked, he said.

To treat contagious abortion, Nzaba gave his cows *amayeza* that he got from a man called Jovose in a nearby village. He did not provide any further information about this medicine and, typical of him and others, seemed to be uninterested in it except for its efficacy. When Nzaba's cattle were sick, he would ask Xolani for help with providing the medicine and injections. He mentioned getting Hi-tet (Bayer) from Xolani. He also gave Xolani money to buy acaricide (Triatix, CKL) for him in a nearby town and he used Xolani's knapsack and pump to spray his cattle.

Xolani told me he lost six cattle in 2004–2005. There was plenty of *idakada* (liverfluke) in the village at that time. Other people were also losing cattle, so it was reassuring not to be the only person to lose so many cattle and he did not suspect anything 'bad' directed at him personally.

Going to a traditional healer to ascertain the supernatural cause was not an option for Xolani, because 'you don't know which of their many stories to believe: they will tell you about things that have happened and some things that will [read 'might'] happen to you. You are easily confused by such nonsense'. When his six cattle died, Xolani was away on police deployment. He asked his uncle, Nqoyi, to come around to check the cattle and to do post-mortems on the animals. His younger adult brother Zamuxolo was still in the village, so he was relaxed as he knew Zamuxolo would be part of this procedure.

For a difficult birth, Fakati, a man with nearly 100 head of cattle, always tried to assist with the birth himself, rather than call Mr Gobani, the accepted expert in the village for dealing with difficult births. In any case, he claimed that Mr Gobani was 'an old man now and can't go running all over the village to sort these things out'. Actually, Fakati did not trust anyone other than his four sons to help him with his especially large herd that had caused some disquiet in a village where only half the homesteads had any cattle and average cattle holdings were around five head.

Another herder, Themba, was 38 years old and an agricultural science graduate from the University of Fort Hare and had an MSc in Rangeland Science from another highly rated university. He held a senior management position in a government agency in the large regional city of East London. He was a very successful and very keen 'scientific' cattle farmer in the often trying communal environment that is the rural Ngqushwa Municipality. By this, I mean that his herd management was underpinned by his academic training, sustained empirical observation and science-based experimentation.

This background made Themba quick to dismiss as 'ignorance' much of what constituted the veterinary regimes of the other cattle farmers around him. Interestingly, he was as quick to dismiss scientific 'textbook stuff' as nonsense if it did not square with his experience and local knowledge of village husbandry conditions. While he acknowledged the efficacy of local remedies for some livestock ailments, he had sufficient money to consistently buy veterinary medicines for tick-borne and other serious livestock diseases. Since he felt highly aggrieved when any of his animals succumbed to disease, he maintained a rigorous and thus expensive animal health regime that was driven by a strict timetable from which he hardly deviated. But this was only for the benefit of his own substantial herd of 40 cattle and did not benefit the cattle owned by others in the village.

To maintain this regime, he visited his home in rural Peddie every weekend to oversee, inspect and personally dose his animals. Over recent years, he had tried to impart some of his knowledge to the other stock owners in the village, particularly in respect of tick management, which needs a collective effort if it is to have any lasting impact. He had had very limited success in sharing his formal agricultural knowledge and, indeed, practical experience with the older men of the village. He reasoned that they resented his runaway success with expanding his herd and were jealous of his successful career. Interviews with the older men in the village suggested that they did not like his independent-minded nature, which they took to be highly disrespectful of them and their seniority in terms of age. They would also have valued a financial contribution from him for the purchase of the expensive acaricide treatments that were used in the communal diptank. Despite frequent sanctions against him for not acting in unison with the rest of the livestock owners, Themba had consistently not dipped his cattle with the other villagers at the village diptank, because he regarded this as 'a waste of time' as the facility was poorly managed. Instead, for control of tick-borne diseases he preferred to spray his cattle himself.

Xolani (mentioned above) was an age-mate of Themba and was employed as a policeman outside the village. Although he had grown up in the village and had his own store of bovine knowledge, he said he had discussions with his trusted friend Themba about how to get the best from his herd of cattle. From Themba, Xolani learnt that it was good to have a tick-load on the animals so as to build up their natural immunities to tick-borne diseases. He admitted that he did not quite understand how this worked, but Themba convinced him that if you dip and spray all the time, you do not allow the animals to become strong on their own. So even though the sight of ticks on the animals was worrying, Xolani now tried to fight the urge to keep them totally clean. When he sprayed his animals in the cattle enclosure, the ticks fell off right there, but no sooner were the animals back in the fields than they were covered with ticks again. This was a stressful situation that called for a strong nerve. Nevertheless, Xolani's herd was growing steadily, not least because he had the wherewithal to purchase veterinary medicines as and when required.

What this section has tried to demonstrate is the complex and unevenly distributed hybrid knowledge that exists in relation to bovine diseases and their treatment. Trust and thus certification of knowledge claims emerge as particularly challenging for herders.

The Political Economy and Contestations around 'Local Knowledge'

So much for the endogenous parameters of people's local knowledge and its gaps. At the level of political-economy, these rural African livestock keepers have long had a generalized mistrust of the 'white' (i.e. European) proponents of veterinary science. This is historically linked to government programmes to control livestock numbers through culling and taxation that used the cattle dipping and inoculation programmes as the government's main sources of contact with African farmers (Bundy, 1987). In fact, their scepticism extends to veterinarians, cattle speculators, veterinary medicine retailers, 'white' farmers selling cattle to them and indeed textbooks and pamphlets on animal health that are aimed at them. What this mistrust does is help to underpin a Xhosa agropastoral identity and to feed an intentional disregard ('ignorance') for the official 'scientific' view of the causality of bovine diseases. By largely discounting the husbandry and veterinary advice from educated others who are typically regarded as untrustworthy and exploitative, most African cattle farmers in the Eastern Cape constantly replay a script where the critical ingredient in all successful knowledge exchanges, i.e. trust, remains at a near-permanent low level (Freire, 1970; Sillitoe, 2010).

It might well be asked if it really matters what these livestock farmers know or do about tick-borne animal diseases. In fact, what they know is particularly important in present-day South Africa, because African herders now hold the country's largest and still steadily increasing proportion of the 'national herds' of livestock, i.e. cattle, goats and sheep (Ainslie, 2002, 2013; Palmer and Ainslie, 2007). In the Eastern Cape Province alone, they are estimated to hold over 3 million head of cattle (ECDA, 2005). Through state land reform and private land purchase, African farmers are slowly acquiring more land on which to run their livestock (Beinart and Brown, 2013).

Ticks pose a particular challenge to successful animal husbandry in these areas, not only because they transmit debilitating bovine diseases, but also because heavy tick-loads damage the reproductive capacity of cows (by damaging their urinary tracts, udders and teats) and bulls (damaging their genitals) (Masika *et al.*, 1997; Ndhlovu *et al.*, 2009). Animals infected by diseases, whether overtly or sub-clinically, exhibit greater morbidity, are listless and therefore both less productive and less fertile (Tice *et al.*, 1998; Minjauw and McLeod, 2003). Farmers whose objective is to grow the size of their individual herds know full well that animals in poor health are not fertile and they thus implement practices to counter this situation (Hlatshwayo and Mbati, 2005).

For the reasons recited above, both the knowledge and practices of rural African livestock farmers are characterized by high levels of uncertainty (Ainslie, 2013). In essence, this uncertainty is around who might be trusted to provide local veterinary knowledge and, furthermore, which (expensive) biomedical treatments actually work in the local context at the current conjuncture (see below) to keep animals in reasonable reproductive health.

Somewhat counter-intuitively, it is the case that this uncertainty is more pronounced now under a more benevolent African National Congress government than in the past when the apartheid state intervened more definitively in this sector, but also carried most of the financial and institutional burden of animal disease control, including the costly dipping programme to control tick-borne diseases (Beinart, 2007; Brown and Gilfoyle, 2007; Beinart and Brown, 2013). This is because the political transition from apartheid to a democratic order in 1994 was followed by a period of fiscal belt-tightening. In the process, the provincial administrations of the nine new provinces (including the Eastern

Cape Province) had to bear the financial and institutional costs of the provision of veterinary/animal health services. When this system crumbled in the late 1990s, African livestock owners themselves had to step into the breach and take on far greater responsibility, both individually and jointly for dealing with livestock diseases than at any time in the past 100 years (Brown *et al.*, 2013).

What has happened in the past 20 years is that farmers either practise very minimalist tick control or are seemingly indifferent to this aspect of bovine management and adopt a 'do-nothing' approach. However, because all their herds of cattle run together on the common rangelands, and are dipped with communally purchased acaricides in common dipping facilities at village level, all farmers 'share' the tick problem and hence the constant burden of tick infestation. To be effective, managing the ticks and the TBDs must be tackled cooperatively, but cattle farmers are individually and collectively uncertain about control strategies and struggle to find adequate and affordable solutions to the challenges that ticks pose.

It is now well known that very limited scientific research was conducted among cattle herds in the reserve/bantustan regions over the decades (Spickett *et al.*, 2007; Beinart and Brown, 2013). This is especially so for the latter half of the 20th century and in relation to animal health. Nevertheless, scientific prescriptions for state intervention in these areas have been required in support of the financial outlay necessary to support the dipping programme. These prescriptions have of course been subject to revisions over time and such revisions have come essentially from two sources: (i) scientific advances and refinements that are driven by the findings of new and innovative veterinary research in the service of the nearby 'white' commercial farming sector; and (ii) the application and modification of scientific advice by farmers in this same sector, in response to a range of exigencies and changes in the nature of their farming enterprises. Scientific prescriptions for the control of TBDs have become increasingly more fluid and complex in the light of, among other things, tick resistance to acaricides and changing tick distributions – in part driven by the increased circulation of wild ungulates as nature reserves were established in a number of sites over the past 50 years – and thus of bovine disease patterns in across the countryside (Allsopp, 2009).

Significantly, every revision/advance of scientific knowledge and the related management prescriptions filters imperfectly through to rural African livestock farmers via a variety of formal and informal channels. These include the state veterinary service with its animal health technicians and until the late 1990s when they were retrenched, the dipping foremen at each village diptank. The channels of information also include the mostly ad hoc interactions African livestock farmers have with commercial retailers of veterinary products, through their dealings with white farmers from whom they buy heifers and young bullocks, and with cattle buyers and speculators who visit the villages or whom they encounter at the intermittent district livestock sales. Each of these channels has specific interests and tensions which adhere to it on the part of the provider of information and of the receiver.

The key point is that trust features centrally in the practices of cattle farmers in rural Peddie. Farmers are often reluctant to share knowledge about specific medicines and remedies with each other. A fear of bewitchment by others features in at least some people's animal health routines. This is the realm of more specialist knowledge. When Makhulundaba was asked whether he believes in securing supernatural assistance to keep his cattle healthy, he replied that, since like others here he believes in his ancestors, he tells his ancestors about any new animals that he put in the kraal. 'The ancestors will look after them for us, because that is what the *amaXhosa* believe,' he explained.

Tellingly, diagnosing the cause of an animal's death begins with (frequently vague) descriptions of the animal's behaviour in the lead up to the decision to slaughter it. It continues during the slaughter, with the close examination of various organs, particularly the gall-bladder, spleen, lungs, liver and heart. The most senior men present must then reach a decision, based on experience, about what caused the death and whether it is safe to consume the meat or not. In the case of anthrax or bovine tuberculosis they might suggest desisting, but this is not a certain outcome. In any event, the diagnosis when performed so publicly can never involve a full account of the suspected causes of the animal's demise.

Conclusions

Relations between cattle farmers and the state (and its various agents) have involved a dynamic but essentially hostile interplay over more than a century, manifest in the tensions and feigned indifference in relation to the trustworthiness and efficacy of scientific and local knowledge. None of these various knowledges in isolation holds the 'answer' to the problem of managing tick-borne diseases. Both 'sides' face serious constraints. Financially constrained African farmers experimenting with all manner of local remedies and thinking through hybrid cultural notions of disease aetiology continue to mistrust state veterinary officials and scientists, who make very occasional, isolated and thus largely unsystematic forays into rural areas to do sporadic surveys into these complex questions.

Given the state's fluctuating but still significant financial outlay on dipping materials, expert veterinary input remains important to plot just where Ngqushwa Municipality and neighbouring areas lie along a gradient of tick resistance to acaricides, in respect of 'endemic stability', the longer-term prevalence/incidence of TBDs and indeed the other major ailments of cattle (cf. Orzech and Nichter, 2008). But it seems this will only be possible after intensive and sustained veterinary research in the area, which, given the present resource constraints, seems unlikely to happen.

Nor would 'improving the science' be a simple undertaking; the seasonal and inter-annual variations in climate (and specifically rainfall) are serious considerations when it comes to understanding any changes in the condition of cattle, which in turn influences their health status and disease susceptibility (Allsopp, 2009). Such variability, coupled with both upswings and downturns in the economic cycle and the on-again, off-again nature of the state's animal health programme has an impact on what aspects of animal health are (i) absolutely necessary in an 'objective', practical sense and (ii) affordable for the socio-economically differentiated cattle owners in the villages of Ngqushwa Municipality.

The decline of the state-sponsored dipping programme has meant that for people who do not have much disposable income and who, on the whole, are getting poorer, the privatization of animal health has been challenging. For the moment, livestock keepers rely on what they have learned from their fathers and elders, what they can glean from their friends and kinsmen and what they can learn from the suppliers of veterinary medicines.

To conclude, local knowledge about animal health and specifically tick-borne diseases is not shared across the population of rural cattle owners. People have asymmetrical access to this knowledge and the nature of this asymmetry relates directly to the social and political-economic context in which cattle farmers are located. As Sillitoe has identified in another context, trust emerges as the key ingredient (and in this case, deficit) permeating all the relationships related to the management of animal health. In the absence of trust, the certification of knowledge that is critical to learning, and thus to effective action, remains undeveloped.

Acknowledgements

Fieldwork was conducted between 2009 and 2012, as part of a case study contributed in support of an ESRC-funded project, *Social History of Veterinary Medicine in South Africa since 1930*, led by Karen Brown and William Beinart at the University of Oxford. Financial support from this grant and the collegial support of Brown and Beinart are gratefully acknowledged.

References

Ainslie, A. (2002) Cattle numbers. In: Ainslie, A. (ed.) *Cattle Ownership and Production in the Communal Areas of the Eastern Cape, South Africa. Research Report no.10*. Programme for Land and Agrarian Studies. University of the Western Cape, Cape Town, pp. 38–45.

Ainslie, A. (2005) Farming cattle, cultivating relationships. Cattle ownership and cultural politics in Peddie District, Eastern Cape. *Social Dynamics* 31(1), 129–156.

Ainslie, A. (2013) The socio-cultural contexts and meanings associated with livestock keeping in rural South Africa. *African Journal of Range and Forage Science* 30(1), 1–4.

Ainslie, A. (2014) Harnessing the ancestors: mutuality, uncertainty and ritual practice in the Eastern Cape Province, South Africa. *Africa* 84(4), 530–552.

Agrawal, A. (1995) Dismantling the divide between indigenous and scientific knowledge. *Development and Change* 26, 413–439.

Agrawal, A. (2002) Indigenous knowledge and the politics of classification. *International Social Science Journal* 54(173), 287–297.

Akerlof, G.A. and Dickens, W.T. (1982) The economic consequences of cognitive dissonance. *The American Economic Review* 72(3), 307–319.

Allsopp, B.A. (2009) Trends in the control of heartwater. *Onderstepoort Journal of Veterinary Research* 76, 81–88.

Barth, F. (2002) An anthropology of knowledge. *Current Anthropology* 43(1), 1–18.

Beck, U. (1999) *World Risk Society*. Blackwell, Oxford, UK.

Beinart, W. (2007) Transhumance, animal diseases and environment in the Cape, South Africa. *South African Historical Journal* 58(1), 17–41.

Beinart, W. and Brown, K. (2013) *African Local Knowledge and Livestock Health. Diseases and Treatments in South Africa.* James Currey, Oxford, and Wits University Press, Johannesburg.

Bennett, J., Ainslie, A. and Davis, J. (2010) Fenced in. Common property struggles in the management of communal rangelands in Central Eastern Cape Province, South Africa. *Land Use Policy* 27(2), 340–350.

Bennett, J., Ainslie, A. and Davis, J. (2013) Contested institutions? Traditional authorities and land access and control in communal areas of the Eastern Cape Province, South Africa. *Land Use Policy* 32, 27–38.

Brown, K. and Gilfoyle, D. (2007) Introduction: Livestock diseases and veterinary science in Southern Africa. *South African Historical Journal* 58(1), 2–16.

Brown, K., Ainslie, A. and Beinart W. (2013) Animal disease and the limits of local knowledge: dealing with ticks and tick-borne diseases in South Africa. *Journal of the Royal Anthropological Institute* 19(2), 319–337.

Bundy, C. (1987) 'We don't want your rain, we won't dip': popular opposition, collaboration and social control in the anti-dipping movement 1908–16. In: Beinart, W. and Bundy, C. *Hidden Struggles in Rural South Africa*. James Currey, London, pp. 191–122.

Cash, D.W., Clark, W.C., Alcock, F., Dickson, N.M., Eckley, N. *et al.* (2003) Knowledge systems for sustainable development. *Proceedings of the National Academy of Sciences* 100(14), 8086–8091.

Dilley, R. (2010) Reflections on knowledge practices and the problem of ignorance. *Special issue of the Journal of the Royal Anthropological Institute (JRAI)*, 'Making Knowledge', S176–S192.

Dold, A.P. and Cocks, M.L. (2001) Traditional veterinary medicine in the Alice district of the Eastern Cape Province, South Africa. *South African Journal of Science* 97, 375–379.

ECDA (2005) *ECDA Strategic Plan for 2006*. Eastern Cape Department of Agriculture, Bhisho.

Freire, P. (1970) *Pedagogy of the Oppressed*. Penguin Books, London [1996 revised edition, translated by Myra Bergman Ramos].

Harris, M. (1997) *Culture, People, Nature. An Introduction to General Anthropology*. Longman, New York.

Hlatshwayo, M. and Mbati, P.A. (2005) A survey of tick control methods used by resource-poor farmers in the Qwa-Qwa area of the Eastern free state province, South Africa. *Onderstepoort Journal of Veterinary Research* 72, 245–249.

Kirsch, T.G. and Dilley, R. (2015) Regimes of ignorance. An introduction. In: Kirsch, T.G. and Dilley, R. (eds) *Regimes of Ignorance. Anthropological Perspectives on the Production and Reproduction of Non-knowledge*. Berghahn, New York, pp. 1–28.

Marchand, T.H.J. (ed.) (2010) Making knowledge: explorations of the indissoluble relation between mind, bodies, and environment. *Journal of the Royal Anthropological Institute* (N.S.) Special Issue, S1–S21.

Masika, P.J., Sonandi, A. and Van Averbeke, W. (1997) Tick control by small-scale farmers in the central Eastern Cape Province, South Africa. *Journal of the South African Veterinary Association* 68, 45–48.

Minjauw, B. and McLeod, A. (2003) Tick-borne diseases and poverty. The impact of ticks and tick-borne diseases on the livelihood of small-scale and marginal livestock owners in India and Eastern and Southern Africa. Research report, DFID Animal Health Programme, Centre for Tropical Veterinary Medicine, University of Edinburgh, UK.

Modisane, B.M. (2009) Field services: eradication and control of animal diseases. *Onderstepoort Journal of Veterinary Research* 76, 115–121.

Moyo, B. and Masika, P.J. (2009) Tick control methods used by resource-limited farmers and the effect of ticks on cattle in the rural areas of the Eastern Cape Province, South Africa. *Tropical Animal Health and Production* 41, 517–523.

Nasson, B. and Samuel, J. (eds) (1990) *Education: From Poverty to Liberty*. David Philip, Cape Town.

Ndhlovu, D.N., Makaya, P.V. and Penzhorn B.L. (2009) Tick infestation, and udder and teat damage in selected cattle herds of Matabeleland South, Zimbabwe. *Onderstepoort Journal of Veterinary Research* 76, 235–248.

Ngqushwa IDP (2002) Ngqushwa Municipality Integrated Development Plan. Unpublished Report. Peddie, Ngqushwa Municipality, Eastern Cape, South Africa.

Orzech, K.M. and Nichter, M. (2008) From resilience to resistance: political ecological lessons from antibiotic and pesticide resistance. *Annual Review of Anthropology* 37, 267–282.

Palmer, A.R. and Ainslie, A. (2007) Using rain-use efficiency to explore livestock production trends in rangelands in the Transkei, South Africa. *African Journal of Range and Forage Science* 24(1), 43–49.

Shepherd, C.J. (2010) Mobilizing local knowledge and asserting culture. The cultural politics of *in situ* conservation of agricultural biodiversity. *Current Anthropology* 51(5), 629–646.

Sillitoe, P. (1998). The development of indigenous knowledge: a new applied anthropology. *Current Anthropology* 39(2), 223–252.

Sillitoe, P. (2010) Trust in development: some implications of knowing in indigenous knowledge. *Journal of the Royal Anthropological Institute* 16(1), 12–30.

Spickett, A.M., Van Der Merwe, D. and Matthee, O. (2007) The effect of orally administered *Aloe marlothii* leaves on *Boophilus decoloratus* tick burdens on cattle. *Experimental and Applied Acarology* 41(1–2), 139–146.

Stone, G.D. (2007) Agricultural deskilling and the spread of genetically modified cotton in Warangal. *Current Anthropology* 48(1), 67–87.

Tice, G.A., Bryson, N.R., Stewart, C.G., du Plessis, B. and de Waal, D.T. (1998) The absence of clinical disease in cattle in communal grazing areas where farmers are changing from an intensive dipping programme to one of endemic stability to tick-borne diseases. *Onderstepoort Journal of Veterinary Research* 65, 169–175.

Warren, D.M. (1998) Comments on Sillitoe's 'the development of indigenous knowledge: a new applied anthropology.' *Current Anthropology* 39(2), 244–245.

6 Integrating Indigenous Knowledge for Technology Adoption in Agriculture

Florencia G. Palis*

Technology adoption is a necessary foundation to achieve impact in agricultural research. Agricultural technologies as seen and practised on the ground by farmers have always been a concern as they remain less realized to this day. The problem regarding the adoption of any technology is its usual tie-up with economics as its key driver. However, to ensure technology adoption it is necessary to consider not only the profitability of the technology but the social and cultural dimensions as well (Vanclay, 2004; Palis *et al.*, 2007). This is because the act of adoption is a deliberate decision made after considering a wide range of issues, done within a social and cultural context in which different individuals with accumulated knowledge and experience may interact and influence the decision. Hence, the associated social and cultural complexities need to be factored into the adoption process.

In the case of the Philippines and most Asian countries, the 2008 food crisis has led many of their political leaders to invest more in rice agriculture programmes, including rice research. This is because rice is the staple food of Asia, a continent that accounts for more than half of the world's population, and its cultivation a common source of livelihood of most Asian farmers (GRiSP, 2013). Aside from the economic, political and nutritional significance of the rice crop in Asia, it has high significance in the social and cultural lives of the people, deeply woven into the fabric of Asian cultures and civilization. This is reflected in their cosmology, language, community structure, rituals, songs, material culture, local knowledge and perception of the landscape, among other aspects (Conklin *et al.*, 1980). Thus, since the 2008 food crisis, the development and extension of agricultural technologies to rice farmers have never been more active. However, adoption of these technologies by rice farmers remains a challenge.

This chapter highlights the importance of integrating indigenous knowledge of rice farmers into the development and extension of agricultural technologies to facilitate widespread farmer adoption and achievement of multidimensional impacts. It illustrates a few case studies where indigenous knowledge of rice farmers in Asia was integrated with that of Western science, resulting in technology adaptation and adoption and policy recommendations for technology adoption to happen, including the aspect of farmer and community safety. This is more evident when we deal with knowledge-intensive technologies – those in the form of knowledge and information that are made accessible to end users in a less tangible form than physical products such as seed or machinery (Price and Balasubramanian, 1998).

The case studies presented here deal mostly with the adoption of natural resource management

* E-mail: florenciagpalis@gmail.com

technologies in rice production such as fertilizer best management practices (FBMPs), integrated pest management (IPM), rodent pest management, pesticides and pesticide safety, and alternate wetting and drying water-saving technology. These are mostly knowledge-intensive technologies where indigenous knowledge had significantly contributed to addressing complex technology adoption issues. These issues revolve around appropriate timing and amount of fertilizer and pesticide application, water scarcity (especially with climate change), right timing and need of community action for rodent control, and the need for personal protective equipment as part of pesticide safety protocol.

In the discourse on innovation systems that tackles evolving roles and existence of multiple sources and utilization of knowledge, the knowledge possessed by farmers is widely acknowledged (Lundy and Gottret, 2007). The integration of indigenous knowledge in development should therefore be encouraged, a shift from the centralized, technically oriented solutions of the past decades that failed to alter life prospects for a majority of the peasants and small farmers of the world (Agrawal, 1995).

Indigenous knowledge in this chapter is interchangeably referred to as farmer knowledge or local knowledge. It is the knowledge that people in a given community have developed over time and continue to develop based on experience and is adapted to the local culture and environment. It is embedded in community practices, institutions, relationships and rituals, held by individuals or communities, dynamic and changing, and unique to each culture and society (Flavier, 1995). It relates to the entire system of concepts, beliefs and perceptions that people hold about the world around them, including the processes whereby knowledge is generated, transmitted and shared within specific social and agroecological contexts (IIRR, 1996; Warburton and Martin, 1999).

Adoption of Best Management Practices for Natural Resource Management in Rice Production

In this section, the importance and utilization of indigenous knowledge for the adoption of best management practices for natural resource management (NRM) in rice production is discussed. The NRM best practices considered here are FBMPs, IPM and rodent pest management.

Building on farmer knowledge for FBMPs

The FBMP discussed here focuses on site-specific nutrient management (SSNM), which is a plant-based approach for optimally supplying rice with essential nutrients (Dobermann and Witt, 2004; Buresh, 2007; IRRI, 2007). It enables farmers to adjust fertilizer use by supplying optimum amounts of nutrients at critical times in the crop's growth cycle to produce high yield. It emphasizes the importance of applying the right type of fertilizer at the right amount and at the right time.

Farmers' knowledge of nutrients is stored in their minds, but this is also reflected in their perceptions about fertilizers, which are embedded in their nutrient management practices in the whole rice production process. Filipino rice farmers normally apply fertilizers on schedule two to three times in a season in accordance with the growth stage of the rice crop (Palis *et al.*, 2007). However, fertilizer management practices are based on the belief that human and plant healthcare are the same, resulting in many overlaps in linguistic terminologies employed for both humans and plants (Palis *et al.*, 2006).

The first application is done at the early stage of the plant, within 15 days after transplanting (DAT) (Palis *et al.*, 2007). The plant is described as a baby or a child, emphasizing the vulnerability of both plants and humans to illnesses. Thus, at this stage, farmers apply more fertilizer (Palis *et al.*, 2007), particularly nitrogen (N), to improve crop growth and enhance plant vigour. In the same way that a baby is given vitamins to make sure that the child is well fed and has good growth and good health, the bulk of the nutrients is applied to the young rice crop.

The second application is, on average, at about 38 DAT, which is near panicle initiation. The third application is at about 55 DAT, when the plant is at the reproductive stage. Farmers normally consider their second and third applications as topdressing. On their second and third applications, the amount of fertilizer is reduced because they view the plant as going towards adulthood, implying that the nutritional requirements of the plant would be less, in the same way that an adult person requires less care.

The science behind the SSNM approach is very much in line with the farmers' logic that considers nutrient management in the context of growth stages. However, SSNM differs in terms of the amount of nutrients needed at specific growth stages. SSNM espouses that farmers need to apply less N early, more N at critical stages of active tillering and panicle initiation, and less or none at all at the later reproductive stages. In the context of human growth, a baby actually needs very little food because it is growing slowly, a teenager needs the most food – such as at active tillering and especially panicle initiation – which is somewhat like adolescence going towards pregnancy.

The way that farmers view fertilizer management practices could be useful in the further refinement and extension of FBMPs such as SSNM. One effective way is through the process of engaging farmers in both research and extension projects called participative research and extension (Percy, 2005) to effect experiential learning by farmers with the use of FBMPs. Here, the farmers are engaged in technology development and validation processes that may lead to local modifications of the technology (Peng *et al.*, 2006). A fusion of farmers' local knowledge and scientific knowledge is necessary to modify any technology and make it more appropriate for end users. Therefore, farmer knowledge, experience and experimental capacity should be used in the design and validation of technology. The challenge, however, is on what and how to build on farmer knowledge in technology development and extension to effect farmer adoption.

Building on farmer knowledge for IPM-FFS

IPM is an ecological approach that builds on biological control as its ecological foundation and the use of pesticides as a last resort. It has been promoted since the 1970s but did not gain wide adoption until the 1990s, when it was disseminated through farmer field schools (FFS) (Palis *et al.*, 1990; Navarro *et al.*, 1998; Matteson, 2000). The FFS is essentially a non-formal school educational approach to IPM extension that is experiential and participatory in nature. All learning activities take place in the field and are based on farmers' experiences through group experimentation during the entire life cycle of the crop.

As in the case of fertilizers, pesticide use is also associated with a belief in the analogy of human and plant healthcare (Palis *et al.*, 2006). To Filipino farmers, rice plants at the early stage are more vulnerable to pest infestations than at ripening stage, like a baby is more susceptible to illnesses and diseases than an adult. This results in farmers applying pesticides in the first 40 days after transplanting, mostly against leaf-feeding insects such as leaf-folders. However, scientific evidence has shown that leaf-folder damage at the vegetative stage could not affect crop yield, because the plants could still recover at that stage (Heong *et al.*, 1994). Hence, a large proportion of insecticide spraying may actually be unnecessary. Farmers commonly remarked that 'if the plants are still young, they have to be taken care of' to prevent pest infestation, ensure good health and eventually good yield. In the same way that a sick baby is given medicine, a young rice crop needs pesticides to combat sickness.

Over the decades, Asian rice farmers have become dependent on agrochemicals to control insect pests. Before IPM-FFS, farmers generally believe that all insects are harmful, which is one of the underlying causes behind the continuous and indiscriminate use of insecticides (Bentley, 1989; Palis, 1998). They also believe that insects will transfer to a neighbouring field when a farm is sprayed. Farmers believe that insecticides effectively control them, aside from being very convenient to use, and they need to spray when neighbouring farmers spray to prevent insects from transferring to their farms. This is further precipitated by the Asian farmers' concept of pesticides as medicine to plants (Heong and Escalada, 1997; Palis *et al.*, 2006). Filipino farmers refer to pesticides as *gamot* (medicine) and they are good for plants because they heal the crop's illness or disease. This blurry perception about these helpful effects is reinforced when chemical salespersons stress the 'medicinal' effects of pesticides on plants.

An important example of an effective integration of farmer knowledge in IPM-FFS was through farmer group experimentation while being involved in FFS, which resulted in IPM adoption (Palis, 2006). Learning theorists have observed that people learn in different ways. Kolb (1984) stated that learning is a process whereby knowledge is created through transformation of experience. Farmer participants used their concrete experiences to test ideas and consequently change their pest management practices through group experimentation. In this context, farmers interpret observations, facts and experiences both individually and as a group and generate a consensus that is culturally enforced. The perceptions of Filipino rice farmers that all insects are harmful changed through experiential learning in the FFS (Palis, 1998, 2006). The FFS participants learned that not all insects harm rice plants through group insect-zoo experimentation where spiders and brown planthoppers were placed in a cage with a rice plant in it. They were able to see how the pests were entrapped with the spider's web after some time. The knowledge they gained from that experiment gave them the courage to ignore insects when they saw them on their respective farms, as long as there were sufficient spiders around.

Building on farmer knowledge for rodent pest management

Rodents are the most important pre-harvest pests in Indonesia for irrigated rice crops and among the three most important pests in Vietnam (Singleton, 2003). In Asia, pre-harvest rice losses are typically 5–10%, with losses of > 20% occurring in some years in some regions. Rodent control is therefore vital to sustain food production, especially for rice. In most Asian countries, rodents are often perceived by farmers to be consistently outsmarting humans – a belief that needs to be overcome (Palis *et al.*, 2011). Thus, ecologically based rodent management (EBRM) was introduced to rice farmers in Vietnam as a possible solution to manage rodent populations effectively, which in turn can help sustain food security in Vietnam and Asia.

Farmers' rodent control practices are generally reactive and rely essentially on chemical and physical methods. EBRM was developed in the late 1990s to manage rodents in rice-based farming systems in Vietnam and other parts of South-east Asia. It combines both cultural and physical rodent management practices such as synchrony of cropping, short 2-week rat campaigns at key periods in key habitats, increasing general hygiene around villages, and use of a community trap-barrier system (Singleton *et al.*, 1999). Although EBRM has been reported to be economically profitable, the successful adoption of this set of technologies requires community participation. Hence, the adoption of EBRM presents a challenge because it is a knowledge-intensive technology. It requires both a solid understanding of the biology of rats and collective action among community members.

The introduction of EBRM in Vietnam was done using a participatory approach through community action. It enabled the farmers to be actively engaged in the technology development and validation process, which promoted the use of indigenous rodent control practices of farmers, especially the digging of burrows and hunting with dogs, and the use of local rat traps. The EBRM did not replace what the farmers were doing, but rather built on their practices and incorporated a scientific basis by encouraging farmers to work together at key times based on knowledge of the dynamics of habitat use and breeding of rodents (Brown *et al.*, 2006). The common practices of digging and hunting, either as a small group or individually, at arbitrary times was consequently transformed into the community working together at key times before rodent damage occurred in their crops. This elucidates the importance of the fusion of local knowledge with scientific knowledge brought about by experiential learning to achieve farmer adaptation and adoption (Palis *et al.*, 2011).

Collective action is not a novel concept in Vietnam. Coordinated community action is the norm that emanated from their history and culture of collective farming in the past and the Chinese influence of Confucianism. Confucianism is viewed as both a philosophy of life and a religion, which emphasizes the importance of loyalty, respect for authority and peacefulness (Quang, 2003). Respect for social hierarchies is therefore basic to Vietnamese families and society.

By far the most important of these values are those associated with family and community, where individual interest is subordinate, if not irrelevant, to the welfare of the whole group (Muoi, 2002). Thus, in the implementation of EBRM, the free-rider problem, which can constrain collective action for rodent management (Palis *et al.*, 2004a), was less of a problem because of the embedded trust among community members, especially among members of the agricultural cooperatives. The norm of working in small groups intermittently for rodent management was easily transformed into community action at key times of the rice growth stage, particularly in the first 2 weeks of transplanting until maximum tillering, which is around 30–40 DAT.

Pesticide Safety

Although pesticides have played a major role in food production since the Green Revolution in the 1970s, they also caused adverse health effects. An estimated 1.3 billion workers are active in agricultural production worldwide and 80% of these are found in Asia (Rice, 2000). According to the International Labour Organization (ILO, 2014), from more than 2.3 million fatalities that take place annually, over 2 million are caused by work-related diseases. In the Philippines, studies had shown that human health, especially of the farmers, is at risk due to pesticide exposure (Pingali *et al.*, 1994; Kedia and Palis, 2008; Lu *et al.*, 2010). Poisoning cases were attributed to lack of protective equipment and use of defective equipment (Jeyaratnam, 1995; Andreatta, 1998).

Personal protective equipment, in the form of gloves and masks, was promoted among 162 Filipino farmers and pesticide applicators to minimize the adverse health effects of pesticides (Palis *et al.*, 2006). Around 70% of the farmers and labourers were not willing to pay for the PPE, particularly the gloves, for the following reasons: gloves were too stifling, uncomfortable, and can cause a condition called *pasma*, which literally means 'spasm' or 'exposure illness'. This concept of *pasma* is generally believed to be caused by hot/cold syndrome and is characterized by weakness or trembling muscles, as in symptoms seen in arthritis, numbness and paralysis (Hart, 1969). The effects of *pasma*, like the effects of overheating when wearing gloves or taking a cold shower right after spraying, are thought, by Filipino farmers, to be worse than the effects of pesticides (Palis *et al.*, 2006).

The inadequate protection of rice farmers from pesticide hazards can also be gleaned from their perceptions and beliefs about illness and pesticides. The farmers perceive illness in terms of inability to function. They think that pesticides may not be a threat because they are immune to them and because these are regarded as medicine that is needed by the plants. They assume that exposure is only through inhalation and ingestion and not through dermal contact (Palis *et al.*, 2006).

Immunity or non-susceptibility to pesticides was seen as inherent to a person and not as the result of precautions taken. Inherent immunity to pesticides is associated with 'strong blood' or *malakas na dugo* (good health and youth), thus explaining the farmers' notion that pesticides only harm certain types of people (i.e. the old and the weak). This belief has led to farmers employing their sons or hiring young people as pesticide applicators, as soon as they are old enough to spray, in the belief that young men are less susceptible to pesticide poisoning because they are younger and stronger or *kaya ng dugo*.

The dual concept of pesticides – as poison to pests and as medicine to plants – has led to divergent views about their health effects on humans, that pesticides are either harmless or harmful (Palis *et al.*, 2006). According to farmers, pesticides are harmful because they kill not only the pests, but also people and animals. This is evident in their re-interpretation of the pesticide dichotomy that, if pesticides were people, they are viewed as insincere people because they show good but have bad intentions, politicians who hide their true identity but do something different from what they say, and usurers who give money but charge high interest. Pesticides, to farmers, come in as both good and bad in the same package. There is the belief that pesticides help in killing insects but other lives are sacrificed in return.

The implication here is that health education programmes should tap the farmers' belief system and cognitive categories when stressing the need for precaution in using pesticides. These programmes should likewise stress the poison angle of pesticides, as the proper choice of words is critical in educational campaigns for safety practices in pesticide use. Furthermore, the practice of IPM that discourages the use of pesticides should be widely and continuously promoted.

Water-Saving Technologies

Water plays a critical role in rice production, especially in the irrigated ecosystem wherein 75% of global rice is produced (GRiSP, 2013). It takes some 3000–5000 l of water to produce 1 kg of irrigated rice (IRRI, 1995). An inadequate supply of water during crop establishment and at vegetative and reproductive stages of the crop would mean a significant yield reduction, while an oversupply of water is an avoidable waste. But water is becoming scarce because of the increase in multiple competing demands, i.e. industrial, domestic and safe drinking water, and sanitation. This is further exacerbated by the increasing adverse effects of global climate change such as the El Niño phenomenon, where rising temperatures translate into increased crop water demand (FAO, 2011). This continuous wasteful reduction in water supply threatens food security in Asia and increases the cost of irrigation development as well as any kind of water use in general.

Water-saving technologies are important to safeguard food security, economic security and water supply for multiple uses of Asian farmers and the Asian population in general. One such water-saving technology is alternate wetting and drying (AWD), which was developed by the International Rice Research Institute (IRRI) to reduce water consumption in irrigated rice and at the same time produce more rice with the same amount or less water (Bouman *et al.*, 2007). It entails an irrigation schedule where water is applied to the field a number of days after the disappearance of ponded water, in contrast to the common practice of continuous flooding. Although there is no standing water in the field, under AWD, rice plants may not be under stress because roots are adequately supplied with water. With AWD, it is estimated that at least 25–30% water reduction can be achieved in irrigated rice production, including those with pump irrigation systems (Palis *et al.*, 2004).

The AWD was validated and promoted among farmers in the Philippines, Vietnam and Bangladesh through farmer participatory action-research using a multi-stakeholder partnership platform. There were some variations though, depending on the socio-political organization and culture of respective societies.

Philippines

The AWD was introduced and promoted in a deep-well pump rice cultivation system among farmers who were members of an irrigation service cooperative in Tarlac, Philippines. Two dry-season experiments were conducted by 12–15 farmers. Each farmer had two plots with 500–1000 m^2: one plot for farmers' practice (FP) and the other one for AWD or controlled irrigation (Palis *et al.*, 2004). The selection of farmers was based on motivation and willingness to participate in the field trials and on on-site criteria such as accessibility to the farm, spread of farmers across the site, position in the toposequence and nearness to the pump. The farmer participatory experiment was done in partnership with rice farmers, scientists, extension agents, irrigation engineers and staff of the National Irrigation Administration (NIA) (the government agency responsible for the irrigation systems in the country), Bureau of Soils and Water Management (BSWM), the Philippine Rice Research Institute (PhilRice) and local government units.

The AWD approach uses a tool, a perched plastic tube inserted into the soil, to monitor the water table in the field. Serving as a decision tool, it tells one when to irrigate the farm. Thus, during the 2002 experiment in the dry (DS) and wet seasons (WS), farmers used this tube to monitor the water level below the surface to determine if irrigation was needed. When water is 15 cm below the surface during DS and 20 cm during WS, the farmer needs to irrigate the AWD plot. It also requires a 3–5 cm depth only, in contrast to the common practice of gauging 10–12 cm depths when farmers irrigate their rice fields.

After the experiment with 12–15 farmers, scientists and other partner stakeholders, AWD was implemented in the P-38 deep-well system level with 63 farmers. After a season-long implementation, farmers did not use the setup any more. Instead, farmers replaced these with a perched bamboo tube and a meter stick (Fig. 6.1).

At the same time, instead of monitoring the irrigation water and measuring the 5 cm water depth, which is required for AWD, they just maintained a 5 cm water depth by reducing a certain portion of the paddy bunds (Fig. 6.2). So, they do not need to actually measure water in their respective farms, because the water automatically flows to the next paddy until the farm is fully irrigated. After 1 year of implementation, farmers no longer used the tube; they just stomped on the soil to see if irrigation was needed.

AWD indeed saved water significantly, by 21% and 26% in the 2002 and 2003 DS, respectively, with a maximum saving of 33%. Likewise, irrigation cost was reduced by 25% in the 2002 DS and by 20% in the 2003 DS. Furthermore, there was no significant yield difference between AWD and FP plots in both years (Palis *et al.*, 2004b). In like manner, profitability of AWD was significantly higher in the 2002 DS, but it was not significant in the 2003 DS. This indicates that the profitability of AWD may be higher than or similar to that of FP. But if water savings are considered as an added benefit that FP cannot provide, AWD proved to be the better alternative irrigation scheme.

Farmers were also partners in the dissemination of AWD. Farmer participatory extension was done through field days attended by farmers from several places, especially from nearby towns and provinces, in the country. The farmers do the talking and the scientists and local partners do the facilitation. Through this arrangement, understanding between fellow farmers is achieved and local knowledge is brought to the forefront, enhancing farmer acceptance and adoption of AWD technology.

From this pilot farmer participatory experiment on AWD, where farmer knowledge was efficiently integrated, a policy emerged through Administrative Order 25 on 11 September, 2009, which stipulated that AWD should be the main water-saving technology (Sibayan *et al.*, 2010). Through this policy, funds were made available for massive dissemination by NIA, BSWM and PhilRice. Since then, AWD has been used as the

Fig. 6.1. Bamboo perched tube and a metre stick (photo: IRRI).

Fig. 6.2. Maintaining the water depth through bund reduction (photo: IRRI).

banner programme of NIA for nationwide implementation through the active participation of irrigator associations (IAs). Likewise, the BSWM implemented AWD on pump-irrigated rice production areas largely through irrigated service cooperatives.

Bangladesh

It was in 2004 that AWD was introduced in Bangladesh by IRRI to the national agricultural research and extension system (NARES) of the country, starting with the key partner: the Bangladesh Rice Research Institute (BRRI) (Palis *et al.*, 2016). AWD was presented to farmers through a simple tool – a perforated water tube, now widely known as *pani* pipe (*pani* means water in Bangla, the national language of Bangladesh). The same principle was followed for AWD practice.

Farmer participatory research was also done through demonstration plots in farmers' fields. It focused on a process of sequential reflection and action by farmers where local knowledge is appreciated and utilized for research planning (Cornwall and Jewkes, 1995). Farmer field days were likewise organized to showcase the AWD technology to a larger audience for out-scaling.

Demonstrations were coursed through many channels: individual pump owners, farmer field schools, and by group or block. Through demonstration plots, farmers were able to have practical experience with the technology, while other farmers had the opportunity to observe the performance of the standing crops. In the beginning, farmers used the perforated perched tube or *pani* pipe to determine the appropriate time to irrigate their respective rice fields. However, after the first-hand experience with AWD technology, farmers started using locally available materials to substitute for the *pani* pipes, particularly the cola bottles; farmers cut the upper portion of the cola bottles (Fig. 6.3). At harvest time farmer field days were organized in key demonstration trials/sites. Invited to these events were farmers within and among the neighbouring villages, as well as private and public organizations and the mass media.

Farmer participatory training on AWD provided to farmers played a key role in the

Fig. 6.3. Plastic cola bottles used to monitor water level (photo: IRRI).

successful dissemination of AWD (Kürschner *et al.*, 2010). The active participation of farmers in the validation and demonstration trials induced farmer-to-farmer learning and sharing. From the 121 farmers surveyed, around 21% of those who did not adopt said that they did not receive adequate training or did not understand AWD. The timing in the conduct of AWD training, however, is important; it should be given during the early part of the DS.

A number of perceived advantages and risks with AWD were reported by farmers and pump owners (Kürschner *et al.*, 2010; Palis *et al.*, 2016). Some of the advantages are that AWD saves water, fuel and electricity for irrigation, thereby resulting in less irrigation costs, and it increases yield as well. Regardless of the perceived advantages reported by farmers and pump owners, AWD adoption can still be considered low. The key factors that were identified to influence AWD adoption by farmers included unreliable water and energy supply, farmer organization and collective action, type of irrigation system, incentives for pump operators (deep tube wells) and pump owners (shallow tube wells), and the payment arrangement for irrigation services (Kürschner *et al.*, 2010; Rahman and Angelsen, 2011; Palis *et al.*, 2016).

Conclusion

From much of the work presented, it is clear that indigenous knowledge is central to the adaptation and adoption of best practices for natural resource management in rice production. In turn, the adoption of these best practices for natural resource management (such as AWD and IPM-FFS) has paved the way to higher yield and income, which implies increasing food production and improving farmers' income.

The way local knowledge was elicited and integrated with Western knowledge could serve as impetus in fostering the utilization of indigenous knowledge for technology extension and adoption. Its use in this area should therefore be encouraged to realize multiple impacts on farmers' lives, be they economic, social, cultural, political, or environmental. This blending of local and scientific knowledge would significantly contribute to the goal of reducing hunger, alleviating poverty and securing environmental integrity.

References

Agrawal, A. (1995) Dismantling the divide between indigenous and scientific knowledge. *Development and Change* 26(3), 413–439.

Andreatta, S.L. (1998) Agrochemical exposure and farmworker health in the Caribbean: a local/global perspective. *Human Organization* 57, 350–358.

Bentley, J.W. (1989) What farmers don't know can't help them: the strengths and weaknesses of indigenous technical knowledge in Honduras. *Agriculture and Human Values* 6, 25–31.

Bouman, B.A.M., Lampayan, R.M. and Tuong, T.P. (2007) *Water Management in Irrigated Rice: Coping with Water Scarcity*. International Rice Research Institute, Los Banos, Laguna, Philippines.

Brown, P.R., Tuan, N.P., Singleton, G.R., Ha, P.T.T., Hao, P.T. *et al.* (2006) Ecologically-based management of rodents in the real world: application to a mixed agro-ecosystem in Vietnam. *Ecological Applications* 16, 2000–2010.

Buresh, R.J. (2007) *Site-specific nutrient management (SSNM) in rice*. Paper presented at the workshop on 'Balanced Fertilization for Optimizing Plant Nutrition' sponsored by the Arab Fertilizer Association, the International Potash Institute and the World Phosphate Institute, 8 February 2007, Sharmel-Shiekh, Egypt.

Conklin, H.C., Lupãih, P. and Pinther, M. (1980) *Ethnographic Atlas of Ifugao: a Study of Environment, Culture and Society in Northern Luzon*. Yale University Press, New Haven, Connecticut.

Cornwall, A. and Jewkes, R. (1995) What is participatory research? *Social Science Medicine* 41(12), 1667–1676.

Dobermann, A. and Witt, C. (2004) The evolution of site-specific nutrient management in irrigated rice systems of Asia. In: Dobermann, A., Witt, C. and Dawe, D. (eds) *Increasing Productivity of Intensive Rice Systems through Site-Specific Nutrient Management*. International Rice Research Institute, Los Baños, Philippines, pp. 75–100.

Flavier, J.M. (1995) The regional program for the promotion of indigenous knowledge in Asia. In: Warren, D.M., Slikkerveer, L.J. and Brokensha, D. (eds) *The Cultural Dimension of Development: Indigenous Knowledge Systems*. Intermediate Technology Publications, London, pp. 479–487.

FAO (2011) *Climate Change, Water and Food Security*. Food and Agriculture Organization of the United Nations, Rome.

GRiSP (2013) *Rice Almanac*, 4th edn. Global Rice Science Partnership, International Rice Research Institute, Los Baños, Philippines.

Hart, D.V. (1969) *Bisayan Filipino and Malayan Humoral Pathologies: Folk Medicine and Ethnohistory in Southeast Asia*. Southeast Asia Program Data Paper No. 76. Cornell University, Ithaca, New York.

Heong K.L. and Escalada M.M. (eds) (1997) *Pest Management of Rice Farmers in Asia*. International Rice Research Institute, Manila.

Heong, K.L., Escalada, M.M. and Mais, V.O. (1994) An analysis of insecticide use in rice: case studies in the Philippines and Vietnam. *International Journal of Pest Management* 40(2) 173–178.

ILO (2014) *Creating Safe and Healthy Workplaces for All*. International Labor Organization, Geneva, Switzerland, pp. 1–18.

IIRR (1996) *Recording and Using Indigenous Knowledge: a Manual*. International Institute of Rural Reconstruction, Silang, Cavite, Philippines.

IRRI (1995) *Water, a Looming Crisis. Program Report*. International Rice Research Institute, Los Banos, Laguna, Philippines.

IRRI (2007) Site-specific nutrient management. Available at: www.irri.org/ssnm (accessed 12 March 2016).

Jeyaratnam, J. (1995) *Acute pesticide poisoning in Asia: the problem and its prevention*. Paper presented at a meeting of the World Health Organization Western Pacific Working Group on Health and the Use of Pesticides, Manila, Philippines, 5–8 December 1995.

Kedia, S.K. and Palis, F.G. (2008) Health effects of pesticide exposure among Filipino rice farmers. *The Applied Anthropologist* 28(1), 40–59.

Kolb, D.A. (1984) *Experiential Learning*. Prentice Hall, Englewood Cliffs, New Jersey.

Kürschner, E., Henschel, C., Hildebrandt, T., Jülich, E., Leineweber, M. and Paul, C. (2010) *Water Saving in Rice Production – Dissemination, Adoption and Short Term Impacts of Alternate Wetting and Drying (AWD) in Bangladesh*. SLE Publication series S241. SLE Postgraduate Studies on International Cooperation, Humboldt-Universität, Berlin.

Lu, J.L., Cosca, K.Z., and Del Mundo, J. (2010) Trends of pesticide exposure and related cases in the Philippines. *Journal of Rural Medicine* 5(2), 153–164.

Lundy, M. and Gottret, M.V. (2007) Learning alliances: building multi-stakeholder innovation systems in agroenterprise development. In: Smits, S.J., Moriarty, P.B. and Sijbesma, C. (eds) *Learning Alliances: Scaling Up Innovations in Water, Sanitation and Hygiene*. Technical Paper No. 47. IRC International Water and Sanitation Centre, Delft, Netherlands, pp. 37–57.

Matteson, P.C. (2000) Insect pest management in tropical Asian irrigated rice. *Annual Review of Entomology* 45, 549–574.

Muoi, N.V. (2002) *A Glimpse into Vietnam's Culture*. Nha Xuat Ban. Dai Hoc Gia TP, Ho Chi Minh City, Vietnam.

Navarro, R.L. Medina J.R. and Callo, D.P. (1998) *Empowering Farmers: The Philippine National Integrated Pest Management Program*. Southeast Asian Regional Center for Graduate Study and Research in Agriculture, Los Banos, Laguna, Philippines.

Palis, F.G. (1998) Changing farmers' perceptions and practices: the case of insect pest control in Central Luzon, Philippines. *Crop Protection* 17(7), 599–607.

Palis, F.G. (2006) The role of culture in farmer learning and technology adoption: a case study of farmer field schools among rice farmers in Central Luzon, Philippines. *Agriculture and Human Values* 23(4), 491–500.

Palis, F.G., Pingali P.L. and Litsinger, J.A. (1990) A multipest economic threshold for rice production (a case study from the Philippines). In: Teng, P. (ed.) *Crop Loss Assessment in Rice*. International Rice Institute, Los Banos, Laguna, Philippines, pp. 229–242.

Palis, F.G., Sumalde, Z.M. and Hossain, M (2004a) Assessment of Rodent Control Projects in Vietnam: Adoption and Impact. Impact Assessment Series Report No. 24. ACIAR, Australia.

Palis, F.G., Cenas, P.A., Bouman, B.A.M., Hossain, M., Lampayan, R.M. *et al.* (2004b) Farmer adoption of controlled irrigation in rice: a case study in Canarem, Victoria, Tarlac. *Philippine Journal of Crop Science* 29(3), 3–12.

Palis, F.G., Flor R., Warburton, H. and Hossain, M. (2006) Our farmers at risk: behavior and belief system in pesticide safety. *Journal of Public Health* 28(1), 43–48.

Palis, F.G., Buresh, R.G., Singleton G.R. and Flor, R.J. (2007) Adoption of fertilizer best management practices: the need for a sociological approach. In: *Fertilizer Best Management Practices: General Principles, Strategy for their Adoption and Voluntary Initiatives vs. Regulations*. Proceedings of the IFA International Workshop on Fertilizer Best Management Practices, 7–9 March 2007, Brussels, Belgium. International Fertilizer Industry Association, Paris.

Palis, F.G., Singleton, G.R., Brown, P.R., Huan, N.H., Umali, C. *et al.* (2011) Can humans outsmart rodents? learning to work collectively and strategically. *Wildlife Research* 38, 568–578.

Palis, F.G., Lampayan, R.M. and Bouman, B.A.M. (2016) Adoption and dissemination of alternate wetting and drying technology for boro rice cultivation in Bangladesh. In: *ADB Technical Assistance Consultant's Report, Regional: Development and Dissemination of Climate-Resilient Rice Varieties for Water-Short Areas of South Asia and Southeast Asia.* Prepared by IRRI for ADB. Project Number: 47163-001, June 2016. Asian Development Bank, Manila.

Peng S., Buresh, R.J., Huang, J., Yang, J., Zou, Y. *et al.* (2006) Strategies for overcoming low agronomic nitrogen use efficiency in irrigated rice systems in China. *Field Crops Research* 96, 37–47.

Percy, R. (2005) The contribution of transformative learning theory to the practice of participatory research and extension: theoretical reflections. *Agriculture and Human Values* 22, 127–136.

Pingali, P.L., Marquez, C.B. and Palis, F.G. (1994) Pesticides and Philippine rice farmer health – a medical and economic analysis of impact. *American Journal of Agricultural Economics* 76, 587–592.

Price, L.M.L. and Balasubramanian, V. (1998) Securing the future of intensive rice systems: a knowledge-intensive resource management and technology approach. In: Dowling, N.G., Greenfield, S.M. and Fischer, K.S. (eds) *Sustainability of Rice in the Global Food System*. Pacific Basin Study Center, Davis, California, and International Rice Research Institute, Los Baños, Philippines.

Quang, M.L. (2003) *Glimpses of Vietnam*. The Gioi Publishers, Hanoi.

Rahman, M.S. and Angelsen, A. (2011) Comparison between irrigation payment systems and probability of using water-saving technology. *Journal of Economics and Sustainable Development* 2(10), ISSN 2222-1700 (Paper) ISSN 2222-2855. Available at: www.iiste.org (accessed 10 October 2016).

Rice, A. (2000) Pesticides in agriculture: the extent of the problem in Asia. In: *Top on the Agenda: Health and Safety in Agriculture*. International Labour Organization (ILO), Geneva, pp. 1–2.

Sibayan, E.B., de Dios, J.L. and Lampayan, R.M. (2010) *Policy support in nationwide dissemination of alternate wetting and drying irrigation management in the Philippines*. Paper presented at the 3rd International Rice Congress, 8–12 November 2010, Hanoi, Vietnam.

Singleton, G.R., Leirs, H., Hinds, L.A. and Zhang, Z. (1999) Ecologically based management of rodent pests – re-evaluating our approach to an old problem. In: Hinds, G.R., Leirs, H., Hinds, L.A. and Zhang, Z. (eds) *Ecologically-Based Management of Rodent Pests*. ACIAR Monograph No. 59. Australian Centre for International Agricultural Research, Canberra, pp. 17–29.

Singleton, G.R., Hinds, L.A., Krebs, C.J. and Spratt, D.M. (2003) *Rats, Mice and People: Rodent Biology and Management*. ACIAR Monograph No. 96, Australian Centre for International Agricultural Research, Canberra.

Vanclay, F. (2004) Social principles for agricultural extension to assist in the promotion of natural resource management. *Australian Journal of Experimental Agriculture* 44(3), 213–222.

Warburton, H. and Martin A. (1999) Local peoples' knowledge: its contribution to natural resource research and development. In: Grant, I.F. and Sear, C. (eds) *Decision Tools for Sustainable Development*. Natural Resources Institute, Chatham, UK, pp. 66–96.

7 Seeds of the Devil Weed: Local Knowledge and Learning from Videos in Mali

Jeffery Bentley,* Paul Van Mele, Sidi Touré, Tom Van Mourik, Samuel Guindo and Gérard Zoundji

Since 2006, the International Crops Research Institute for the Semi-Arid Tropics (ICRISAT) and partners have conducted farmer field schools (FFS) to experiment with options to control striga, a weed so difficult that some call it the 'devil weed'. As a result, practical and profitable integrated striga and soil fertility management practices have been developed for pearl millet and sorghum. The farmer field school experiences provided the key building stones for a number of videos related to ISSFM. In 2011, ICRISAT approached Agro-Insight to have their staff and their partners in Niger, Nigeria, Ghana and Mali trained in the production of farmer-to-farmer training videos. When the teams went back to the farmer field schools to film the videos, and the farmers began talking about how they managed striga, ICRISAT realized how much the farmers had learned in the FFS.

The ten 'Fighting Striga' videos comprise a set that gives the background biology and ecology to help the farmers understand how control options work and also gives them clues to adapt those options to their local circumstances. The videos use an animation (a technical cartoon) to show how striga seeds germinate and attach themselves to the roots of a cereal crop (or die, if germination is stimulated by a legume, cotton or tobacco crop).

In 2012, the series of farmer training videos was translated into French, English and six major West African languages (Bambara, Bomu, Hausa, Mooré, Peulh and Zarma). In 2013, the videos became more widely known and popular. The 8–12-minute videos were put on to a single DVD under the title 'Fighting Striga'. On the opening page, people could select any of the eight languages. The videos could be watched in any order and each one was standalone. An overview and short description of each video is given in Box 7.1. To make sure the videos got into the hands of service providers and farmers, in Mali ICRISAT distributed some 10,000 copies of the 'Fighting Striga' DVDs to many organizations and individuals.

With the involvement of 173 extension agents, more than 12,000 farmers watched the 'Fighting Striga' videos through public screenings (Table 7.1). Some of the 13 villages visited in 2014 had FFS and the videos; others had only watched the videos. Most of the villages saw the videos in 2012, soon after they were released, except for two villages that saw them later.

Farmers Learn from Videos

From 10–21 November 2013, we visited 11 groups of farmers in ten villages in three regions in Mali: Sikasso, Koulikoro and Ségou to learn

* Corresponding author; e-mail: jeff@agroinsight.com

 Indigenous Knowledge: Enhancing its Contribution to Natural Resources Management (ed. P. Sillitoe)

Box 7.1. Description of the ten farmer-to-farmer training videos under the 'Fighting Striga' series.

1. Striga biology
How the weed develops from tiny seeds, not from the roots, as many people think. Striga is a true parasite which attaches itself to the host's root, then remains hidden underground for weeks, so it is the last weed to emerge, and escapes the early season weeding.
http://www.accessagriculture.org/node/241/en

2. Integrated approach against striga
Fertilize the soil with composted manure, because striga damage is worse in poor soils. Add small doses of mineral fertilizer to the base of the plants and intercrop with legumes which kill striga. Hand pull remaining striga plants before they bear seed.
http://www.accessagriculture.org/node/243/en

3. Succeed with seeds
How to test crop varieties to find the ones that are the most resistant to striga and adapted to farmers' real conditions.
http://www.accessagriculture.org/node/245/en

4. Composting to beat striga
Tips on making compost from manure and crop residues, especially in an arid climate.
http://www.accessagriculture.org/node/247/en

5. Micro-dosing
Less is more; apply smaller amounts of fertilizer to the base of the plant, instead of spreading it over the whole field. This improves yields and saves money.
http://www.accessagriculture.org/node/249/en

6. Animals and trees for a better crop
Livestock can eat the leaves and seed pods of some trees, leaving manure to fertilize crops, especially if farmers establish friendly relationships with herders.
http://www.accessagriculture.org/node/251/en

7. Storing cowpea seed
Simple ways to keep insect pests out and keep cowpea seed healthy and viable, so the household has enough seed to intercrop cowpea with cereals.
http://www.accessagriculture.org/node/253/en

8. Grow row by row
Cowpeas and other legumes are trap crops which stimulate striga to germinate, but not to attach to the host's roots. Intercropping and rotating with legumes kills the striga seed in the soil.
http://www.accessagriculture.org/node/255/en

9. Joining hands against striga
Weeding is pretty boring and tedious work, but farmers can beat the drudgery by working together.
http://www.accessagriculture.org/node/258/en

10. Let's talk money
A brief view of costs and benefits under farmer practice and integrated striga and soil fertility management. Managing striga costs more, but the bigger yields are worth it.
http://www.accessagriculture.org/node/260/en

about technical innovations and agricultural experiments farmers had undertaken after watching the training videos. We used a simple questionnaire and did follow-up phone interviews in some cases. These interviews were designed to build on quantitative data which ICRISAT is collecting. In 2014 we interviewed farmers in 13 villages (mixed groups of women and men) about social innovations they had tried after watching the videos.

Some technical innovations can be adopted by individuals; for example, a single person, working alone, can change the sowing density of the peas, without worrying about what the neighbours are doing. Other technical innovations are obligatorily social. For example, adopting the cell phone in isolation makes little sense; the experience is greatly enhanced if one's friends and contacts also get phones and exchange numbers.

In this study we wanted to see what sorts of social innovations the villagers were trying in order to facilitate technical innovations; for example, women in Benin joined together in groups to parboil rice (Zossou *et al.*, 2010). We were also

Table 7.1. Screenings of 'Fighting Striga' videos in Mali, 2013 (table compiled by the authors with data provided by ICRISAT).

Number of village screenings	Men	Women	Children	Audience size
156	5309	3423	3490	12,241

looking for socio-structural change not neatly related to technology (new savings and loan groups might qualify as an example). Watching videos is an example of didactic learning, which allows farmers to incorporate outside ideas into local knowledge (Stone, 2016).

The seed growers (village of Siby)

In Siby we met a few men from a seed producers' group CooproSem (Coopérative pour la Promotion de la Filière Semence de Mandé). Cooprosem was started in Siby in 2006 with nine farmers who had previously helped Icrisat with participatory selection of sorghum varieties (for Striga resistance and other traits). Later the farmers asked Icrisat for seeds from the selected varieties, but were told that Icrisat only had foundation seed (i.e. used to produce certified seed). So the farmers organized the cooperative to produce certified sorghum seed, by planting foundation seed from Icrisat. By 2009 CooproSem was producing certified sorghum seed on five ha, and selling it to seed companies, projects and other farmers (Dalohoun et al., 2011). This example shows that contact with research organisations can lead smallholders to make institutional innovations. As we will see later in this chapter, videos can also induce social innovations. The group recalled the content and how the part about striga biology impressed them, to know how the seeds spread and that animals could even be a vector for the striga seeds. They had also innovated. One farmer described a kind of intercropping he had innovated. He put two cowpea seeds and three or four sorghum seeds in one hole. In the next he only put sorghum. The cowpea covered the soil, and he put small doses of nitrogen fertilizer into the soil. They also waited until the ground was soft to hand pull the striga, so as not to break it off at the roots, a technique shown in the 'Joining hands against striga' video. One farmer also covered the crop deeper at the time of ridging, to keep the striga from coming back up.

Innovating with seed storage (village of Sanambelé)

In Sanambelé, Karim Coulibaly (a mature man), Maimouna Coulibaly (peanut farmer and information secretary for a woman's group) and Souleyman Sacko (youth) had adopted some ideas, mainly the use of chilli and native plants to store cowpea seed. Karim used a plant called *benefi* which he saw on the video. After threshing the cowpea, he winnowed the seed and put it in the granary with the *benefi*, one layer at a time. Maimouna used chilli powder to conserve her groundnut seeds. She put in layers of chilli in the groundnut seed, as shown in the cowpea seed storage video, and the insects did not attack. She also used the leaves of the chilli plant, which is an innovation not shown on the video.

Cropping patterns (village of Yorobougoula)

The village had a farmer-trainer from the NGO Mobiom, Bakary Diallo, who showed the videos in the village. The farmers recalled a lot, such as how the video showed working together as a community to control striga. The people put many ideas into practice. For example, Bakary Diallo started making compost out of manure after watching the video, but he made it on top of the ground, instead of in a pit. This innovation saved him the work of digging a pit. He had tried micro-doses of fertilizer in sorghum, as discussed in the video, but also in cotton, which was his own innovation. Sira Diakité, a farmer and small-scale merchant, tried intercropping cowpea with sorghum, but she put in three lines of cowpea for each one of sorghum, because she wanted more fodder for her animals. People adapted the ideas for their own needs, specifically the intercropping technique. Fanta Diallo remembered inter-planting sorghum and millet with groundnut and cowpea, which is a good point, since the video only mentions groundnut in passing, while stressing cowpea, so she was paying attention. She had experimented with a new way of intercropping millet with cowpea, and sorghum with cowpea, and had seen that the cowpea kept the soil humid. Intercropping allowed her to earn more, because besides her cereal harvest, she got the cowpea crop residues to feed to her livestock. She also remembered the integration of livestock and agriculture. They had practised that for a long time here, but now people understood that the two really were compatible and people were improving their arrangements between pastoralists and farmers. Now farmers had more crop residues and so they could give some to people who had animals.

Video and Farmer Field School

Representatives from FFS groups in Makandiana, near Siby, said that:

- the videos allowed them to see how the striga seeds spread, which can be in an animal's dung, making livestock a vector for striga; and
- before, they used to leave striga in the field after weeding it, but now they see that it is necessary to take it out of the field and burn it.

When we asked what they wanted for future videos, local farmer Bamory Camara had invented a new way of intercropping sorghum and cowpea, by planting the cowpea 15 days after the sorghum. He said he wanted to make one on the intercropping style he had invented. That shows a lot of self-confidence and that they know that the farmers in the videos are real, not a sham.

In Gnamana village near San, the FFS farmer-facilitators sat waiting in the shade of a millet storage platform. They were dignified elder men who all had a notebook or a sheaf of papers. They had led an intensive cluster of five FFS for their neighbours. The first year was a conventional FFS on integrated striga management, sponsored by International NGO World Vision and supported technically by ICRISAT. By the second year they had received the striga videos and showed them all.

In Gnamana village, a farmer called Mama Fabe said that he had not realized that striga came from seeds until he saw the videos. Not to have noticed this key bit of bio-ecological background was quite an admission for a person who had already received training and then led an FFS on striga, but not everything can be observed with the naked eye in a field of sorghum. Scientists have the time and use the tools to observe these difficult-to-observe truths. Striga seeds are so small that they look like powder. They easily escape notice. Through the videos, scientists can explain this fact to a large audience.

The videos had also helped the expert farmers, because they all did experiments. In Gnamana village, one farmer named Bouba Tangara tried micro-doses of 6 g of mineral fertilizer per millet hill. He measured the 6 g by taking three pinches with three fingers. Another farmer called Mama Fabe intercropped two lines of millet with three of cowpea, leaving no space for striga and obtaining a good production. He also intercropped his maize with cowpea. Local farmer Brama Tangara mixed urea and mineral fertilizer in a micro-dose with cowpea seed. At only 7 kg per 0.25 ha, he was surprised at how much the yield improved. He mixed the seed with the fertilizer in a small container and planted them together.

Gnagalé Camara, the leader of a women's group called Gneléni (which means 'light'), watched all ten videos at home, twice, by borrowing her brother's laptop. She used a lot of hand pulling, and she had innovated by using a short-handled hoe in her right hand to get under the roots, as she plucked up the plant with her left hand. It is a clever adaptation not shown on the videos. This would help in developing future videos.

Table 7.2 summarizes farmer experiments with new ideas, after watching the videos.

Institutional Change in Villages

Several of the villages had groups of women and youths that did wage labour, and they added hand pulling of striga to the roster of services they offer. Table 7.3 gives a list of institutional changes. The cooperatives in the village of Sirakèlé added hand pulling of striga as a reciprocal service between members. Someone reports striga and the others go to the field to help pull up the weeds, which is a kind of early-warning striga control. In the village

Table 7.2. Farmer experiments after watching 'Fighting Striga' videos.

Village and farmer group	Compost	Micro-dosing	Hand pulling	Intercropping	Storing cowpea seed	Varietal testing
Siby						
Seed producers, Cooprosem	Making good compost in a pile on top of the ground		After hand pulling place soil over the base of crop to keep striga plants from emerging	Several sorghum with cowpea mixes, with nitrogen. Maize with cowpea and with groundnut		
Siby						
Gneléné women's group			Lots of hand pulling. Uses hoe to avoid breaking striga			
Makanjana						
Representatives from several FFS[(a)]	Making compost, but applying it to tomatoes			Experiments with sorghum with cowpea		
Sanambelé						
Villagers (not organized)					Stored cowpea seed with repellent plant. Mixed chilli powder with groundnut seed in storage to discourage insects	
Yorobougoula						
Some people linked to Mobiom and other neighbours	Making compost pit. Also making compost on top of the ground	Micro-doses of organic fertilizer on sorghum and cotton		Experiments with millet and sorghum with cowpea		
Nampossela						
FFS[a] group (they learned these ideas in the FFS; only one of them saw the video)	Good compost controlled striga and raised yields. Less striga germinates		Many use hand pulling and it does control striga	Cowpea with sorghum and with maize lowered the germination of striga	Triple sack to store cowpea seed	

Continued

Table 7.2. Continued.

Village and farmer group	Compost	Micro-dosing	Hand pulling	Intercropping	Storing cowpea seed	Varietal testing
Sirakelé						
Women's group and other groups	Four people have made compost pits	Micro-doses of compost	The community organizes to hand pull striga		Double sack	
Yorosso						
Ben Kadi	Good compost stops striga from germinating	Micro-doses of compost	Hand pulling striga			
Souara						
Farmers who helped make video on compost	Many make compost pits. They know compost helps control striga		The village does hand pulling	Sorghum with cowpea and with groundnut		
Gnamana						
FFS[a] farmer facilitators	They learned about compost in the video	Micro-dose of mineral fertilizer and urea		Millet with cowpea prevents striga from growing		
Wakoro						
People from two FFS[a] and some variety testers	Several made compost, but because of a lack of water some made it once instead of twice during the year	Micro-dose of urea, and planting seed closer together. Micro-dose of mineral fertilizer mixed with seed	Uses hand pulling with micro-dose	Sorghum with cowpea		Testing new sorghum varieties. Watched video to help select seed. In an FFS[a] they chose six varieties. After the film they selected varieties based on yield, resistance to striga and ease of drying

[a]FFS, farmer field school

Table 7.3. Institutional change in the villages after watching the videos.

Region and village	Change	Notes
Mopti		
Promani	Strengthened groups	They formed groups during the FFS[a], with AKF[b], but after watching the videos they made special women's groups; 20 new men and 25 additional women joined the groups. These changes were probably stimulated by AKF
Madiama	Strengthened groups	After the videos their groups were strengthened and they started a large cooperative. Probably due more to the organizational efforts of AKF[b] than to watching the videos. Women's group tends a collective field using striga control and puts the money from the harvest in their *caisse*
Kouna	Video committee	After the first screening, village leaders organized a video committee, which played the videos every night for 2 weeks, and took them to outlying hamlets, so that everyone saw the videos
Torokoro	None	
Orgnon	Women's savings and loan	They claim to have started small groups for women after the videos. The women claim that they organized a savings and loan group as a result of watching the video. The president of the women's association recalled the video 'Let's Talk Money' and claimed that it helped them
Ségou		
Dobo	Hand pulling striga (from FFS)	The groups already existed. Some were organized spontaneously and some with outside help. After the videos, each one also organized to hand pull striga, which is now institutionalized in the village, but the villagers attribute it more to the FFS[a] than to the videos
Hasso	Hand pulling striga and compost	The groups already existed but were strengthened, e.g. adding tasks like hand pulling striga and making compost which are difficult to do alone. Groups that already did tasks like weeding fonio and harvesting groundnut began to cooperate to dig compost pits and transport manure
Daga	Hand pulling striga	Groups of women and youth that already existed to do farm work for pay. After watching the videos, these groups added hand pulling of striga to the list of services they offer, and farmers hired them to do it. They are not always able to meet the demand from farmers (Guindo, 2016)
Togo	Hand pulling striga	No new groups were formed, but existing ones were strengthened. Groups of women who did farm work for wages added hand pulling as one of their services. It is now much in demand
Sikasso		
Nampossela	Slight	The groups that already existed grew crops to sell, bought and sold cereals, or did farm work for cash
Zantiela	None	
N'Tonasso	New women's groups formed	Groups started by AMEDD[c] because of their experience with FFS and filming the videos
Sirakèlé	Hand-pulling striga and counting money	Both cooperatives now pull striga. If a member reports striga, the others help to pull it up. One of the women's groups piles striga and burns it. The women say that the 'Let's talk money' video has helped them to analyse their accounts and tell if they have made money or not

[a]FFS, farmer field school
[b]AKF, Aga Khan Foundation
[c]AMEDD, *Association Malienne d'Eveil au Développement Durable*

of Togo, about 80% of the people in the village had seen the videos. The groups of women who were already organized to work for wages (e.g. doing weeding and harvesting) added pulling up striga to their repertoire. Some people worked in groups to make compost, or to haul it.

Small groups and a long-running field school (village of Promani)

The president of the cooperative (Fig. 7.1) and the president of the FFS and three of the FFS trainers said that, in the videos, they saw that some women were organized into women's groups, so they decided that they could do that too. After creating a special group for women it grew from just the five trainers to include a total of 30 women in 2014. The cooperative grew from 30 to 50 people. That is one of the institutional changes which the villagers themselves attributed expressly to the video. As we have seen in Benin, just seeing people do certain things together may be enough to stimulate thoughts of emulating them (see Zossou *et al.*, 2010).

The FFS graduates are still doing experiments. The head of the cooperative, Bakar Coulibaly, showed us one of his. This millet field had been split into plots. The thriving millet on the right side was fertilized with micro-doses of compost, while on the left side the millet was intercropped with cowpea, but not fertilized (farmer experiments often test two variables at once). (Fig. 7.2). People were still experimenting, years after first learning to manage striga. Gorou Traoré, president of the local farmer field school, said that striga was their worst problem, except for rain, and no one could control the rain. Koumba Daou showed us her field of millet and cowpea: one row of cowpea and two of millet, with some goat manure, so the videos really had inspired technical change. She said she no longer had striga problems. The videos were actually designed to include women-friendly technologies. When the video makers showed the rough edit of the livestock video in northern Ghana, women suggested adding footage of small livestock like goats,

Fig. 7.1. Farmers like field schools and, if given the chance, may participate for years, adding new, creative experiments along the way.

Fig. 7.2. A farmer experimental field, with millet fertilized with a micro dose of compost on the right, and unfertilized millet intercropped with cowpea on the left.

otherwise the film might have inadvertently implied that soil fertility can only be improved with cow manure, which is men's property.

Gorou Traoré showed us his plot. He put 2 kg of mineral fertilizer on a 300 m^2 plot; another field had two rows of millet and one of cowpea (Figs 7.3 and 7.4). The Aga Khan Foundation (AKF) extension agent came about once a month to advise them on their experiments. Gouro was testing the millet varieties Toroniou and Guéfoué 16, with the cowpea variety Dunafana.

Koumba Daou was intercropping two rows of millet with one of cowpea, fertilized with goat manure. It was the fourth time she had done it and the two techniques had reduced the striga infestation. The women had to go through men to get a field. One year they may get a field and manage the striga and then the next year they may get a different field full of striga. Land tenure is key.

The video committee (village of Kouna)

In the village of Kouna they did not have FFS. But when an extensionist brought the striga videos in 2012, the village leaders watched them, and although they invited 40 people, more than 40 came. They liked the videos so much that they decided to show them to everyone; they set up a special video committee and screened them in the *place publique* for 2 weeks, until everyone in this large village had seen them. They got 11 copies of the DVD. People from the nearby hamlets came to watch the videos in the village centre, and people took the DVDs to watch in the further hamlets.

Strengthening cooperatives and groups (village of Sirakèlé)

This village organized a large screening of 500 people, attended by Icrisat, AMEDD (*Association Malienne d'Eveil au Développement Durable*) and Amassa (*Agence Malienne pour la Sécurité et Souveraineté Alimentaire*). Then they watched the videos three times in small groups of 30. They remembered the video contents well. They were organized into cooperatives. One had 26 men who sold grains to the World Food Programme (WFP) and produced millet seed to sell to farmers. A women's cooperative worked a collective

Fig. 7.3. Millet and cowpea.

Fig. 7.4. In another plot, Koumba Daou intercropped two rows of millet with one of cowpea, fertilized with goat manure.

field, and they harvested for other farmers for wages. It was an old group and there were also older groups of women. The cooperatives have had a lot of input from the non-governmental organizations (NGOs). If a member reported striga, the men's group went to his field and pulled it up, because it was tedious to do alone. And they wanted to pull it up before it went to seed. Knowing about striga seed may be the key drive to act speedily, stimulating people to act in groups rather than alone. In two villages (Orgnon and Sirakèlé), young women said that they had paid more attention to their accounting after watching the video 'Let's Talk Money'. It helped them to do their own analysis of costs and benefits when selling food to the WFP.

Farmers continued to experiment with the techniques they learned in the videos and FFS. Many were trying hand pulling, intercropping, micro-dose and crop rotation. The farmers already knew that composted manure enriched the soil, but they rarely had enough compost. From the videos, people learned that they could add vegetal waste, especially cereal stalks, to the compost and they were pleased that they could now make more organic fertilizer.

Conclusion

In village after village, even if people had only watched the videos once, they always had something intelligent to say about the contents, even 2 years after the screening. In each of the villages visited, farmers tried out new ideas they had learned from the videos. Farmers were not passive observers and were sensitive to observe social dynamics presented in videos. Farmers also made social innovations, even though the videos were not intended to trigger social change. In a previous study, women in Benin watched a video on parboiling rice, which showed a small group of women doing the job together. This had a great impact on women in the audience, who decided that they would no longer process the rice individually but rather in a group (Zossou *et al.*, 2010). In this study in Mali, farmers were quick to notice that striga weeding can be effectively done in groups and started organizing themselves to do so. There are several accounts from Niger where the 'Joining hands against striga' video sparked similar dynamics to the extent that even special village days were organized to hand pull striga. All farmers learned from the videos that the battle against striga can only be won when they join forces and apply many technologies together.

All of the villages were changed in some ways, either by solving their striga problems, or by changing their organizations, or both. Like FFS, the videos do more than just explain technology. Farmers learn background biological and ecological information, and then use that to conduct their own changes.

The main technical changes included: hand pulling of striga; making compost; micro-dosing fertilizer and intercropping with legumes. Organizational changes included: strengthening women's groups; groups adding striga pulling to their repertoire of services and organizing to watch videos.

Acknowledgement

This study was generously supported by the Swiss Agency for Development and Cooperation (SDC).

References

Dalohoun, D.N., Van Mele, P., Weltzien, E., Diallo, D., Guindo, H. and vom Brocke, K. (2011) Mali: when government gives entrepreneurs room to grow. In: Van Mele, P., Bentley, J.W. and Guéi, R. (eds) *African Seed Enterprises: Sowing the Seeds of Food Security,* CAB International, Wallingford, UK, pp. 65–88.

Guindo, S. (2016) Earning money from weeds. In: Bentley, J., Boa, E. and Salm, M. (eds) *A Passion for Video. 25 Stories about Making, Translating, Sharing and Using Videos on Farmer Innovation*. Access Agriculture, Nairobi and CTA, Wageningen, Netherlands, pp. 35–36, 56.

Stone, G.D. (2016) Towards a general theory of agricultural knowledge production: environmental, social, and didactic learning. *Culture, Agriculture, Food and Environment* 38(1), 5–17.

Zossou, E., Van Mele P., Vodouhe, S.D. and Wanvoeke, J. (2010) Women groups formed in response to public video screenings on rice processing in Benin. *International Journal of Agricultural Sustainability* 8(4), 270–277.

8 'I Will Continue to Fight Them': Local Knowledge, Everyday Resistance and Adaptation to Climate Change in Semi-Arid Tanzania

Lars Otto Naess*

Local knowledge[1] (LK) has over recent years gained a prominent place in research, policy and practice on adaptation to climate change. A growing body of literature has demonstrated the merits of local knowledge, both for instrumental or social justice reasons, to support adaptation (Crate and Nutall, 2009; Reid *et al.*, 2009; Eriksen *et al.*, 2011; Castro *et al.*, 2012). On the policy side, the reference to traditional knowledge in the Paris Agreement[2] to the United Nations Framework Convention on Climate Change (UNFCCC) is significant, if long overdue.

Yet, this near-universal endorsement of local place-based knowledge in adaptation literature hides a level of ambivalence over its role and value, and a number of tensions remain. Some of these tensions are well known, and arguments well-rehearsed also in the broader literature on local knowledge and development. These include the concern and critique over documenting and 'packaging' of context-specific information (Ellen *et al.*, 2000) and the epistemological debates over science and local knowledge (Pepin, 1996; Berkes, 2002). Another related concern is that local knowledge, as embedded in worldviews and cultures, may restrict options available to people and hence increase rather than reduce vulnerability to climate change (Patt and Schroeter, 2008). Yet others relate to the fact that some local and traditional practices, in a modern cultural context and in the face of climate change, may support short-term coping, but will be maladaptive in the long run (Ford *et al.*, 2013). Thus, what prevails is often what may be seen as a cautionary approach to local knowledge, illustrated in the Paris Agreement's formulation that adaptation action should be guided by 'the best available science and, as appropriate, traditional knowledge.' (Ford, 2013, p. 25).

Arguably, the tensions over the role of local knowledge in part reflect the fact that much of mainstream adaptation literature has a focus on technology-driven incremental adjustments to climate scenarios and climate impacts, amidst a still emerging focus on root causes – social, political, economic – of vulnerability and, with it, critiques of mainstream interventions of resilience and adaptation (Brown, 2015; Watts, 2015). However, in this chapter it will be argued that the tensions also represent a bigger challenge, namely that while studies of local knowledge and adaptation have been good at bringing out what can potentially be achieved by integration of local knowledge, there has been less focus and attention in the adaptation literature on how

* E-mail: naess@ids.ac.uk

local knowledge is situated within the everyday lives of people who tackle climate change amidst a range of other livelihood shocks and stressors.

Illustrated by a case study in semi-arid Tanzania, this chapter shows how individuals and households use LK and science, intermixing and also circumventing rules and regulations, much of which has the nature of everyday resistance – challenging of internal and external norms, and using scripts from local knowledge to justify their actions that (as they are often aware) are not sustainable, nor likely to help them in the long run. The chapter explores how farmers, faced with what people perceived as declining and more unpredictable rainfall, little or no external support and less access to land, are able to mobilize resources to navigate droughts. In turn, what do the strategies tell us about the role and importance of local knowledge for adaptation? It will be argued that a focus on everyday resistance is useful in that it helps to explain the persistence of local knowledge-based practices and, secondly, it helps to reflect on the often-repeated call to integrate and promote local knowledge for adaptation.

Local Knowledge, Everyday Resistance and Adaptation

The contributions of local knowledge to climate change adaptation have been well articulated and documented over the past 10–15 years (e.g. Riedlinger and Berkes, 2001; ACIA, 2004; Byg and Salick, 2009; Petheram *et al.*, 2010; Lebel, 2013; Leonard *et al.*, 2013). However, climate-related shocks and stressors have been central in much of the work on local, indigenous and traditional knowledge, which goes back many decades (Brokensha *et al.*, 1980; Chambers, 1983; Richards, 1985; Thrupp, 1989; Scoones and Thompson, 1994; Warren *et al.*, 1995; Sillitoe, 1998). Local knowledge was central to the argument about social responses to the droughts and famines of the 1960s and 1970s, for example. In this context, adaptation has had a chequered history. Originally an ecological term, the concept of adaptation first rose to prominence – and contestation – during the 1970s. Over the 1970s and 1980s, it became increasingly associated with the notion of individuals choosing to adapt (or not) to hazards (Burton *et al.*, 1978), increasingly challenged by work focusing on unequal power relationships and structures that underpin vulnerability (Torry, 1978; Hewitt, 1983; Watts, 1983). Adaptation re-emerged in the context of climate change with the establishment of the Intergovernmental Panel on Climate Change in 1988 and the UN Framework Convention on Climate Change (UNFCCC) in 1992. Since then, the field of adaptation has spanned a range of disciplines and approaches, from a focus on responses to discrete projected climate changes ('impacts' focus) on the one hand, to a focus on adaptation as a socio-political process ('vulnerability' focus) on the other (O'Brien *et al.*, 2007).

Following Eriksen *et al.* (2015), adaptation to climate change is considered here as a fundamentally socio-political process. Adaptation decisions and actions are outcomes of complex processes, steered by historical trajectories of power that guide how changes are understood and acted upon, and the range of options available. This does not mean ignoring or downplaying the physical manifestations of climate change as adding and changing climate-related shocks and stressors, but rather the view that those manifestations are determined by the hazards as they converge with multiple interrelated historical social, political and economic processes. And this is where local knowledge comes in as a central concern. As Watts (1983) demonstrated, people's responses are conditioned by the coming together of historical processes of domination and marginalization with the cultural frames through which people see and act on changes. Based on Brown (2015), the focus here will be in particular on the role of local knowledge as affecting, and reflecting, rootedness and resourcefulness, but notably also everyday forms of resistance, the latter based on Scott (1985).

Eriksen *et al.* (2015) usefully distinguished three aspects of adaptation to climate change as a socio-political process: authorities, knowledges and subjectivities. Authorities reflect the way adaptation is framed, and by whom; knowledges relate to how discourses come into being; and subjectivities to how people place themselves within this. In each, it is thus about how adaptation is being promoted, but also how it is being opposed. The concept of resistance, and in particular everyday resistance (Scott, 1985), is useful here, as it cuts across all three aspects, i.e. why and how authorities, knowledges and subjectivities encompass resistance – to formal adaptation interventions, to what types of knowledge count and do not count, and from whose perspective adaptation is conceived and implemented. As Katrina Brown argued in her recent book (Brown, 2015), adaptation (or resilience building) is ultimately about how people make sense of and respond to climate change. Brown (2015) focused on resistance – as in everyday resistance (Scott, 1985) – along with rootedness and resourcefulness.

Arguably, the literature on local knowledge and adaptation to climate change has no shortage of focus on resourcefulness, nor rootedness: this is the core of the argument, namely that while it is context specific, it demonstrates just how resourceful people are, the knowledge they have and how adaptation work can only advance, indeed succeed, if local knowledge is considered and duly integrated. There have been significant advances over recent years in creating dialogues and 'level playing fields' between scientists and local people (e.g. Guthiga and Newsham, 2011). However, there has been less focus on how certain strategies are rejected or opposed. The starting point here is that a focus on everyday resistance may help better formulate the role of local knowledge in resolving some of the tensions outlined above. As Kalland (2000) argued, resistance may be to formal rules but also to informal, culturally embedded rules: people do not necessarily adhere to traditional rules, especially in situations where knowledge, values and worldviews are increasingly 'hybridized'. We also know that local knowledge and the norms and values they express reflect local power structures and gender inequalities, but like other aspects of local knowledge, these are dynamic and changeable. Thus while local knowledge has, rightfully, been seen as crucial for understanding and opposing long-standing marginalization and exclusion of local perspectives, views and capacities (most visibly in relation to indigenous peoples and climate change, e.g. in small island developing states (SIDS) and the Arctic[3]), actions taken to support local knowledge may also affect pathways for adaptation, for example through intersecting with power relations. This chapter will argue that attention to resistance may help to contribute to uncovering nuances in the role of local knowledge for adaptation, which in turn may help towards understanding why and how some interventions aimed at helping individuals and households work while others fail.

Everyday Resistance and Coping with Drought in Semi-Arid Tanzania

Background

Central semi-arid parts of Tanzania, as other dryland areas of Africa, have a long history of droughts and famines and people have developed a range of skills, practices, management systems and institutions to tackle drought-related shocks and stresses (Brooke, 1967; Rigby, 1969; Maddox, 1988). This section will attempt to show how many of the actions to cope with drought may be considered as acts of everyday resistance to authority, mobilizing resources in the face of what is perceived as an increasingly challenging social and environmental context.

Authority here comes in two main forms: internal authority as reflected in traditional norms and values; and external norms as expressed through policies and regulations on resource use. This section is based on ethnographic research carried out by the author in 2006 in two villages in Dodoma Region, central Tanzania. Data comprised 140 survey interviewees, 80 semi-structured interviews and eight focus groups, along with transect walks and a range of informal conversations with villagers. Key informant interviews were conducted at district, regional and national levels. 2006 was a drought year in the study area and while interviewees and discussions were based on recall of historical (mainly) drought-related shocks and stressors, they were also related to perceptions of change and actions taken during the time of fieldwork.

Dodoma Region is located in the semi-arid areas of central Tanzania. The rural areas around the municipality of Dodoma broadly coincide with the historical area of Ugogo, named after the dominant ethnic group in the area, the Gogo. Rural parts of Dodoma are among the poorest and most food-insecure parts of Tanzania. The Gogo are described as agropastoralists or 'cultivating pastoralists' (Rigby, 1969).

Cattle are historically a core part of their culture, but sedentary agriculture has become an increasingly important part of Gogo livelihoods over recent decades.

The area has a low and erratic rainfall pattern, with a long record of devastating famines (Rigby, 1969; Maddox, 1988) which are linked to, albeit often not caused by, droughts. For example, it is no coincidence that the worst famine in oral history happened after the First World War, during which colonial powers confiscated food and required labour to support the war effort, leaving fields uncultivated (Maddox, 1988). The following quote typifies the Gogo people (*Wagogo* in Swahili) and their relationship with the environment: 'It is a way of life that has defined Ugogo and the Wagogo; one that ensured the survival of the society as a whole, if not all its members, in an environment with little water and uncertain grain supplies' (Maddox, 1988, p. 4). The quote embodies the ability of the Gogo to adapt in a harsh and unpredictable natural environment, in many cases using collective institutions, but also that these processes produce winners and losers.

There was a strong perception in the study area that the weather (used interchangeably with the concept of climate) was changing. Of particular concern were what people saw as less overall rainfall, less reliable rainfall and warmer temperatures, which all affected crops negatively. A wide range of responses was documented. Local knowledge played an important role in identifying drought signals and their causes, as well as responses to these. Examples include the use of traditional signs to monitor rainfall (such as trees, animals and rainstone rituals), oral history of past famines and their significance in relation to the recent famine, perceptions of cycles of rainfall and causes of these changes, and observations of the effects of changes on resources that support livelihoods. There was concern in the two study villages about what they saw as reduced rainfall overall and changing rainfall patterns. There was no marked difference among social groups with regard to observations, and only two respondents in each village (who had returned from living in the urban centre of Dodoma and Dar es Salaam, respectively) linked their observations of change to human-induced emissions. Secondly, farmers had made, or were in the process of making, a number of changes to their livelihood strategies in response to what they saw as a reduction in rainfall, overall and at crucial times during the year. The main ways in which they were tackling droughts in the short term included: (i) local resource use; (ii) drawing on social networks and informal structures; and (iii) (but limited) formalized structures, notably distribution of emergency seeds and food aid. A number of people had migrated, many temporarily, to either Dodoma or Dar es Salaam, to look for work. Despite this, remittances played only a minor role in people's incomes. On the contrary, a common complaint was that able-bodied men 'came back to eat' after harvest, contributing little to the household labour or incomes.

The following will focus on resistance to authority in relation to two key livelihood resources: regulation of access to natural forests; and crop regulations, including emergency seed support. As will be seen, both of these were subject to everyday resistance, but in different ways and with different outcomes.

Forests: regulations and resistance

Forest reserves adjacent to the study villages (Chigongwe and Chinyami forest reserves,[4] respectively) were monitored by a forest officer at the Division level, each reporting to the District Office in Dodoma. Charcoal making and cultivation were major challenges in both areas, but more so for Chigongwe than Chinyami. Many farmers in both villages supported protection of forest and trees. Much of this support was connected to a widely held perception that trees 'held' rain and hence secured local rainfall. Loss of large trees was seen as a major reason for the recent drought and the perceived reduction in rainfall over recent years. The awareness of this was more widespread in Nzali than in Kigwe. In Nzali, protection of large trees of particular species were sanctioned in informal rules set and enforced by the traditional leader (*mtemi*). Traditionally, if villagers were caught while cutting a protected tree, they were required to give a black sheep to the *mtemi* to be slaughtered at the site where the tree was cut, and the person would have to confess in front of the *mtemi*. Informants said this rule was still enforced, and had last happened only a few years ago. In contrast, while respondents in Kigwe also referred to drought as linked to forest loss, references to protected trees were fewer, and many were unaware of any traditional restrictions on tree cutting.[5]

While respondents were aware of forest protection, and many expressed their support for it, many were also clear that if they had the chance and were able to, they would clear the forest to cultivate or to make charcoal. Justifications for people's contravention of forest regulations fell in two main groups of responses. One was that the forest was their only option locally to secure incomes and food during droughts, such as the one they were experiencing at the time of fieldwork. The rains were perceived as starting in the forest areas and they could hence be sure to get a harvest in the forest, unlike in the village, also because of better soil fertility, and less problem with weeds. Charcoal and sesame were both secure sources of income for those able to produce them. The other main group of responses was that the current government regime was unfair to those who had lost land or were barred from using the forest. Several respondents in Nzali had been evicted from the forest reserve and had had to move to other areas in the village with lower soil quality and less secure rainfall. In Kigwe it was the forest officer who caught people illegally making charcoal or cutting down trees in the forest reserve, also taking action if forest products were being sold in the market. Perhaps because of his prominent role in forest management (the forest officer in Nzali was further away and so not able to visit Nzali very often), respondents saw the forest in Kigwe as 'belonging' to the forest officer. A quote from a male farmer in Kigwe illustrates this: '... the whole forest belongs to him.... If you have cut trees from Chigongwe and he finds you, you are going to pay for everything you take from the forest.'

Further explanations can be found in traditional rules. Aside from large trees, there had traditionally been few restrictions on the clearing of land for cultivation in the forests, which was generally allowed as long as they kept the protected trees. In Nzali, some areas were protected for their ritual importance. Particular hilltop areas were used in relation to rainstone rituals and sacrifices. Some elders said that in the past they were allowed to do farming in the valleys, and one informant noted that he could remember the traditional leader telling them that they were allowed to cultivate in the forest as long as it was not 'in his face', meaning that they should hide the areas they cleared from his view.[6] Only a few respondents in Kigwe knew of any protected forest areas in the past, and if mentioned it was related to hilltops and in connection with forest protection rules by the colonial authorities.

Responses also revealed differences in how people respected traditional rules. An internal conflict between those cultivating in the forests and producing charcoal and other villagers was illustrated by many (especially elderly) villagers' references to the youth having lost respect for traditions, being 'thieves' and 'stealing'. By clearing not only land but also large trees, they brought drought on everybody. Respondents also had little faith in the protection systems, seeing them as inefficient and prone to misuse. In Nzali, farmers who were themselves barred from going to the forest had seen other villagers who were able to clear forest get high yields and earn large incomes. They could observe the same people now building bigger houses roofed with corrugated iron sheets, which were worth much more than the traditional Gogo houses and could, for example, be used as collateral.[7]

Some pointed out that the increasingly strict formal enforcement of forest conservation meant that some farmers who earlier had farms in the forest were now moving to the village, resulting in higher prices and increased land shortages. While neither charcoal nor sesame were new income sources in the area, the production of both was considered to have increased rapidly over the past 5–10 years due to better market conditions. Responses suggested that it was possible to pay a fine to some of the guards in order to continue farming in the forest. One male informant in Nzali claimed that, after paying the fine, he could continue working in the forest: 'The government cannot send the guards there because I have already paid the fine.' He said he had paid 50,000 Tanzania shillings (TSh) (about £20) to the guards to be able to continue farming.

Younger generations justified their actions by saying there was little or no land available in the village, in effect forcing them to make a choice of whether or not to remain farmers. Those who could take advantage of the situation had high incomes, but for the majority of informants the strict forest regulations meant less access to resources. The forest regulation called HADO (after the government's Hifadhi Ardhi Doroma soil conservation project under which reserves had been managed) was seen as the organization taking 'their' forest away from villagers (Nzali).[8] Along with farmers' stated views that forest clearing and tree felling bring drought, they generally support measures for stricter enforcement of regulations, but with an ambivalent attitude. Some add that they 'should set aside some areas for cultivation'. Perhaps the clearest illustration of this ambivalence is the following statement from a Kigwe farmer (male, 30–39 years old): 'If there is a proper guarding system maybe the cutting down of trees will be reduced ... but because I am a charcoal maker, I will continue to fight [the guards].'

'They are confusing us Gogo': crop advice and regulations

Other key government regulations in Kigwe and Nzali related to crop varieties and distribution of drought-tolerant emergency seeds, supplied to the government by the UN's Food and Agriculture Organization. For many years, the government had discouraged farmers from growing traditional sorghum and maize varieties, and encouraged the replacement of them with improved drought-resistant varieties. Drought-resistant varieties had long been a key part of government strategies to increase farmers' ability to cope with drought in Tanzania, and the need to teach farmers to plant drought-resistant crops featured in government policy documents, such as the government's disaster reduction strategy (URT, 2003). It was also central to government thinking on climate change adaptation as set out in the National Adaptation Programme of Action, as a way to combat drought and hunger (URT, 2007). The seriousness with which the government regarded this goal is illustrated by the fact that agricultural officials had at times in the past gone to farmers' fields and cut their traditional crop plants.[9] In recent years the official line had become less confrontational. At the time of fieldwork, the Regional Commissioner for Dodoma had issued a directive saying that farmers should set aside 2 acres for drought-resistant varieties, but that beyond this they could plant the varieties they wanted.[10]

Farmers in both villages traditionally grow a number of local sorghum varieties, often referred to as *lugugu*.[11] The local varieties were preferred for many reasons, echoing findings elsewhere in Africa (e.g. Friis-Hansen, 1999): their resistance or attractiveness to pests (birds, wild boars), their taste and suitability for cooking as well as their higher tolerance to variable rain patterns. Village leaders and agriculture officials, however, made little secret of their negative view of local crop varieties, along with other traditional agricultural practices, because in their view they were unsuitable in a drying climate. One elected village leader called the cultivation of local crop varieties 'encouraging the enemy hunger'. Many considered a main obstacle to be that farmers were not willing to accept or be responsive to new knowledge, such as taking up improved varieties, because they were 'ignorant', 'slow to take up new knowledge' or because of cultural and social barriers.

The implementation of the ban on local varieties differed in the two villages, however. In Nzali, the local agricultural officer (*Bwana shamba*) was of the opinion that, to prevent hunger, there was a need to be strict and not allow them to be grown even in areas with good access to water during drought: 'We cannot start letting them [grow these varieties], therefore we ban them all.' He stressed that actions would be taken especially against those growing maize and the most common local sorghum variety *lugugu*, because 'they are the big enemy'. In Kigwe, there was a strong recommendation from the village leadership not to grow maize or the local sorghum varieties, but the directive was less strictly enforced by the leadership than in Nzali.

Interviews with farmers revealed a more complex picture. While farmers valued their local varieties and expressed a strong desire to continue growing them, many had also stopped growing *lugugu* and other local sorghum varieties, or only grew them in a small part of their farm. The overall pattern was a move away from maize and the tall local sorghum varieties, notably *lugugu*, and towards more millet of the local variety as well as improved sorghum varieties. The main reason given was decline in rainfall. Many said that *lugugu* takes too long to mature (6 months) and that they no longer give enough food because of the low rainfall. It was further held that rains today were so short in duration that farmers needed crops that matured quickly. Seeing that local varieties did not give good yields any longer, they grew the largest area with improved varieties. A typical quote, from an elderly farmer (> 60 years, poor category) in Kigwe, included: 'We really like [*lugugu*], but these days the rain has been reduced, we are afraid of growing [it].' The local variety of millet was one of the few that could be counted on with the changes: 'We trust *uwele*, that's why we grow it.' Similarly, fruit trees produced lower yields because of low rainfall, and water wells that used to last until December now dried up much earlier.

The government's alternative was to encourage the use of improved drought-resistant seeds. Because of the ongoing drought, emergency seeds were being distributed in both villages. The seed distribution was observed at close hand during the time of fieldwork in Nzali. A village official, through orders from the District Commissioner, announced the distribution of seeds during a group discussion meeting held with all sub-village leaders in Nzali.[12] The official announced that nobody was allowed to start planting before 15 November,[13] otherwise they would be punished.[14] The criteria for receiving seeds were as follows: first, the farmers would have to plough their fields, and they would have to apply manure. Then, the seeds had to be planted in straight rows, and a rope was to be used for that purpose. Each household was given 2.5 kg. Those receiving seeds would would have to return twice the amount (5 kg) to the village office after harvesting, to be redistributed within the village the coming year. If someone was caught not sowing the seeds in the prescribed way, they would be taken to the village office and would have to pay a fine (TSh5000 as quoted by informants, TSh1500–2000 by village leaders).[15,16] Informants said that there had been no prior consultation on which seeds would be distributed. They had just appeared one day and farmers had no choice of seeds other than the one variety that had been chosen. The Agricultural Officer in Nzali explained that he had requested that the district provide other varieties, but that selection was decided at the district level.

The seed support embodied what village and ward leaders saw as part of a major challenge to improve food security in the area, namely to make farmers change their behaviour, including 'to teach them to plant in straight rows'. Farmers, on their side, considered the improved varieties to give better yields than their local varieties during drought years. Probing further into how they considered improved varieties in comparison with local ones, respondents held that they planted them to save them from hunger, out of a desire to get food or cash or because they were told to. As expressed by a female head of household in Nzali: '[We] are given the seeds for free . . . that's why [we] are growing them but not because [we] like them very much.' They matured early, which meant that farmers could obtain harvest while waiting for slower maturing varieties to yield, especially millet.

Views on the criteria of ploughing, applying manure and planting in straight rows differed among the informants, but some common themes emerged. The value of ploughing and manure per se was not questioned by any informants, but it was largely a matter of affordability and practicality. Farmers showed less understanding of the demand to plant seeds in straight rows, which differs from the traditional Gogo method of scattered sowing. Officials justified it with the need for appropriate spacing, and that it would increase yields. Discussions with people engaged in sowing in straight rows revealed that it was not clear to them what the benefits were. They reported they were doing it only because the government told them to. The typical response in interviews and informal discussions were that 'the experts have told us to do it', with some respondents adding that they saw 'no benefit', or 'maybe that it looks nice'. One young female farmer in Kigwe (18–29 years, poor wealth group) said she 'did not know, the government hadn't told them why'. Some expressed confusion over the messages from the government and did not know what to believe. The claim of bigger harvest was also disputed, typified by one farmer saying: 'We used to get big harvests by planting traditionally, but now we don't know anymore. They are confusing us Gogo people.'

A number of farmers could be observed preparing their lands and sowing in straight rows after receiving seeds (Fig. 8.1). It became clear, however, that not everybody was following the criteria that came with the provided seeds and, importantly, that farmers had their own

Fig. 8.1. Farmer preparing land for sowing in accordance with government directives.

strategies for handling the directive. Informants said that while they planted the farms along the main road in straight lines, they did not do so in the farms further away. Informants justified this by the fact that it was more time consuming and labour intensive than their traditional ways. Even access to ropes to guide the planting was a constraint to many. With labour and time having a high opportunity cost, they could ill afford any delays. Some said they planted in traditional ways because they did not see any benefits; they had big harvests in the past while planting in the traditional way. An interviewee said that since their fields were far away, they were out of reach of the government: 'It is far, they can't see there, they can't catch us there.'

Views from farmers contrasted with the views of village leaders. While acknowledging the problems many farmers had with access to ox-ploughing, they also saw farmers' responses and non-adherence to the rules in terms of lack of knowledge, willingness to learn and physical ability, as illustrated by one official in Kigwe, saying, 'Some are ignorant, others are old and can't plant with a rope to get the spacing right.'

Discussion and Conclusions

The preceding sections have shown how many actions and strategies for coping with drought as they play out in practice in Dodoma can be considered as forms of everyday resistance; circumvention of rules and regulations farmers do not understand, or consider unfair or impossible to uphold. Local knowledge – in the form of people's reference to past practices, knowledge of crop varieties and perceptions of linkages between trees and rainfall – play a key role in shaping the responses and the motivations and justification for the resistance.

Interestingly, in terms of the two examples given above, namely the need for drought-resistant varieties and forest conservation, farmers' views and perceptions did not differ markedly from those of the local leaders and the externally imposed regulations they implemented. Farmers desired improved crop varieties, and they expressed many of the same types of concerns over forest loss as the leaders and officials did. Many farmers also referred to their fellow villagers' practices as a key problem in why farmers still depended on food aid. For example, there were complaints that many villagers, and especially the youth, were no longer adhering to Gogo traditions, that people were 'stealing' when they cut down large trees and that the people in general had become more 'cunning'. Thus, the same broad narrative from leaders was found among farmers, namely that farmers' practices were breaking rules and causing problems for the community.

However, a more detailed look reveals some significant differences in views and understanding. Firstly, it is clear that the practical options presented through external interventions were at odds with farmers' knowledge, understanding and preferences. The drought-tolerant seeds they received were considered low-quality food, were easily attacked by pests and did not keep well in storage. This is in line with other studies of farmers' views of modern crops compared with their own varieties (Friis-Hansen, 1999; Mwanga *et al.*, 2005). Farmers' responses also demonstrate that they did not understand the reason for many of the interventions. While they agreed with forest conservation, they also saw the forest as a necessary source of land for agriculture, which it was in the past. By tradition they were allowed to cultivate in the forest, as long as they did not cut large trees of specified species. Such conflict between traditional rules, specifying particular sacred areas or particular trees, and modern legislation, protecting whole areas, is well known in Africa (e.g. Serra, 2001).

Secondly, there was a clear mistrust towards the government's ability to provide assistance, and farmers saw the government as the cause of many of their problems. The distribution of seeds, for example, was seen as coming too late, and in too small amounts, and there was a perception that only leaders were receiving seeds. A typical response was that 'the leaders say the seeds will arrive, but they never do'. Farmers also stressed that they did not know beforehand what type of seeds they would receive and if they would receive them. Neither did they know exactly why they had to plant them in straight rows. As shown, there were also examples of allegations of mismanagement and corruption, further strengthening the atmosphere of mistrust.

Thirdly, the implementation of seed supply and forest regulations narrowed farmers' livelihood options, especially for the poorest, who could not access the labour and inputs needed to sow the seeds. The use of government-provided seeds could be seen as something farmers did because they had to, something that was out of their control, but which they nevertheless adapted to as best they could. Through this process, farmers' own knowledge and understanding were at best ignored, or at worst undermined. Farmers were not allowed to plant seeds before a certain date; they had to plant in a way that was not traditional for the Gogo, and they were not allowed to use all the varieties they had wanted. In case of conflict between what they themselves thought and the external regulations, farmers thus found their way around regulations, particularly in cases where the government was not able to reach. A regional agricultural official even acknowledged that she knew the farmers made 'one farm for the leaders, and one for themselves'.

But where does this leave us on the role of local knowledge and adaptation? To answer this, we shall have to look at whether the actions address the root causes of (current) vulnerability to climate risks, and in turn, whether the outcomes are adaptive in a way that increases flexibility and options for the future, and avoiding any particular 'lock-ins' that are difficult or costly to reverse. There is also an issue of scale: Eriksen *et al.* (2011) showed how one person's adaptation may be another one's maladaptation. Roncoli *et al.* (2001) demonstrated that while people may adapt, there are 'painful trade-offs' along the way that mean that they may survive in the short term, but which will undermine their livelihoods in the long term (see also Pelling 2011; Ford *et al.*, 2013).

The picture emerging is one of farmers trying to make do with an increasingly narrow range of farming options. Farmers' responses give insights into why they seem to be slow to take up new practices, and the paradox of why certain traditional practices persist: rather than being motivated by a wish to continue using their traditional practices (which they see as insufficient), or not understanding the benefits of new technologies, responses suggest that the lack of uptake often has more to do with households' asset and labour limitations, and a view that the alternative options they are being presented with are not going to support them. These findings may also go some way towards explaining the apparent contradiction between considering current external assistance as a hindrance, but still wanting more of it, which must be seen in relation to the way external interventions are undertaken at present. The authoritarian way of implementing government regulations in Nzali is a case in point.

The findings reinforce the need for considering the local level as an arena of negotiation between many different interests, embedded in formal as well as informal institutions, and where the relative interests, powers and capabilities are shaped by social and historical legacies. As shown, local knowledge is socially embedded and not necessarily a reflection of socially just or equitable structures (Scoones and Thompson, 1994; Agrawal, 2002), and, as pointed out in a study from Burkina Faso by Roncoli *et al.* (2001), farmers' resourcefulness in coping with climate variability is accompanied by significant and unequally distributed costs. In this perspective, understanding local knowledge can be seen as an entry point for addressing and understanding historical aspects of coping and adaptation, people's agency in responding to changing circumstances, their preferences, as well as the constraints and barriers experienced by different social groups in a particular location.

The argument advanced in this chapter is that the role of local knowledge needs to be understood in the context of adaptation as a political and power-laden process; and that adaptation responses and outcomes emanate from an interplay between the knowledge systems, the ability to use this knowledge, and a situated understanding of the knowledge systems. In conclusion, the primary role of local knowledge may be seen as increasing understanding of, and directing attention to, the constraints and enablers to addressing the root causes of vulnerability to climate change. This chapter has shown how resistance to formal or informal authority is closely linked to people's understanding of their environment. The question is perhaps not whether or not local knowledge should be considered or integrated, but understanding these complex and 'messy' processes and responding to how livelihoods are produced and sustained.

Notes

[1] Local knowledge is used here as a term that brings together local, indigenous and traditional knowledge, as an embedded knowledge–practice–belief complex (Berkes, 1999) and their many derivations and sub-categories (Antweiler, 1998). There are important distinctions among these terms and this is an imperfect approximation, but one that is suitable for the scope of this chapter.

[2] Annex in FCCC/CP/2015/L.9/Rev.1, available at https://unfccc.int/resource/docs/2015/cop21/eng/l09r01.pdf

[3] See e.g. www.manystrongvoices.org

[4] Chinyami was classified as a HADO reserve in 1989 (District Natural Resources Office, Dodoma Rural District).

[5] In-depth interviews and informal interviews in Kigwe.

[6] Interestingly, most of the clear-felled areas in the forests in Nzali were out of view from the road through the village. Rather than being a reflection of traditional rules, it is not clear whether or not this was to respect traditional rules. Another possible explanation is that it was to hide forest clearing from particularly District and Region level officials when they were passing through the village.

[7] Corrugated-roofed houses were, however, deemed less safe than traditional houses, especially during the rainy season, when the noise of the rain on the roof meant that the occupants were not able to hear thieves and burglars.

[8] At the time of fieldwork, there had been discussions in Nzali about the possibility of moving the HADO boundary to allow cultivation in some areas and alleviate some of the land shortages.

[9] Agricultural official, Dodoma Regional Office, Dodoma.

[10] Officials, Dodoma District Office, Dodoma.

[11] *Lugugu* refers to a tall local sorghum variety, but was also used as a collective term for local sorghum varieties.

[12] The official came to the meeting and asked for permission to make an announcement towards the end of the meeting (group discussion in Nzali, November 2006).

[13] The official forecast was that the rain would start the week of 19 November.

[14] Farmers were being pursued by the village leadership for starting to plant before this date. At the same meeting where the announcement was made, some of the leaders were summoned to pursue a farmer who had started to plant before he was allowed. On a later occasion, a village official expressed dismay of many farmers not following the rules: 'We are going to punish them, we are tired of farmers [not following the rules]' (field observation, Nzali, November 2006).

[15] Head of village emergency committee

[16] At the time of fieldwork, the exchange rate was roughly TSh2000 to UK£1.

References

ACIA (2004) *Impacts of a Warming Arctic: Arctic Climate Impact Assessment*. ACIA Overview report. Cambridge University Press, Cambridge, UK.

Agrawal, A. (2002) Indigenous knowledge and the politics of classification. *International Social Science Journal* 54(173), 287–297.

Antweiler, C. (1998) Local knowledge and local knowing: an anthropological analysis of contested 'Cultural Products' in the context of development. *Anthropos* 93(4–6), 469–494.

Berkes, F. (1999) *Sacred Ecology: Traditional Ecological Knowledge and Resource Management*. Taylor & Francis, London.

Berkes, F. (2002) Epilogue: making sense of Arctic environmental change? In: Krupnik, I. and Jolly, D. (eds) *The Earth is Faster Now: Indigenous Observations of Arctic Environmental Change*. Arctic Research Consortium of the US (ARCUS), Fairbanks, Alaska, pp. 335–349.

Brokensha, D.W., Warren D.M. and Werner, O. (eds) (1980) *Indigenous Knowledge Systems and Development*. University Press of America, Washington, DC.

Brooke, C. (1967) Types of food shortages in Tanzania. *Geographical Review* 57(3), 333–357.

Brown, K. (2015) *Resilience, Development and Global Change*. Routledge, London.

Burton, I., Kates, R.W. and White, G.P. (1978). *The Environment as Hazard*. Oxford University Press, Oxford, UK.

Byg, A. and Salick, J. (2009) Local perspectives on a global phenomenon – climate change in Eastern Tibetan villages. *Global Environmental Change* 19(2), 156–166.

Castro, A.P., Taylor, D. and Brockensha, D.W. (2012). *Climate Change and Threatened Communities: Vulnerability, Capacity and Action*. Practical Action Publishing, Rugby, UK.

Chambers, R. (1983) *Rural Development: Putting the Last First*. Longman, Harlow, UK.

Crate, S.A. and Nuttall, M. (eds) (2009) *Anthropology and Climate Change: From Encounters to Actions*. Left Coast Press, Walnut Creek, California.

Ellen, R.F., Parkes, P. and Bicker A. (eds) (2000) *Indigenous Environmental Knowledge and Its Transformations: Indigenous Critical Anthropological Perspectives*. Routledge, London.

Eriksen, S., Aldunce, P., Bahinipati, C.S., Martins, R.D.A., Molefe, J.I. et al. (2011) When not every response to climate change is a good one: identifying principles for sustainable adaptation. *Climate and Development* 3(1), 7–20.

Eriksen, S.H., Nightingale, A.J. and Eakin, H. (2015) Reframing adaptation: the political nature of climate change adaptation. *Global Environmental Change* 35, 523–533.
Ford, J.D., McDowell, G., Shirley, J., Pitre, M., Siewierski, R. et al. (2013) The dynamic multiscale nature of climate change vulnerability: an Inuit harvesting example. *Annals of the Association of American Geographers* 103(5), 1193–1211.
Friis-Hansen, E. (1999) *The Socio-Economic Dynamics of Farmers' Management of Local Plant Genetic Resources – A Framework for Analysis with Examples from a Tanzanian Case Study*. Centre for Development Research, Copenhagen.
Guthiga, P. and Newsham, A. (2011) Meteorologists meeting rainmakers: indigenous knowledge and climate policy processes in Kenya. *IDS Bulletin* 42(3), 104–109.
Hewitt, K. (ed.) (1983) *Interpretations of Calamity from the Viewpoint of Human Ecology*. Allen and Unwin, London.
Kalland, A. (2000) Indigenous knowledge: prospects and limitations. In: Ellen, R., Parkes, P. and Bicker, A. (eds) *Indigenous Environmental Knowledge and its Transformations: Critical Anthropological Perspectives*. Harwood Academic Publishers, Amsterdam.
Lebel, L. (2013) Local knowledge and adaptation to climate change in natural resource-based societies of the Asia-Pacific. *Mitigation and Adaptation Strategies for Global Change* 18(7), 1057–1076.
Leonard, S., Parsons, M., Olawsky, K. and Kofod, F. (2013) The role of culture and traditional knowledge in climate change adaptation: insights from East Kimberley, Australia. *Global Environmental Change* 23(3), 623–632.
Maddox, G.H. (1988) Leave Wagogo, you have no food: famine and survival in Ugogo, Tanzania, 1916–1961. PhD thesis, Northwestern University, Evanston, Illinois.
Mwanga, J., Letayo, E., Nonga, D., Senyagwa, A., Lugeye, E. *et al.* (2005). *A Study of Local Knowledge in Relation to Management of Agro-Biodiversity and Food Security in Semi-arid Central Tanzania*. Consolidated Intermediate Report. LPRI-Mpwapwa, and FAO LinKS Project, Dar es Salaam, Tanzania.
O'Brien, K., Eriksen, S., Nygaard, L.P. and Schjolden, A. (2007) Why different interpretations of vulnerability matter in climate change discourses. *Climate Policy* 7(1), 73–88.
Patt, A.G. and Schröter, D. (2008) Perceptions of climate risk in Mozambique: implications for the success of adaptation strategies. *Global Environmental Change* 18(3), 458–467.
Pelling, M. (2011) *Adaptation to Climate Change: from Resilience to Transformation*. Routledge, London.
Pepin, N. (1996) Indigenous knowledge concerning weather: the example of Lesotho. *Weather* 51(7), 343–348.
Petheram, L., Zander, K.K., Campbell, B.M., High, C. and Stacey, N. (2010) 'Strange changes': indigenous perspectives of climate change and adaptation in NE Arnhem Land (Australia). *Global Environmental Change* 20(4), 681–692.
Reid, H., Alam, M., Berger, R., Cannon, T., Huq, S. and Milligan, A. (2009) Community-based adaptation to climate change: an overview. *Participatory Learning and Action* 60(1), 11–33.
Richards, P. (1985) *Indigenous Agricultural Revolution: Ecology and Food Production in West Africa*. Hutchinson Education, London.
Riedlinger, D. and Berkes, F. (2001) Contributions of traditional knowledge to understanding climate change in the Canadian Arctic. *Polar Record* 37, 315–328.
Rigby, P. (1969) *Cattle and Kinship among the Gogo: a Semi-pastoral Society of Central Tanzania*. Cornell University Press, Ithaca, New York.
Roncoli, C., Ingram, K. and Kirshen, P. (2001). The costs and risks of coping with drought: livelihood impacts and farmers' responses in Burkina Faso. *Climate Research* 19(2), 119–132.
Scoones, I. and Thompson, J. (eds) (1994) *Beyond Farmer First: Rural Peoples' Knowledge, Agricultural Research and Extension Practice*. Intermediate Technology Publications, London.
Scott, J.C. (1985) *Weapons of the Weak: Everyday Forms of Peasant Resistance*. Yale University Press, New Haven, Connecticut.
Serra, A. (2001) *Legitimacy of Local Institutions for Natural Resource Management: the Case of M'Punga, Mozambique*. Marena Research Project Working Paper no. 3. School of African and Asian Studies, University of Sussex, UK, and Centro de Experimentação Florestal, Sussundenga, Mozambique.
Sillitoe, P. (1998) The development of indigenous knowledge – a new applied anthropology. *Current Anthropology* 39(2), 223–252.
Thrupp, L.A. (1989) Legitimising local knowledge: from displacement to empowerment for third world people. *Agriculture and Human Values* 6(3), 13–24.
Torry, W.I. (1978) Natural disasters, social structure and change in traditional societies. *Journal of Asian and African Studies* 13(3–4), 167–183.

URT (2003) *Disaster Vulnerability Assessment (Phase II)*. Prime Minister's Office, Dar es Salaam, United Republic of Tanzania.

URT (2007) *National Adaptation Programme of Action (NAPA) for Tanzania*. Vice President's Office, Division of Environment, Dar es Salaam, United Republic of Tanzania.

Warren, D.M., Slikkerveer, L.J. and Brokensha, D. (eds) (1995) *The Cultural Dimension of Development: Indigenous Knowledge Systems*. Indigenous Knowledge and Development Series, Intermediate Technology, London.

Watts, M. (1983) On the poverty of theory: natural hazards research in context. In: Hewitt, K. (ed.) *Interpretations of Calamity from the Viewpoint of Human Ecology*. Allen and Unwin, London, pp. 231–262.

Watts, M.J. (2015) The origins of political ecology and the rebirth of adaptation as a form of thought. In: Perreault, T., Bridge, G. and McCarthy, J. (eds) *The Routledge Handbook of Political Ecology*. Routledge, London, pp. 19–50.

9 Indigenous Soil Enrichment for Food Security and Climate Change in Africa and Asia: A Review

James Fairhead,* James Fraser, Kojo Amanor, Dawit Solomon, Johannes Lehmann and Melissa Leach

Studies of indigenous soil knowledge or 'ethnopedology' identify how farmers differentiate and understand their soils and reveal their sophisticated appreciation of the biophysical characteristics of the landscapes that they inhabit (Barrera-Bassols and Zinck, 2003). Local ethnopedological frameworks share differences and similarities to scientific systems of categorization. Farmers often categorize soils according to colour, texture (i.e. sand/clay/ stone/silt content), geographical or topographical position, potential, and wider ecological interactions. Such knowledge informs soil fertility management practices (Adewole Osunade, 1995; Krogh and Paarup-Laursen, 1997; Birmingham, 2003; Gray and Morant, 2003; Niemeijer and Mazzucato, 2003; Osbahr and Allan, 2003; Ngailo and Nortcliff, 2007; Duvall, 2008). Yet whilst these studies show a complex appreciation of existing soils, none describes local people recognizing anthropogenic soils in their environments, or their formation. They are silent on the creation of enduringly enriched soils through endogenous practices and biological inputs. Soil science is also silent on this issue. We have searched almost in vain for studies that discern improved soils in Africa, let alone those that examine the formation, qualities and use of such soils (but see Fairhead and Leach, 1995; Solomon and Lehmann, 2000; Lobe *et al.*, 2001; Solomon *et al.*, 2007; Zingore *et al.*, 2007).

Many works show how farmers nurture and appreciate high-fertility patches within fields and wider landscapes (Scoones, 1991, 2001), practices of manuring and how its intensity is frequently greater near compounds and villages, establishing rings of heightened fertility in accordance with Von Thunen's location theory (Leigh, 1946; Pelissier, 1966; Ruthenberg, 1971; Prudencio, 1993). Yet these works speak of soil fertility 'maintenance' rather than enduring improvement and attribute any patches of improved soils to the ephemeral inputs of everyday life (manure, crop residues, ash, excreta, fish bones, etc.). Although this research appreciates the huge variety of 'fertilizing' practices developed by African farmers, the improvements are represented either as transient or as requiring the continuous import of fertilizing materials. There is virtually no attention to durable improvements to the soil in Africa, in strong contrast to the burgeoning literature concerning the upgraded and anthropogenic 'dark earth' or *terra preta* soils of Amazonia (e.g. Glaser and Birk, 2012).

Literature concerning 'landesque' capital has been more appreciative of enduring positive legacies of past land users (Håkansson and Widgren, 2014). Typically these works highlight

* Corresponding author; e-mail: j.r.fairhead@sussex.ac.uk

improvements to physical properties through terracing, irrigation and drainage structures, but the concept of landesque capital also embraces improvements to soil biological and chemical properties. Brookfield draws attention to the 'manufactured' soils of Europe (plaggen) and those in Papua New Guinea and the Amazon (Brookfield, 2001). Accordingly soils should be understood within the balance between investment and disinvestment (as distinct from the balance between the natural history of soil formation and social history of soil degradation). Yet as Widgren rightly remarks on Brookfield's assessment of the global distribution of landesque capital, 'the most difficult estimation of distribution concerns anthropogenic soils' (Widgren, 2007). Brookfield does draw on one paper from Africa (Strømgaard, 1992) which acknowledges long-term improvements in soil cation exchange capacity at depth in soils under the Zambian *chitemene* system in which soils are enhanced by wood ash, but basically Africa does not come out well in the survey of landesque capital. It has 'rich occurrences of terraced agriculture and irrigation as well as farming systems that have used different kinds of mounds, ridges, etc.' but this is nothing like other regions of the world (Widgren, 2007).

Our recent 'discovery' of African Dark Earths (AfDE) demonstrates, however, that West African farmers can durably improve what were previously thought to be 'inherently' infertile tropical soils (Solomon *et al.*, 2016). Anthropogenic AfDE sampled in Ghana and Liberia have 200–300% more organic carbon than the soils from which they are derived, with 2–26 times more pyrogenic carbon (PyC), which is more effective in promoting nutrient retention than other non-pyrogenic organic carbon. PyC is also highly persistent in soil and therefore not only important for long-term improvements of soil fertility; it also possesses potential for carbon sequestration to mitigate climate change. Compared with the strongly acidic (pH 4.3–5.3) soils in which they form, anthropogenic AfDE also exhibit higher pH (5.6–6.4), 1.4–3.6 times greater cation exchange capacity and more plant-available nutrients (1.3–2.2 and 5–270 times more nitrogen and phosphorus, respectively) (Solomon *et al.*, 2016).

This chapter reveals that practices leading to enduring improvements to African soils are much more widespread in sub-Saharan Africa (SSA) but have been overlooked in ethnopedology and soil science (Fairhead *et al.*, 2012). Curiously these soils have been noticed by ecologists and archaeologists who identify the sites of past settlements by their luxuriant and specific vegetation, hundreds if not thousands of years after their abandonment (Keay, 1947; Ramsay and Rose Innes, 1963; Blackmore *et al.*, 1990; White and Oates, 1999). Archaeologists have frequently been frustrated by farmers who preferentially seek out the sites of ruined settlements for their fields and muddy their horizons. This chapter considers their agricultural significance.

A Typology of Indigenous Soil Enrichment

The two areas of practice that are considered here which produce enriched soils concern: (i) ruined settlements (rendered rich in organic material and biochar) that permit intensified production; and (ii) techniques of slow anaerobic burning of vegetation under a soil covering, sometimes called soil burning, burn beating, denshiring or (in French) *écobuage*, including the impact of charcoal manufacture on soils. Permutations of these practices that have been identified from the literature are probably only a small fraction of extant soil-enhancing practices that may even include regular shifting cultivation in humid conditions when incomplete combustion produces back carbon (BC) or PyC, but this is not considered in this chapter.

Former settlement sites

To assess the agricultural significance of former settlement sites we contacted a network of Africanist archaeologists. This revealed evidence from Central, West and North East Africa. The best documented cases concerning the importance of ruined settlements to African farming are the mining of the midden sites of Pharaonic Egypt (Bailey, 1999). The mixture of mud-brick remains (of nutrient-laden Nile silt) and ancient organic waste of the middens in ancient Egyptian cities and villages is referred to as *sebakh* and is valued by farmers. In the cities, household rubbish and animal excreta were constantly dumped over hundreds (even thousands) of years, causing the streets to rise ever higher and the ground floors of houses to become basements due to the increasing depth of the rubbish (Bailey, 1999). The importance of these middens (some covering six square miles or more) as a fertilizer was recognized in the early 19th century when they were mined on an industrial scale. Large companies built rail tracks to transport and sell it as fertilizer throughout the Nile Valley. Fields throughout Egypt became 'archaeologically polluted' by the ceramics that had been distributed. *Sebakh* digging was banned in the 1930s but still persists locally.

Cultivating old settlement sites is common to many regions in Ethiopia. Temesgen Burka (personal communication) reported that among Oromo of western Ethiopia, '*ona*' refers to ruined or abandoned settlements which are appreciated for their rich soils. They are used especially by women for their homegarden crops. Fertile soils form as ruined houses collapse and decompose, and mix with the soils enriched by the ash and human and animal manure from the years of settlement. Domestic animals were kept in the houses. Threshing places acquire similar properties and are used by women for vegetable planting. This practice apparently has a spiritual or symbolic function of associating oneself with ancestors. T. Kathryn Arthur (personal communication) reported that Gamo speakers of southern Ethiopia cultivate soils of ruins, as they are rich and fertile due to the decomposition of animal and plant remains.

Pierre Kinyock (personal communication) reported that throughout Cameroon, farmers use old settlement sites for new farms. The Cameroonian Agricultural Research Institute used to collect such soil to fill the plastic bags for their seedlings. On the borders between Nigeria and Cameroon, Nicholas David and Judy Sterner (personal communication) reported that inhabitants of the Mandara Mountains appreciate the fertility of abandoned compounds – so much so that farming is the prime agent of their destruction and the dispersal of potential archaeological sites. Once abandoned, farmers quickly take advantage of their rich soils. Old owners live nearby to work the sites. They either do not revere these sites, or, if they do, farming is not anathema to reverence. Demanding crops such as tobacco and maize that are usually associated with older men are often the first crops to be grown there. Gradually buildings disappear as they are assimilated into the larger system of terraces or fields. Paul Basu (personal communication) reported that in northern Sierra Leone at a site called 'Old Yagala,' which is a deserted hilltop settlement not far from the town of Kabala, villagers intensively cultivate areas among ruined buildings. They focus particularly on the slopes where household waste was clearly thrown, and where there is plenty of midden material. It can be noted in passing that the improved soils of ruined settlements were also observed in Asia by German botanist Wilhelm Sulpiz Kurz, including around the ancient town east of Rajmehal, Bengal, and on the ruined pagodas of Myanmar (Kurz, 1875). Just as in Africa, but with one notable exception of East Kalimantan (Sheil *et al.*, 2012), these have not been the focus of inquiry to date, but warrant further research.

Charring vegetation in anaerobic conditions under soil

Agronomists have long observed a second set of practices that more clearly enriches the soil in BC that involves charring vegetation under a covering of soil. Grass or bush/tree vegetation is cut, dried and then covered with 5–10 cm of soil and ignited. Farmers craft the conditions so that the vegetation burns very slowly 'without a flame', often for a week or two. The biomass is not thermally oxidized completely to ash so a significant proportion remains as BC-rich char or 'biochar' if used in the context of a soil amendment (Lehmann, 2007). Afterwards, the soil, biochar and ash mix can be farmed in situ on the mound or is distributed over the field. These practices long attracted the negative attention of colonial agronomists from the 1880s, and were reviewed extensively by Portères (1972). This technique is not restricted to Africa and the tropics. It was popular in Europe from the 17th to the 19th century when farming clay and acidic soils. Many effects of this practice have been examined by agronomists, but none has attributed any importance to its most obvious effect: the production of biochar, which largely remains to be researched.

The technique is variously called soil burning, burn-beating, sod burning (suggesting that the main effect is in the mineralization by heat of organic matter/roots in the soil), paring (Rennie, 1834), or denshiring (after Devonshire in England, where it was practised). The prevalence of this technique has remained obscure, partly due to its varied nomenclature. It is labour-intensive, requiring about 100 person-days/ha, as opposed to other techniques of field clearance and preparation (clearing and burning) that require perhaps 20–40 person-days/ha (Portères, 1972). That people put in so much extra labour suggests the importance of its soil transformations beyond the release of nutrients by fire. To date these have been understood in relation to soil structure. The advent of interest in biochar as a soil conditioner now suggests, however, that the BC-rich char itself may be important. Thus whether this extra labour is simply the labour of obtaining a crop, or can also be considered a longer-term 'investment' in soil improvements (since the biochar will remain in soil for much longer than the ash), remains an open question.

Research into these techniques has been sporadic. Portères compiled a thorough review of the practice across Africa, suggesting that the technique was used: (i) as a last resort on poor soils when there was no other way to achieve a satisfactory crop; (ii) to farm previously uncultivated or uncultivable land; or (iii) to restore land that had degraded to pasture (Portères, 1972, p. 153). Yet this is not the whole story. As summarized in Table 9.1, which collates descriptions of these practices in Africa and Asia and what is known of the stated reasons for their use, the technique is used variously: on highland grass soils (e.g. in Ethiopia, Guinea's Futa Djallon and Assam in India) to render them suitable for grain crops (e.g. barley and rice); on clay and compacted soils to render them friable and suitable for root/tuber crops; and on unspecified soils to render them suitable for rice. Whilst the technique has been understood as driven by the short-term benefits that it produces, these have generally been understood as balanced by the long-term 'degradation' that 'soil burning' has. That short-term benefits might also dovetail with long-term benefits associated with biochar production has not been appreciated or investigated.

Impacts on Soil

The practices are diverse but all include PyC into the soil from slow-burning fire hearths and from the regular burning of trash. Accumulation of charred material has been found to: (i) transform soil fauna, encouraging ants and earthworms (over caterpillars, millipedes, springtails) that ingest, powder and defecate biochar, mixing it down the soil profile (Topoliantz and Ponge, 2005; Topoliantz *et al.*, 2006; Ponge, 2013); (ii) change soil microbial community (O'Neill *et al.*, 2009; Whitman *et al.*, 2016); (iii) augment phosphorus and calcium (from excrements and bone residues) and improve its plant availability (Sato *et al.*, 2009; Zwetsloot *et al.*, 2015); (iv) reduce acidity; and (v) build up plant-available N through improved retention (Lehmann *et al.*, 2003). It is likely that there are also positive feedbacks within an increasingly productive ecosystem, apparently effecting a durable transformation, 'ratcheting up' the soil ecosystem into a more productive state.

Considerable research explores the impact of these techniques on soil structure and chemistry but no research considers either their impact on biochar or the long-term impact of these techniques more generally. Such analysis as there is only picks up the short-term effects, for which results are strongly consistent across all cases. Here we review research on the direct effect of infield, under-soil charring – and we would be the first to admit that this raises as many questions as it answers.

Firstly, the heat causes the clay fraction in these soils to aggregate into pseudo-sands, helping to make friable, lighter soils. Mboukou-Kimbatsa (1997), Praquin (1976) and Kassapu (1979) all showed that clay is destroyed (from 19% to 20% in one case study), soluble phosphates are released and clay particles coagulate into coherent aggregates that gradually decay back to clays. There are also results showing that the reversal does not happen but that there are permanent changes in mineralogy to more amorphous minerals that fix more phosphorus which is unavailable to plants (Ketterings *et al.*, 2000, 2002).

Mesfin (1981) found that the sand fraction increased (temporarily) from 34% to 78% and the clay fraction fell from 33% to 4%. This leads to improved surface porosity and infiltration.

Table 9.1. Exemplars of indigenous soil enrichment.

Place and Name Locality	Practice	Agroecological zone	Agricultural system	References
Guinea				
Futa Djallon Plateau and westwards towards the coast among Fula, Susu, Dialonke and Baga speakers	Locally called *muki* or *moki*, meaning 'to be consumed slowly by fire', derived from Malinke *mugu-mugu*, 'to be consumed slowly just until forming ash'. On grasslands, grass root balls ('plaques') are hoed up, dried and, as the rainy season approaches, are piled 75–200 cm upside down. A hole is made to the bottom of the pile and filled with grass, and flammable wood or dried cattle dung is inserted. The whole is covered with soil 5–6 cm deep, 'just right' for the necessary combustion. The base of the ignition hole/chimney is then lit. Farmers thus make a field of small 'volcanoes' out of which initially emerges thick smoke. After several days of charring, the piles cool and the soil is raked out, and sown with rice	(a) Kollade: the almost uncultivable, poorly drained soils that are hydromorphic in wet season and dry and hard in the dry season, usually with short grass vegetation (*Loudetia*) (b) Bowal: equally poor soils around ferruginous outcrops (c) Don'kere (valley soils: more or less inundated, eluvial, by runoff or alluvial by water courses	Used for one cropping cycle of West African rice (not eleusine millet) or tubers followed by 10–50 years of fallow. In some regions this was practised only in certain '*muki* years' (perhaps 1 in 9). The regularity may depend on cattle dung so its diminution may link to cattle scarcity	Arcin, 1907; Anonymous, 1923; Levare, 1925; Richard-Mollard, 1947; Sudres, 1947; Adam, 1958; Adames, 1962
Chad				
Banks of Logone river, near Bogor	In the dry season, tall grasses are hoed up with their roots, put in ridges, dried, covered with soil and burnt. When cool, once the rains come, millet (eleusine, fonio) is planted on the mound sides and covered over with a little of the ridge soil. The ridge itself is planted with four rows of rice. Rows are 4–6 m apart. Before the grasses are fired by pastoralists, the grass root plaque is uprooted and laid in rows, dried, covered with soil and fired at one end, the fire eventually travelling to the other. In other locations, the ridges are opened and distributed over part of the intervening furrow. Rice is sown on this and on the opened ridge (doing better on the ridge). This technique needs as much grass as possible, which can be achieved in 1–4 years	On seasonally inundated river plains	For rice, millet (eleusine, fonio) and taro (*Colocasia*)	Bouteyre, 1955; Gaide, 1958; Bezot, 1966

Continued

Table 9.1. Continued.

Place and Name Locality	Practice	Agroecological zone	Agricultural system	References
Cameroon				
	Farmers bury and burn vegetation in raised beds. Thus vegetation growing on a previous ridge is removed, gathered in the intervening furrows and dried. The furrow is then covered with soil from the previous bed, making a ridge over the furrow, and a furrow where the ridge had been. At regular intervals on each ridge the farmer makes a grass chimney down to the bottom. The vegetation is lit and burns slowly This technique appears to create improved soil structure such as penetration for the yam root crop. Where land prepared for maize and *Colocasia*, grasses are buried in a similar way, but not burnt	Bamileke grass fields	For yams but ridges initially planted with squash and an indigenous vegetable (*Solanum nodiflorum*), with yams planted a month later	Jacques-Felix, 1947; Jacques-Felix and Betremieux, 1949; Inter-African Bureau for Soils and Rural Economy, 1958; Praquin, 1976; Kassapu, 1979; Fotsing, 1992
Cameroon				
North, in flood zones of the Lagone River	Towards the end of the dry season the vegetation, basically of *Andropogon* grass, is uprooted, placed in swaths 4 m apart and covered with earth 50–100 mm deep with occasional gaps. It is ignited and chars slowly for several days. The swaths subside gradually but retain ash and biochar residues			Inter-African Bureau for Soils and Rural Economy, 1958
Ghana				
Asuano (a single innovative farmer in Brong Ahafo)	Five-year bush fallow is felled, dried, piled, covered with earth and fired. The charred product is spread over the plot(s). The motive for repeated charring is not the maintenance of benefits, but to build them gradually. Periodic inputs by bush fallow charring are enhanced by annual charring of crop residues		Maize, plantain, cocoyam	Oguntunde *et al.*, 2004, 2008; Sohi and Yeboah, 2009

Ethiopia				
Highlands	*Guie* (*gye*) or 'soil burning.' Grass sods (root masses to which soil adheres and the grass itself) that have grown during several years of fallow are uprooted at the end of the rains with the traditional 'mixing' hoe, are dried and piled up to 80 cm high (diameter 160 cm) with a density of about 800–1600 piles/ha). Whether some soil is added on top is unclear. Piles are ignited with some charcoal and burn slowly for up to 2 weeks. They produce a burnt layer in the centre and then a carbonized layer, a heated layer and an outer cover layer. When cooled, these are distributed over the field		Barley for two successive seasons.	Legesse, 1974; Abebe, 1981; Mesfin, 1982; Roorda, 1984, 1988; Sertsu, 1987; Pulschen and Koch, 1990; Mertens *et al.*, 2015
Congo and Gabon				
Niari valley among Babembe, Bakamba, Bakougnis, Bateke and Pool (Bahangalas) speakers	*Maala* burn-beating system. Grass is uprooted, and piled in parallel mounds about 2 m × 0.9 m, and dried. Branches are placed on top and then all is covered with soil. When lit they burn for several days. There are many permutations (rectangular and round)	On clay soils and drained ferrallitic soils (Ultisols). This technique restores soils degraded during mechanization	First season: squash, maize, courgette, yams, sweet potatoes, onion manioc, groundnut. Oseille de Guinée (*Hibiscus sabdarifa*), baselle (*Solanum negrum*), amaranthus, and tomato. Ground cover is always maintained. Second season: yam or sweet potato. Second year: opened up, and sown with groundnuts, perhaps intercropped with manioc. Third and fourth years: manioc, then fallow for 7–10 years	Sautter, 1953, 1955; Guilloteau, 1957; Martin, 1958, 1970; Soret, 1959; Jurion and Henry, 1967; Nyete, 1991; Mapangui, 1992; Mboukou-Kimbatsa, 1997

Continued

Table 9.1. Continued.

Place and Name Locality	Practice	Agroecological zone	Agricultural system	References
Malawi				
Chewa plains south-west of Lake Chilwa	Tall grass is collected together, tied in a knot and root tufts are struck to sever them. The bunches are left standing (like pre-mechanised shocks of corn at harvest) to dry. These are heaped, covered with earth, and burnt		Mapire and Egyptian dura (*Holcus sorghum*), pumpkins, cucumbers, cassava, sweet potatoes, tobacco and hemp	Livingstone, 1875
Malawi				
South of Lake Malawi	Grass and weeds are uprooted, dried and placed in flat heaps. Soil is placed over them making round mounds 2 m × 2 m × 0.8 m high. The burning is slow. In a separate practice, sods are separated from soil beneath and collected into flattened heaps, grass uppermost. When dry, fire is applied and slow combustion goes on; most of the products of the burning are retained in the ground and much of the soil is incinerated. Finally, subsoil round the mound is dug, is pulverized by hand and then thrown on to the heap. New soil is thus distributed over the ashes and burnt ground of the original heap and is very clear of weeds		Beans and maize and squash intercrops	Livingstone, 1866
DR Congo				
Uele, Eastern Azande, Katanga, on seasonally flooded plains	Grass is partially burnt, hoed up, dried, piled, covered with soil and fired			Goffinet, 1913; Dubois, 1957; Humblet, 1958

India				
South-west Gujarat, Western Maharashtra (Thana district); these are districts of heavy and continuous rainfall (2.5–4.25 m/year)	*Ráb* (meaning 'cultivation'): successive layers of cow-dung, tree loppings and brushwood (brought from the forest), shrubs, leaves and grass are piled up to about 90 cm, with earth placed on top to keep it all down, and then set on fire. It is not seeded until the rain actually comes. Among Guzrathi, more soil is added to temper the fire		Rice (finest kinds of transplanted varieties) and millet (*Eleusine coracana*, locally called *nagli*)	Voelcker, 1893; Mollison, 1901; Saldanha, 1990
India				
Assam, Shillong Plateau, and Khasi Hills, on pine forest highlands	Fields laid out in beds 2 m long × 1 m broad and separated from one another by about their own width. Each is overlaid with boughs from the nearest pine wood. A strip of earth 5 cm thick or more is skimmed from the space between the beds and laid on top of the boughs. When the boughs and twigs are really dry they are set alight and after they have burned out the earth is mixed with the ashes		Potatoes. In the following year the intervening spaces become the beds and so on	Bor, 1942; Awasthi *et al.*, 1981
India				
Assam, Shillong Plateau, Khasi Hills, on pine forest highlands	In a second version, grass sods are turned with a hoe early in the winter. They are dried over 2 months and then piled up, but with bunches of dried grass inserted. These grasses are lit and a slow fire is kept up. This is rather similar to European 'paring and burning.' The ashes are distributed		To grow a crop of rice, millet or 'Job's tears'. This appears to be similar in ecology, practice and intent to the practices in the Ethiopian highlands (above)	Gurdon, 1914
China				
	Crop residues are covered with soil and charred			Liebig, 1878

Improved porosity and infiltration also arise from other indirect effects. In particular, Mboukou-Kimbatsa (1997) revealed in the Niari valley system a change in biological activity generally, and specifically a change in the balance of macrofauna from termites to earthworms, and, relative to a control, an eventual increase in earthworm activity. This leads to improved aeration (see Legesse, 1974). The importance of the heat effect on soil biology is limited, given that it only affects a small portion of the soil at any one time, permitting the recolonization by soil macro- and microflora and fauna from neighbouring soil.

All studies suggest a short-term improvement in soil structure (see also Clinnick and Willatt, 1981; Nzila and Nyete, 1996), rendering clay soil suitable for tuber and other crops such as groundnuts – an improvement that several studies suggest will be temporary and outweighed by the loss in soil organic matter.

Secondly, the heating associated with this practice reduces the weed load and changes the proportion of remaining weeds from grasses (monocots) to broadleaf weeds (dicots), some of which are also 'indigenous vegetables' (Legesse, 1974). Pulschen (1987) found that, in Ethiopia, weed densities were only one-third of similar but unburnt controls, and there was a lower proportion of grass weeds. This reduction in weeds and indeed of insect pests is suggested by most authors.

Thirdly, all studies find that the heat effect produces a reduction in soil organic matter as against a control (Mesfin, 1981; Roorda, 1984; Mboukou-Kimbatsa, 1997) and biological activity is initially reduced due to disappearance of organic material and the burn (Mboukou-Kimbatsa, 1997). Yet although some of these researchers take these negative effects at face value, others find that a reduction in organic matter in the short term might be balanced in the long term by the way the transformed soils permit the more rapid regeneration of organic matter. Moreover, although Mboukou-Kimbatsa (1997) found that total carbon fell from 2.8% to 0.3% (cf. Mesfin, 1981) this loss seems to be limited in the soil burning by the presence of vegetation that is spared from complete combustion (Nzila and Nyete, 1996; Mboukou-Kimbatsa, 1997). This vegetal debris consists in part of vegetal charcoal, which is also not metabolized by the microorganisms and remains for a long time in the soil. As Lehmann and Joseph (2009, p. 200) described, the effect of carbon loss by charring is outweighed at a certain timescale by the greater persistence of the biochar. Other uncharred residues are quickly decomposed by soil microorganisms. Mboukou-Kimbatsa (1997) found that the overall effect is to reduce carbon on the surface but to increase carbon at depth. This suggests that biochar limits the diminution of overall soil carbon and that these techniques might lead, over time, to the development of dark earths at depth. In addition to leaching with percolating water, the processes of vertical movement of biochar in the soil are presumably also linked to mounding practices and to the earth-moving effects of soil fauna. This echoes the findings from studies of *chitemene* farming in Zambia, where piles of pollarded branches are burnt to generate ash, but also produce charred remains. Strømgaard (1992) suggested that the enduring increased cation-exchange capacity in the subsoil derived from this practice improved the efficiency of nutrient use in subsequent burn cycles.

Fourthly, the technique also leads to changes in soil chemistry as a result of ash and biochar accumulation. Reduced acidity (increased pH) has been shown by Mboukou-Kimbatsa (1997), Mesfin (1981) and Nzila and Nyete (1996). The last of these authors found that pH increased from 5.1 to 6.5. Reduced exchangeable aluminium (and thus reduced aluminium toxicity) is linked to this reduced acidity and was shown by Mboukou-Kimbatsa (1997). Increases in available potassium were shown by Bouteyre (1955) and Mboukou-Kimbatsa (1997). Increases in available phosphates have been shown by Mesfin (1981) (from 40 to 180 mg/100g), Nzila and Nyete (1996) and by IRAT (Kassapu, 1979). Care needs to be taken in interpreting this result, however, because the production of pseudo-sands through the heat effect can lock up phosphates (see, for example, Ketterings *et al.*, 2000, 2002)

Finally, the impact of these techniques on soil nutrients (such as nitrate, NO_3^-) is complex. Mboukou-Kimbatsa (1997) found overall increases in nitrates. Roorda (1984) found that total nitrogen (N) was lost in the burn layer but increased in other layers. Nzila and Nyete (1996) found the N loss as only marginal. Given that yields on these highly weathered soils are often critically dependent on potassium and phosphates and N can be procured through biological fixation of atmospheric N_2, this improvement might outweigh the slightly reduced nitrate availability. Associated with the loss of organic matter is a reduction in the cation-exchange capacity (Roorda, 1984; Mboukou-Kimbatsa, 1997). Yet this was not considered in relation to the full soil profile.

Research into these practices has examined the effect of 'soil burning' but not the effect of slow and often 'cool' charring on char production and the qualities that biochar brings to the soil. Where the impact of these techniques on biochar levels has been noted, it is so only in passing. There has been no significance attributed to it and to its long-term impact on the soil. Indeed, given that this technique has been understood to be about heating or burning the soil, some studies have attempted to examine the effects of soil burning simply by heating soils – thus missing out entirely the potential effect of the biochar which these techniques generate (e.g. Sertsu and Sánchez, 1978).

All those who have studied these systems appreciate the labour input and realize the gains in yields that they bring. Yet colonial and modern regimes have usually understood this as a short-term gain, in a system that they have assumed leads to long-term soil degradation (not investment). The contribution of the fire to soil degradation would appear to be visible in its short-term effects on soil organic matter and on cation-exchange capacity. No one has looked at the long term. Because these are understood to be degrading techniques, national policies have discouraged the practice and have sought to find alternatives, at least in Ethiopia, northern Cameroon and in Guinea. Only in the most recent studies of the practice in the Congo has the potential for this practice actually to rehabilitate degraded soils been considered – and this study itself does not explore the significance of biochar.

Authorities in Cameroon, Guinea and Ethiopia condemned the practice of *écobuage* and 'soil burning'. Curiously, however, other colonial foresters nevertheless adopted this technique as the basis for (re-)wooding savannas (Schmitz and Delvaux, 1958). Portères (1972) considered explicitly why this technique had not attracted more research attention, arguing that the technique had already been considered as 'very primitive' for a century and that the image of a practice on the path to extinction discouraged modern researchers from researching it. Colonial writers contributed to the idea of primitivity. The technique was used in Europe 'before the generalized use of fertilizer' (Jacques-Felix and Betremieux, 1949). Moreover, the concepts of 'burn-beating' and 'denshiring' sound old-fashioned in English, as does *écobuage* in French. These are the kinds of words to be found in obscure antiquarian books that discuss long-superseded ideas and practices that were perhaps best left to historians of science.

However, whilst most works consider this to be a technique in decline, the cases from Congo and Ghana suggests that it is also a technique that is being innovated and acquiring new importance. Certainly in Congo it remains important for farmers restoring land degraded by mechanization. Whilst the practices have been researched for their short-term effects, it is important for our argument to realize that no attention has been given in any of this research to their long-term effects and to the potential importance of the accumulation of biochar itself in the soil. This remains a potentially important research agenda. Given the potential importance of biochar, how does the char effect that has been left out of this calculus matter?

Charcoal production and iron-smelting sites

A third, associated set of practices enhancing soils through the addition of biochar concerns the kilns used to make charcoal. In several accounts farmers assert that areas around charcoal kilns and areas in the vicinity of ancient iron-smelting sites are enduringly improved. Peter Schmidt (personal communication) reported that in Buhaya, Tanzania (Bukoba District), farmers sometimes seek out features such as iron-smelting furnaces for their farming, explaining that the charcoal and ash improve the soil.

Oguntunde *et al.* (2004, 2008) examined the effects of charcoal production on soil physical properties in the forest–savanna transition zone of Ghana. As in 'soil burning practices', water conductivity, porosity and infiltration improved, soil pH increased, exchangeable minerals were much increased (with available potassium up by 329%) but organic C and total N decreased by 9.8% and 12.8%, respectively. Sand (and pseudo-sand) content increased as the clay fraction decreased. Overall, maize yields increased by about 50% relative to a control.

Within the yam farming areas around Kintampo in the Brong Ahafo region, in recent years many farmers have taken up charcoal production as a subsidiary activity, using the deadwood from trees preserved as stakes as a raw material. The circular patches on which charcoal is burnt are recognized by farmers as being highly fertile spots within the farm landscape, particularly for maize and vegetables. However, good yields are only achieved in these areas 2 years after burning, which suggests that there might be some microbial activities associated with charcoal remains which impart positive fertility characteristics to the soil.

Conclusion

Growing recognition of the need for more sustainable alternatives to industrial farming, such as agroecology (De Schutter, 2011), conservation agriculture (Hobbs *et al.*, 2008), organic farming (Foley *et al.*, 2011, p. 339) and 'ecological intensification' (Tittonell, 2014) has led to a revival of interest in the potential of traditional agroecosystems and soil enrichment practices. These have inspired many of the alternatives because they exploit locally available and renewable resources, emphasize nutrient recycling and biological maintenance, rather than inputs of exogenous and finite agro-chemicals to enhance crop production and food security. They also sequester a large amount of carbon in soils and could contribute to mitigation of climate change. In addition, these systems often rely on and conserve both local biodiversity and indigenous culture (Altieri, 2002; Gliessman, 2007; Perfecto *et al.*, 2009). Their study may reveal sustainable food production practices whose key traits can be transplanted to regions that are climatologically similar (or becoming similar) (Altieri and Nicholls, 2013).

Recent policy statements show a renewed interest in possible contributions of small-scale farmers and their knowledge to models or pathways towards sustainability. For example, Oliver de Shutter, United Nations Special Rapporteur on the right to food, is a prominent advocate for agroecology (http://www.srfood.org/en/agroecology). The Montpellier panel of international agricultural experts recommends 'building on and sharing the expertise of African smallholder farmers' and the 'scaling up and out of appropriate and effective technologies and processes' derived from these systems (Montpellier Panel, 2013, p. 4).

This chapter has shown how indigenous soil enrichment practice is a fundamentally overlooked aspect of traditional agroecosytems. Contemporary interest in biochar-based systems that inherently mimic indigenous soil enrichment practices that lead to the formation of enduringly fertile and carbon-rich AfDE is now stimulating emerging research and development in Africa, and some agricultural and soil scientists are involved in this. But the emphasis is on introducing biochar systems as a technology for Africa, transferred from elsewhere (e.g. from the Amazon). The initial emphasis is thus more on induced rather than indigenous innovation. A number of trials are in progress involving partnerships between European and American institutions (Amazonian *terra preta* researchers amongst them) and African ones. Some claim farmer-participatory methods. But none attend to indigenous soil management practices or acknowledge the possibility that biochar might already be part of local farming repertoires. This review alongside our research on African Dark Earths (Fraser *et al.*, 2014; Frausin *et al.*, 2014; Solomon *et al.*, 2016) has drawn attention to the potential importance of widespread existing soil-enriching practices in diverse regions of Africa for sustainable agricultural development strategies.

References

Abebe, M. (1981) Soil burning in Ethiopia. *Ethiopian Journal of Agricultural Science* 3, 57–74.

Adam, J.G. (1958) *Elements pour l'Etude de la Vegetation des Hauts Plateaux du Fouta Djallon (Secteur des Timbis), Guinée Francaise. I. La Flore et ses Groupements*. Bureau des Sols, Dakar.

Adames, F. (1962) *Rapport de la Mission CCTA-FAMA sur les Hauts Plateaux du Fouta Djallon (Guinée) 1961–1962. Vol III (Agriculture)*. Fondation pour l'Assistance Mutuelle en Afrique, Lagos, Nigeria.

Adewole Osunade, M.A. (1995) Identification of crop soils by small farmers of south-western Nigeria. *Journal of Environmental Management* 35, 193–203.

Altieri, M.A. (2002) Agroecology: the science of natural resource management for poor farmers in marginal environments. *Agriculture, Ecosystems, Environment* 93(1–3), 1–24.

Altieri, M. and Nicholls, C. (2013) The adaptation and mitigation potential of traditional agriculture in a changing climate. *Climatic Change*. doi: 10.1007/s10584-013-0909-y.

Anonymous (1923) La situation agricôle dans la Fouta-Djallon (Guinée Française). *Bulletin Mensuel de l'Agence Economique de l'A.O.F. (Paris)* 4(30), 163–166.

Arcin, A. (1907) *La Guinée Francaise*. Challamel, Paris.

Awasthi, R.P., Kori, S. and Grewal, J.S. (1981) Effect of soil burning on the growth and yield of potato in Khasi Hills, Meghalaya [India] (Central Potato Research Inst., Simla (India)). *Indian Journal of Agricultural Science* 51(5), 315.

Bailey, D.M. (1999) Sebakh, sherds and survey. *The Journal of Egyptian Archaeology* 85, 211–218.

Barrera-Bassols, N. and Zinck, J.A. (2003) Ethnopedology: a worldwide view on the soil knowledge of local people. *Geoderma* 111(3–4), 171–195.

Bezot, P. (1966) Les recherches rizicoles au Tchad. *L'Agronomie Tropicale* 1, 70–92.

Birmingham, D.M. (2003) Local knowledge of soils: the case of contrast in Côte d'Ivoire. *Geoderma* 111(3–4), 481–502.

Blackmore, A.C., Mentis, M.T. and Scholes, R.J. (1990) The origin and extent of nutrient-enriched patches within a nutrient-poor savanna in South Africa. *Journal of Biogeography* 17, 463–470.

Bor, N.L. (1942) The relict vegetation of the Shillong Plateau, Assam. *Indian Forestry Recorder* 3(6), 152–195.

Bouteyre, G. (1955) L'ecobuage en culture rizicole dans la Région de Kim et Boumo (Région du Mayo Kebbi et du Logone), summarized in *L'Agronomie Tropicale* 12 (1957) (3), 283.

Brookfield, H. (2001) Intensification, and alternative approaches to agricultural change. *Asia Pacific Viewpoint* 42(2–3), 181–192.

Clinnick, P.F. and Willatt, S.T. (1981) Soil physical and chemical properties measured in an 'ashbed' following windrow burning. *Australian Forestry* 44, 185–189.

De Schutter, O. (2011) *Agroecology and the Right to Food*. Report presented at the 16th Session of the United Nations Human Rights Council [A/HRC/16/49], 8 March 2011. Available at http://www.srfood.org/images/stories/pdf/officialreports/20110308_a-hrc-16-49_agroecology_en.pdf (accessed 7 October 2014).

Dubois, J. (1957) Semis forestiers sur buttes incinérés. Leur importance dans les travaux de deforestation des savanes du Bas-Congo. *Bulletin d'Information de l'Institut Nationals et Agronomique du Congo Belge (INEAC)* 6(1), 21–30.

Duvall, C.S. (2008) Classifying physical geographic features: the case of Maninka farmers in southwestern Mali. *Geografiska Annaler: Series B, Human Geography* 90(4), 327–348.

Fairhead, J. and Leach, M. (1995) Ruined settlements and new gardens: gender and soil ripening among Kuranko farmers in the forest–savanna transition zone. *IDS Bulletin* 26(1), 24–32.

Fairhead, J., Leach, M. and Amanor, K. (2012) Anthropogenic dark earths and Africa: a political agronomy of research disjunctures. In: Sumberg J. and Thompson J. (eds) *Contested Agronomy: Agricultural Research in a Changing World*. Routledge, London, pp. 64–85.

Foley, J.A., Ramankutty, N., Brauman, K.A., Cassidy, E.S., Gerber, J.S. *et al.* (2011) Solutions for a cultivated planet. *Nature* 478 (7369), 337–342.

Fotsing, J. (1992) Stratégies paysannes de gestion des terroirs et de lutte anti-érosive en Pays Bamileke (Ouest – Cameroun). *Bulletin – Réseau Erosion* 12, 241–254.

Fraser, J.A., Leach, M. and Fairhead, J. (2014) Anthropogenic dark earths in the landscapes of upper Guinea, West Africa: intentional or inevitable? *Annals of the Association of American Geographers* 104(6), 1222–1238.

Frausin, V., Fraser, J.A. Narmah, W., Lahai, M.K., Winnebah, T., Fairhead, J. and Leach, M. (2014) 'God made the soil, but we made it fertile': gender, knowledge and practice in the formation and use of African dark earths in Liberia and Sierra Leone. *Human Ecology* 42(5), 695–710.

Gaide, M. (1958) Au Tchad, les transformations subies par l'agriculture traditionelle sous l'influence de la culture cotonnière. *L'Agronomie Tripicale* 9(5–6), 728–730.

Glaser, B. and Birk, J. (2012) State of the scientific knowledge on properties and genesis of anthropogenic dark earths in central Amazonia (*terra preta de Índio*). *Geochimic et Cosmochimica Acta* 82, 39–51.

Gliessman, S.R. (2007) *Agroecology: The Ecology of Sustainable Food Systems*. CRC Press, Boca Raton, Florida.
Goffinet, J. (1913) Apercu sur l'agriculture de l'Uele. *Bulletin Agricole du Congo Belge* 4(4), 587–609.
Gray, L.C. and Morant, P. (2003) Reconciling indigenous knowledge with scientific assessment of soil fertility changes in southwestern Burkina Faso. *Geoderma* 111(3–4), 425–437.
Guilloteau, J. (1957). The problem of bush fires and burns in land development and soil conservation in Africa south of the Sahara. *African Soils* 4, 64–102.
Gurdon, P.R.T. (1914) *The Khasis*. Macmillan and Co., London.
Håkansson, N.T. and Widgren, M. (2014) *Landesque Capital: the Historical Ecology of Enduring Landscape Modifications*. Left Coast Press, Walnut Creek, California.
Hobbs, P.R., Sayre, K. and Gupta R. (2008) The role of conservation agriculture in sustainable agriculture. *Philosophical Transactions of the Royal Society B: Biological Sciences* 363(1491), 543–555.
Humblet, L. (1958) La mise en protection des Jachères dans la Haut-Ituri. In: *Deuxième Conférence Forestière Interafricaine, Pointe Noire, 3–11 July 1958*, Vol. 1. Commission de Coopération Technique en Afrique au Sud du Sahara (CCTA), Ministère des affaires étrangères et du commerce extérieur, Brussels, pp. 110–111.
Inter-African Bureau for Soils and Rural Economy (1958) [No article title.] *Monthly Analytical Bulletin* 8–9, 20.
Jacques-Felix, H. (1947) L'agriculture des noirs au Cameroun. Une forme de l'écobuage. *L'Agronomie Tropical* 2(3–4), 180–182.
Jacques-Felix, H. and Betremieux, R. (1949) Etude de quelques Sols tropicaux. (Congres Africaine des Sols, Goma, 1948). *Bulletin Agricole du Congo Belge*, XL(1), 171–192.
Jurion, F. and Henry, J. (1967) *De l'agriculture itinérante à l'agriculture intensifiée*. INEAC (Institut National pour l'Etude Agronomique au Congo), Brussels.
Kassapu, S. (1979). The impact of alien technology. *Ceres* 67(12), 29–33.
Keay, R.W.J. (1947) Notes on the vegetation of the old Oyo forest reserve. *Farm and Forest* 8, 36–47.
Ketterings, Q.M., Bigham, J.M. and Laperche, V. (2000) Changes in soil mineralogy and texture caused by slash-and-burn fires in Sumatra, Indonesia. *Soil Science Society of America Journal* 64(3), 1108–1117.
Ketterings, Q.M., van Noordwijk, M. and Bigham, J.M. (2002) Soil phosphorus availability after slash-and-burn fires of different intensities in rubber agroforests in Sumatra, Indonesia. *Agriculture, Ecosystems and Environment* 92(1), 37–48.
Krogh, L. and Paarup-Laursen, B. (1997) Indigenous soil knowledge among the Fulani of northern Burkina Faso: linking soil science and anthropology in analysis of natural resource management. *Geojournal* 43, 189–197.
Kurz, W.K. (1875) *Preliminary Report on the Forest and Other Vegetation of Pegu (Burma)*. C.B. Lewis Baptist Mission Press, Calcutta.
Legesse, W.Y. (1974) *Das Bodenbrennen 'Guie' in Äthiopien und seine Wirkung auf Boden und Pflanze*. Faculty Gartenbau und Landeskultur, Technical University of Hannover.
Lehmann, J. (2007) Bio-energy in the black. *Frontiers in Ecology and the Environment* 5, 381–387.
Lehmann, J. and Joseph, S (eds) (2009) *Biochar for Environmental Management: Science and Technology*. Earthscan, London.
Lehmann, J., da Silva Jr, J.P., Steiner, C., Nehls, T., Zech, W. and Glaser, B. (2003) Nutrient availability and leaching in an archaeological Anthrosol and a Ferralsol of the Central Amazon basin: fertilizer, manure and charcoal amendments. *Plant and Soil* 249, 343–357.
Leigh, A.H. (1946) Von Thünen's theory of distribution and the advent of marginal analysis. *Journal of Political Economy* 54(6), 481–502.
Levare, A. (1925) En Guinée Française. *Bulletin de la société Géographique d'Alger et de l'Afrique du Nord*.
Liebig, J. von (1878) *Chemische Briefe*, C.F. Winter'sche Verlagshandlung, Leipzig and Heidelberg, Germany.
Livingstone, D. (1866). *Narrative of an Expedition to the Zambezi and Its Tributaries*. Harper and Brothers, New York.
Livingstone, D. (1875) *The Last Journals of David Livingstone in Central Africa*. Jansen McClurg & Co., Chicago, Illinois.
Lobe, I., Amelung, W. and Du Preez, C.C. (2001) Losses of soil carbon and nitrogen with prolonged arable cropping from sandy soils of the South African Highveld. *European Journal of Soil Science* 52, 93–101.
Mapangui, A. (1992) Etude de l'organisation et du comportement de sols ferrallitiques argileux de la vallée du Niari (Congo): conséquences sur l'évolution physique sous culture de manioc en mécanisé depuis quinze ans. Thèses, Paris 6, ORSTOM, Bondy.
Martin, G. (1958) Essai de bilan de quatre années d'études pédologiques dans la vallée du Niari, Station Agronomique de Loudima, Loudima, Republic of Congo.

Martin, G. (1970) Synthèse agropédologique des études ORSTOM dans la vallée du Niari en République du Congo-Brazzaville. *Cahiers ORSTOM série Pédologie* 8(1), 63–69.
Mboukou-Kimbatsa, I.M.C. (1997) Les macroinvertébrés du Sol dans différents systèmes d'agriculture au Congo: cas particulier de deux systèmes traditionnels (écobuage et brûlis) dans la Vallée du Niari. Thèse de doctorat, Université de Paris VI Pierre et Marie-Curie, Paris.
Mertens, K., Tolossa, A., Verdoodt, A. and Van Ranst, E. (2015) Impact of traditional soil burning (guie) on Planosol properties and land-use intensification in south-western Ethiopia. *Soil Use and Management* 31(2), 330–336.
Mesfin, A. (1981) Soil burning in Ethiopia. *Ethiopian Journal of Agricultural Science* 3(1), 57–73.
Mesfin, A. (1982) Uses of improved seed-bed and fertilizers as an alternative to soil burning or 'guie'. *Ethiopian Journal of Agricultural Science (Addis Ababa)* 4(1), 1–9.
Mollison, J. (1901) *A Text Book on Indian Agriculture. I. Soils, Manures, Implements*. Advocate of India Steam Press, Bombay.
Montpellier Panel (2013) *Sustainable Intensification: a New Paradigm for African Agriculture*. Agriculture for Impact, London.
Ngailo, J.A. and Nortcliff, S (2007) Learning from wasukuma ethnopedology: an indigenous well established system for transfer of agro-technology in Tanzania. *Indilinga – African Journal of Indigenous Knowledge Systems* 6(1), 64–75.
Niemeijer, D. and Mazzucato, V. (2003) Moving beyond indigenous soil taxonomies: local theories of soils for sustainable development. *Geoderma* 111(3–4), 403–424.
Nyete, B. (1991) *Quelques aspects agrotechniques de l'écobuage pratiqué dans le secteur de Bouansa (Vallée du Niari)*. DGRST-ORSTOM, Brazzaville.
Nzila, J.d.D. and Nyete, B. (1996). Pratiques culturales paysannes au Congo: influence de l'écobuage sur l'évolution de la fertilité des sols de la vallée du Niari. In: Pichot, J.P., Sibelet, N. and Lacoeuilhe, J.J. (eds) *Fertilité du Milieu et Stratégies Paysannes sous les Tropiques Humides*. CIRAD-SAR, Montpellier, pp. 245–248.
O'Neill, B., Grossman, J., Tsai, M.T., Gomes, J.E., Lehmann, J. *et al.* (2009) Bacterial community composition in Brazilian anthrosols and adjacent soils characterized using culturing and molecular identification. *Microbial Ecology* 58(1), 23–35.
Oguntunde, P.G., Fosu, M., Ajayi, A.E. and van de Giesen, N. (2004) Effects of charcoal production on maize yield, chemical properties and texture of soil. *Biology and Fertility of Soils* 39, 295–299.
Oguntunde, P.G., Abiodun, B.J., Ajayi, A.E. and van de Giesen, N. (2008) Effects of charcoal production on soil physical properties in Ghana. *Journal of Plant Nutrition and Soil Science* 171(4), 591–596.
Osbahr, H. and Allan, C. (2003) Indigenous knowledge of soil fertility management in southwest Niger. *Geoderma* 111(3–4), 457–479.
Pelissier, P. (1966) *Les Paysans du Sénégal. Les Civilisations agraires du Cayor à la Casamance*. Fabrègue, Saint-Yrieix, France.
Perfecto, I., Vandermeer, J. and Wright, A. (2009) *Nature's Matrix: Linking Agriculture, Conservation and Food Sovereignty*. Earthscan, London.
Ponge, J.-F. (2013) Plant–soil feedbacks mediated by humus forms: a review. *Soil Biology and Biochemistry* 57, 1048–1060.
Portères, R. (1972) De l'écobuage comme un système mixte de culture et de production. *Journal d'agriculture tropicale et de botanique appliquée* 19(6/7), 151–207.
Praquin, J.Y. (1976) *Fertilisation phosphatée et écobuage sur culture pure de maïs (Z 290)*. Documentation interne, IRA, Dschang, Cameroun.
Prudencio, C.Y. (1993) Ring management of soils and crops in the west African semi-arid tropics: the case of the mossi farming system in Burkina Faso. *Agriculture, Ecosystems and Environment* 47(3), 237–264.
Pulschen, L. (1987) *Terminal report for the period May 1983–April 1987*. Soils and Plant Nutrition Section, 31. Agricultural Center, Debre Zeit, Ethiopia.
Pulschen, L. and Koch, W. (1990) The significance of soil burning ('guie') in Ethiopia with special reference to its effects on the Agrestal Weed Flora. *Journal of Agronomy and Crop Science* 164, 254–261.
Ramsay, J.M. and Rose Innes, R. (1963) Some quantitative observations on the effect of fire on the Guinea savanna vegetation of northern Ghana over a period of 11 years. *African Soils* 8, 41–86.
Rennie, J. (1834) Paring and burning. In: Baxter, J. (ed.) *The Library of Agricultural and Horticultural Knowledge*. Baxter, Lewes, UK, pp. 502–508.
Richard-Mollard, J. (1947) Une forme particulière d'écobuage au Fouta-Djallon. *Notes Africaines (IFAN) Dakar* 36, 25–26.

Roorda, T.M.M. (1984) *Effects of Soil Burning (Guie) on Physico-chemical Properties of Soils in the Sheno, Debre Berhan and Chacha Areas of Ethiopia*. International Livestock Centre for Africa, Addis Ababa.

Roorda, T.M.M. (1988) Soil burning in Ethiopia: some effects on soil fertility and physics. In: Jutzi, S.C., Haque, I., McIntire, J. and Stares, J.E.S. (eds) *Management of Vertisols in Sub-Saharan Africa: Proceedings of a Conference Held at ILCA, Addis Ababa, Ethiopia, 31 Aug–4 Sep 1987*. International Livestock Centre for Africa, Addis Ababa, p. 124.

Ruthenberg, H. (1971) *Farming Systems in the Tropics*. Clarendon Press, Oxford.

Saldanha, I.M. (1990) The political ecology of traditional farming practices in Thana district, Maharashtra (India). *The Journal of Peasant Studies* 17(3), 433–443.

Sato, S., Neves, E.G., Solomon, D., Liang, B. and Lehmann, J. (2009) Biogenic calcium phosphate transformation in soils over millennium time scales. *Journal of Soils and Sediments* 9, 194–205.

Sautter, G. (1953) *Le Système de Culture des Villageois dans le District de Boko*. Institut d'Etudes Centrafricaines, Brazzaville.

Sautter, G. (1955) Notes sur l'Agriculture des Bakamba de la Vallée du Niari (Gabon). *Bulletin de l'Institut d'Etudes Centrafricaines* 9, 67–105.

Schmitz, A. and Delvaux, J. (1958) Implantation d'Eucalyptus sur brulis. *Bulletin Agricôle de la Congo Belge* 49(4), 1001–1015.

Scoones, I. (1991) Wetlands in drylands: key resources for agricultural and pastoral production in Africa. *Ambio* 20(8), 366–371.

Scoones, I. (2001) *Dynamics and Diversity: Soil Fertility and Farming Livelihoods in Africa: Case Studies from Ethiopia, Mali, and Zimbabwe*. Earthscan, London.

Sertsu, S. (1987) *Investigation on the Effect of a Soil Burning Practice ('Guie') in Ethiopia on Soil Properties and Barley Yield*. Justus Liebig-Universität Giessen, Giessen, Germany.

Sertsu, S. and Sánchez, P.A. (1978) Effects of heating on some changes in soil properties in relation to an Ethiopian land management practice. *Soil Science Society of America Journal* 42(6), 940–944.

Sheil, D., Basuki, I., German, L., Kuyper, T.W., Limberg, G. *et al.* (2012) Do anthropogenic dark earths occur in the interior of Borneo? Some initial observations from East Kalimantan. *Forests* 3(2), 207–229.

Sohi, S. and Yeboah, E. (2009) Traditional biochar-based management of tropical soil in subsistence agriculture. In: Lehmann, J. and Joseph, S. (eds) *Biochar for Environmental Management: Science and Technology*. Earthscan, London, pp. 161–163.

Solomon, D. and Lehmann, J. (2000) Loss of phosphorus from soil in semiarid northern Tanzania as a result of cropping: evidence from sequential extraction and ^{31}P NMR spectroscopy. *European Journal of Soil Science* 51, 699–708.

Solomon, D., Lehmann, J., Kinyangi, J.,Amelung, W., Lobe, I. *et al.* (2007) Long-term impacts of anthropogenic perturbations on dynamics and speciation of organic carbon in tropical forest and subtropical grassland ecosystems. *Global Change Biology* 13(2), 511–530.

Solomon, D., Lehmann, J., Fraser, J.A., Leach, M., Amanor, K. *et al.* (2016) Indigenous African soil enrichment as climate-smart sustainable agriculture alternative. *Frontiers in Ecology and the Environment* 14(2), 71–76.

Soret, M. (1959) *Les Kongo Nord-occidentaux*. Presses Universitaires de France, Paris.

Strømgaard, P. (1992) Immediate and long-term effects of fire and ash-fertilization on a Zambian miombo woodland soil. *Agriculture, Ecosystems and Environment* 41(1), 19–37.

Sudres, A. (1947) La degradation des sols au Fouta-Djallon. *L'Agronomie Tropical* 2(5–6), 227–246.

Tittonell, P. (2014) Ecological intensification of agriculture – sustainable by nature. *Current Opinion in Environmental Sustainability* 8, 53–61.

Topoliantz, S. and Ponge, J.F. (2005) Charcoal consumption and casting activity by *Pontoscolex corethrurus* (Glossoscolecidae). *Applied Soil Ecology* 28(3), 217–224.

Topoliantz, S., Ponge, J.F. and Lavelle, P. (2006) Humus components and biogenic structures under tropical slash-and-burn agriculture. *European Journal of Soil Science* 57, 269–278.

Voelcker, J.A. (1893) *Report on the Improvement of Indian Agriculture*. Eyre and Spottiswoode, London.

White, L.J.T. and Oates, J.F. (1999) New data on the history of the plateau forest of Okomu, southern Nigeria: an insight into how human disturbance has shaped the African rain forest. *Global Ecology and Biogeography* 8, 355–361.

Whitman T., Pepe-Ranney, C., Enders, A., Koechli, C., Campbell, A., Buckley, D. and Lehmann, J. (2016) Dynamics of microbial community composition and soil organic carbon mineralization in soil following addition of pyrogenic and fresh organic matter. *ISME Journal*. doi: 10.1038/ismej.2016.68.

Widgren, M. (2007) Precolonial landesque capital: a global perspective. In: Hornborg, A., McNeill, A.R. and Martinez-Alier, J. (eds) *Rethinking Environmental History: World System History and Global Environmental Change*. AltaMira, Lanham, Maryland, pp. 61–77.

Zingore, S., Murwira, H.K., Delve, R.J. and Giller, K.E. (2007) Influence of nutrient management strategies on variability of soil fertility, crop yields and nutrient balances on smallholder farms in Zimbabwe. *Agriculture, Ecosystems and Environment* 119(1–2), 112–126.

Zwetsloot, M., Lehmann, J. and Solomon D. (2015) Recycling slaughterhouse waste into fertilizer: how do pyrolysis temperature and biomass additions affect phosphorus availability and chemistry? *Journal of the Science of Food and Agriculture* 95, 281–288.

10 Will the Real Raised-Field Agriculture Please Rise? Indigenous Knowledge and the Resolution of Competing Visions of One Way to Farm Wetlands

Doyle McKey,* Delphine Renard and Marion Comptour

Raised fields (RFs) can be defined as 'any prepared land involving the transfer and elevation of earth in order to improve cultivating conditions' (Denevan and Turner, 1974). Drainage is of course not the only way in which elevating earth improves conditions for farming, and wetlands are not the only environments in which raised fields (or 'raised beds', as little-elevated structures are often termed) are found. Nonetheless, the largest and most extensive raised fields are found in seasonally flooded wetlands and it is in these habitats that raised-field agriculture (RFA) has been most studied. RFA is practised today by farmers of wetlands in numerous parts of Africa, Asia and Oceania (Denevan and Turner, 1974; Baveye, 2013). Ironically, however, RFA has attracted most attention in a region where it has been virtually extinct for half a millennium or more, and almost no attention in regions where it is practised today. A review of flood-based farming systems in Africa (Landarte Puertas *et al.*, 2016) did not even mention wetland RFA. Over 30 years ago, Vasey *et al.* (1984) lamented that while the morphology of raised fields in many parts of the world had been well described, very little work had been devoted to the ecological functioning of wetland raised-field systems. This is still largely true. Even for the famed *chinampas* of the Valley of Mexico, Frederick (2007) found only four studies that actually presented primary data, and this appears not to have changed much since.

Studies of RFA in the Neotropics have been conducted mostly by archaeologists and geographers (e.g. Erickson, 1995; Denevan, 2001; Rostain, 2012). They study the vestiges of ancient raised fields and, in some sites, experimental raised fields that have been built, or rehabilitated, in attempts to explore how the extinct agricultural system may have functioned (e.g. Swartley, 2002; Saavedra, 2009). Researchers in these projects have sometimes included soil scientists (e.g. McKey *et al.*, 2010; Rodrigues *et al.*, 2015, 2016a, b), but not agronomists, let alone agronomists with experience in extant RFA in the Old World.

Why has wetland RFA attracted such interest in the Neotropics? One important reason is the nature of the only system that seems to have survived to the present day (albeit in altered form (Torres-Lima *et al.*, 1994)), the famed *chinampas* of the Valley of Mexico (Coe, 1964; Armillas, 1971; Frederick, 2007; Hagmann, 2009). These represent an intensive wetland agricultural system that sustained dense populations of humans over at least several centuries. Although today they produce organic vegetables and cut flowers

* Corresponding author; e-mail:doyle.mckey@cefe.cnrs.fr

for urban markets, rather than maize for subsistence, today's *chinamperos* continue to apply ecological principles in interesting ways in the design of agroecosystems. Their practices in particular show deft coupling of aquatic and terrestrial compartments in the management of nutrients, drainage and irrigation.

Another reason for the interest in RFA in the Neotropics was the discovery by geographers and archaeologists of huge areas of wetlands in South America, and smaller areas in Mesoamerica and Central America (e.g. Martin *et al.*, 2015), covered by vestiges of pre-European raised fields (Denevan, 2001). The findings that extensive areas of seasonally flooded savannas in South America were modified by pre-Columbian populations became one of the principal tributaries to the flood of new evidence that humans had played a role in shaping the landscapes of pre-Conquest Amazonia, evidence summarized in Denevan's (1992) paper on the 'pristine myth' and popularized in the book by Charles Mann (2005).

As they debunked the 'pristine myth', archaeologists and geographers also increasingly challenged the long-influential ideas of Betty Meggers (1971) about environmental limitations to agricultural production and the development of large, complex societies in Amazonia. According to Meggers, the highly weathered soils of savannas and of forests in *terra firme* between the major rivers are too infertile to support dense human populations. In forested areas, however, new research on *terra preta* ('black earth'), and related soils now grouped as Anthropogenic Dark Earths (ADE), showed that even highly weathered, infertile Oxisols could be modified by humans – through the addition of nutrients, but most importantly through the addition of charcoal that increased nutrient retention – to produce very fertile soils that could sustain intensive agriculture (e.g. Lehmann *et al.*, 2003; Glaser and Woods, 2004; Woods *et al.*, 2009). Wetland RFA seemed to fit into this same vision. The building of raised fields overcame drainage problems, and the *chinampas* model – to which archaeologists frequently turned in efforts to compare their findings with observations about the functioning of a putative present-day analogue – seemed to suggest that nutrient limitations could similarly be overcome. By such inventions, humans could overcome environmental constraints, to the point that the latter became 'negligible analytic phenomena' (Balée and Erickson, 2006, p. 4).

The notion thus grew of wetland RFA as a highly intensive and productive form of wetland agriculture. Based (explicitly or implicitly) on their conceptions of the *chinampas* system, archaeologists often assumed that the almost continuous cultivation of *chinampas* would characterize RFA in general (Erickson, 1994; Lee, 1997; Barba, 2003; Saavedra, 2006). However, the *chinampas* differ in many ways from RFA systems in tropical lowlands. The *chinampas* were built in lakeshore environments where water was continually present, allowing crop growth throughout the year; their soils were relatively rich, with both lacustrine and volcanic inputs, and with the relatively high organic matter contents that characterize mid-elevation tropical settings; and the impact of pests and pathogens of crop plants was moderate compared with lowland sites. In contrast, environments where lowland RFA was conducted in Amazonia are characterized by highly seasonal water availability, with a long and severe dry season. Agricultural activities certainly followed strong seasonal cycles, with pronounced seasonal breaks in cultivation. Soils in most of these areas are much poorer than in the valley of Mexico.

The enthusiasm of proponents of the vision of raised fields as a kind of 'pre-Columbian green revolution' was thus soon met by criticism (Chapin, 1988; Swartley, 2002; Lombardo *et al.*, 2011; Baveye, 2013). These critics pointed to the important ecological differences between the region where *chinampas* are farmed and the regions where vestiges of raised fields occur in South America, and identified important limitations in the experiments aimed at reproducing or rehabilitating raised fields. They argued that RFA was not intensive, but rather extensive, in the sense that cultivation was not permanent but likely required long fallow periods to allow reconstitution of nutrient stocks in the poor, weathered soils where most vestiges of lowland South American RFA occur. They argued that RFA was adopted less because of the opportunities it may have presented than as a matter of necessity: demographic pressure, conflicts with other groups, climate change or some other factor forced some groups to find ways to conduct agriculture in these marginal environments.

Pushing their arguments to the limit, some critics argued that the only function of raised fields was to adapt to the constraint of flooding (Lombardo *et al.*, 2011). Adding to the scepticism about the agronomic potential of RFA is the apparent failure of experimental RFs in all regions where this approach has been tried, in the sense that it was never adopted by local people following the experiments (Renard *et al.*, 2012; McKey *et al.*, 2014). Critics declared these experiments in reintroduction of RFA to be failures (e.g. Lombardo, 2013). However, it is debatable whether their failure can be ascribed to agroecological limitations (Baveye, 2013), to lack of knowledge about wetland agriculture on the part of the persons recruited for these experiments (Renard *et al.*, 2012), or to factors that make raised-field agriculture unable to be adapted to today's social, cultural, or economic contexts in Neotropical regions (Erickson, 1994).

We suggest that both these visions are inadequate and that the real face of RFA has yet to be revealed. Archaeologists have over-generalized from one system in a highly favourable environment, a system that is certainly not representative and perhaps even unique; and have inferred or even 'invented' practices (and thereby also indigenous knowledge) from necessarily imperfect attempts to reconstruct RFA experimentally. It is not surprising that RFA fails to live up to the unrealistic expectations based on these misunderstandings. At the same time, researchers have not taken advantage of the opportunities for comparison with a perhaps more informative set of present-day analogues: RFA as practised by farmers in a range of environments in the Old World tropics today. For these reasons, the real nature of wetland RFA has been misunderstood.

Although two pioneers of the study of RFA in the Neotropics turned to extant Old World systems for inspiration 40 years ago (Denevan and Turner, 1974), this lead has not been followed up. With the many pressures favouring regional specialization – and the frequent reluctance among social scientists to posit that a society's cultural behaviour can be predicted, even partially, based on environmental constraints faced by the society – few archaeologists and human geographers are prepared to attempt broad cross-regional comparative perspectives. For the ecologist, however, using a comparative approach to search for convergent practices in similar environments is second nature. That this approach can bear interesting fruit is shown by a recent comparison of savanna floodplain fisheries in the Bangweulu basin of Zambia and the Beni llanos of Bolivia (McKey *et al.*, 2016).

What Can We Learn From Present-Day RFA as Practised in the Old World?

Despite the long-divergent histories of their peoples and cultures, ecological similarities make present-day RFA in the Old World a more informative analogue of RFA in pre-Columbian South America than are the *chinampas*. Observing Old World practitioners of RFA also avoids the methodological limitations of experimental reconstructions of RFs (McKey *et al.*, 2014), in which the 'farmers' recruited for these experiments have usually lacked any prior knowledge of wetland agriculture and whose practices may not resemble those that prevailed in pre-Columbian times. The practices of present-day Old World RFA farmers, whose livelihood depends on knowing how to farm the soils of these environments, are likely to be much better indicators of how pre-Columbian inhabitants of South America farmed wetlands.

Among the questions about present-day RFA suggested by inferences that archaeologists have made about pre-Columbian RFA – and by controversies about these inferences – are the following.

Why do people construct RFs? All agree that drainage is an important reason why soil surfaces are raised, but some argue that this is their only function (Lombardo *et al.*, 2011), discounting any role of nutrient management. Others argue that vegetation and sediments added to RFs from the surrounding area (for example, in the aquatic component of RF agroecosystems) furnish nutrients to crops (e.g. Erickson, 2008; Renard *et al.*, 2012). Could the supply of such nutrients sustain intensive, continued cultivation of these systems, or are fallow periods required to restore nutrient stocks (or to limit pests and pathogens)? If fallows are integrated, how are they managed? Is fallow vegetation burned or mulched?

The local knowledge of present-day RFA farmers in the Old World tropics has hardly been plumbed, so a synthesis is impossible. The next section offers observations from our field work about local knowledge and practices of present-day RFA farmers concerning a few key themes. This work has been conducted in two sites in Africa: the Bangweulu basin in Zambia and the cuvette region of the Congo Republic. Wetland RFA is also practised in tropical Asia (e.g. Mollard, 1999; Bai *et al.*, 2014) and Oceania (e.g. Kirch, 1978). The most in-depth account of RFA in the Old World is provided by Sillitoe (1996). Although that study was conducted in a non-wetland setting (the New Guinea highlands), it will be shown that its findings are highly pertinent to our African studies.

Resource concentration as an overriding function of raised fields

Drainage is certainly a major function of raised fields. Studying the vestiges of RFA in several sites across the Beni Llanos of Bolivia, Rodrigues *et al.* (2015, 2016a, b) found that the height of vestiges of raised fields varied among sites in relation to the depth of flooding typical of the site. A similar relationship is found between the height of present-day raised fields in Zambia and the high-water level characteristic of the site.

However, drainage can be seen as just one part of an overriding function of RFs that has been largely overlooked: the spatial concentration of resources in a resource-poor environment. The acid, infertile, seasonally flooded savannas in which wetland RFs usually occur are characterized by the scarcity of several resources important for plants; mounds or ridges concentrate these resources at levels that crop plants can use. The most obvious example of resource concentration in these flooded environments is the concentration of soil into well-drained islands, but other scarce resources, such as mineral nutrients, are also concentrated. In present-day RFA in Africa, RFs serve to concentrate on a limited planting surface the topsoil and nutrient-containing plant biomass (e.g. weeds, fallow vegetation) collected from a considerably larger surface around the mound or ridge. The concentration of resources into relatively rich patches may be the key to viability of agriculture in an environment where mean resource levels would support only limited production.

We observed the building of raised fields near Mossaka, in the cuvette region of the Congo Republic (Fig. 10.1; for description of sites see McKey *et al.*, 2014). Where the field is to be built, grass and other herbaceous vegetation is uprooted from a large area and left to dry for a few days (Fig. 10.1a). The partly dried plants are then stacked to form a round mound or an elongated platform, up to 1.5 m tall and 20 m long or more (Fig. 10.1b). The new RF is composed almost entirely of plant biomass, with the exception of small amounts of mineral soil clinging to the roots of each plant (Fig. 10.1c). RFs are long-lived structures, persisting for decades. Farmers told us that it takes 1–2 years of decomposition of these large amounts of organic matter for fertility to reach its highest levels (Fig. 10.1d).

With decomposition RFs lose height, and more material of the same nature (i.e. mostly plant biomass) is added from the surrounding area each year the field is put into use, to maintain the desired height (Fig. 10.2). The concentration of topsoil and litter to make a raised planting surface and the concentration of nutrient-rich biomass on a fraction of the total surface are both essential to the agronomic quality of RF soils.

In the Bangweulu basin, Zambia, historical and comparative information suggests that the initial function of building RFs was to concentrate nutrients and that drainage is a function that was acquired later. In this region, wetland RFA appears to be an extension of the Mambwe mound cultivation system in savannas (Siame, 2006), which is itself in turn derived from large-circle *chitimene* slash-and-burn cultivation in miombo woodlands (Stromgaard, 1989). Soil fertility is low in miombo woodlands, requiring nutrient concentration to support crop growth. In large-circle *chitimene*, 'woodland is not simply felled and burned, as is normal under slash-and-burn agriculture, but trees are chopped in a larger area, the outfield, and the slash piled in a smaller area, the infield, and burned' (Stromgaard, 1989).

Nutrients that can support crops are in even shorter supply in savanna, where plant biomass is lower. Agriculture around the basin was initially conducted primarily in woodland.

Fig. 10.1. Platform raised fields, cuvette region of the Congo, near Mossaka, Congo Republic. (a) To make the field, grass is uprooted from a large surface. After a few days' drying, grass and soil clinging to roots will be piled up and concentrated on a small area to make the raised field. (b) A newly constructed raised field 20 m long. (c) Detail of the surface of the raised field showing that it comprises plant biomass and soil clinging to roots. (d) A platform raised field in its second year, bearing young manioc (cassava) plants. (All photographs by the first author.)

Fig. 10.2. Round raised fields, cuvette region of the Congo, near Mossaka, Congo Republic. (a) Round raised field showing the large surrounding area from which biomass and soil are gathered to add to the field. (b) Plant biomass recently added to the top of the field. (All photographs by the first author.)

Plots were slashed and burned, cultivated for one or two crops, and then left in fallow. However, as demographic pressure increased owing to population growth and inter-ethnic conflict, forest land became ever scarcer and farmers began to open plots in savanna. In finding ways to take advantage of even scarcer resources, farmers moved from slash-and-burn of woodlands to slash-and-mulch and in-field composting in mounds made in savanna, in a system known locally as *fundikila* (Stromgaard, 1989). During the dry season, grassy vegetation is cut or uprooted, concentrated into piles and then covered by dirt to form mounds. These are left fallow through the rest of the dry season. Crop growth in the next rainy season depends on the release of nutrients by the decaying grasses buried in the mounds. Some crops (beans, maize, cassava) are planted on the mounds. Mounds are then flattened and the humic soil is spread and planted with important crops such as finger millet, then cassava and sorghum, which are left during the dry season. During the following rains the mounds are thrown up again for a second cropping season and then levelled before being left to fallow. According to farmers, plots regenerate faster when levelled than in the mounded state (Stromgaard, 1989). In *fundikila*, the primary function of mounds is thus clearly the concentration and composting of nutrients. Stromgaard (1990) showed that this highly labour-intensive system was sustainable. After 4 years of legume–cereal crop rotation on mounds alternating with flat cultivation, more nutrients had accumulated in the soil of the mound than were present before in the fallow.

As demographic pressure continued, populations were pushed to open their fields in ever more challenging environments, from upland savanna to seasonally waterlogged to seasonally flooded savanna. In these environments, RFs acquired a second important function: drainage. In these poorly drained environments, mounds are not flattened before planting; rather, crops are planted on the mounds. And as in the Neotropics, the higher the water level reached during flooding, the higher (and broader) the mounds are constructed (Fig. 10.3a). The extension of agriculture from forest to upland savanna to seasonally waterlogged and seasonally flooded savanna appears to support the hypothesis proposed by some students of archaeological RFA in South America that people adopted RFA more out of necessity than because of the opportunities it offered. However, it is important to recognize that during this transition, nutrient scarcity became a major constraint for agriculture even before drainage. Management of scarce nutrients, via the management of organic matter, has probably always been important for the agronomic viability of RFs (Fig.10.3b).

The importance of nutrient management as a function of RFs was amply documented by Sillitoe (1996), who worked among the Wola in the New Guinea highlands. Here, flooding is not a factor. Farmers plant their crops in large circular mounds (2–3 m in diameter and 1 m or so high), in whose centres weeds and fallow vegetation

Fig. 10.3. Raised fields in the Lunga Bank, Bangweulu basin, Zambia. (a) Large raised fields in a deeply flooded area. Fields are about 1.7 m high. All except the lowermost 70 cm of each mound is planted with manioc (cassava) stakes. (b) Construction of a new set of raised fields. The large piles of plant biomass will be covered with soil. (All photographs by the first author.)

are composted. Nutrients from the decomposing vegetation were shown to be essential for the maintenance of yields of sweet potato, the staple crop; and yields increased with longer fallow periods that supplied more plant biomass and nutrients. Mound building also increased the depth of topsoil and its friability, enhancing root penetration. Although soils in Sillitoe's study site were characterized as being acid (mean topsoil pH 5.13) and of low fertility (mean topsoil cation exchange capacity (CEC) 27.8 mEq/100 g), those of our African sites were much poorer. In the Congo site, pH ranged from 3.58 to 4.02 and CEC from 3.9 to 14.1 mEq/100 g (unpublished data). In Zambia, Stromgaard (1990) reported (for unmounded topsoil in savanna) values for pH of 5.1 (comparable to the New Guinea site), but for CEC values of only 2.6 mEq/100 g. In adjacent mounds, both values were increased, and the concentrations of critical nutrients, including phosphorus and calcium, were doubled (Stromgaard, 1990). In the highly infertile soils of both African sites, the positive effect of mounding on nutrient status is even more crucial than in New Guinea.

In Zambia, as we have seen, the adoption of wetland RFA appears to have been preceded by farming of less constraining upland environments, first forest, then upland savanna and then seasonally flooded savannas, entailing shifts to ever more labour-intensive practices to effectively recycle scarce nutrients. Did the adoption of wetland RFA by pre-Columbian farmers in South America sometimes follow a similar series? Did mound building continue to have an important nutrient-recycling function in wetlands? Whitney *et al.* (2014) found evidence for maize cultivation in RFs. They suggested (as we have argued for our African sites) that in the region's acid, infertile soils this would have been impossible without some kind of soil improvement. However, there appears to be no evidence for practices that could have accomplished this. It might be very difficult to detect evidence in archaeological vestiges of RFA in South America of nutrient-recycling practices like those employed by present-day Zambian farmers. Present-day African farmers concentrate nutrients by concentrating rapidly decomposable herbaceous plant matter. Nutrient concentration is likely to be ephemeral, and since vegetation is usually slashed and mulched, rather than burned, durable signals such as deposits of charcoal – which enable detection of nutrient concentration in *terra preta* anthroposols centuries after their abandonment – are absent. The most suggestive evidence of the importance of amendments to organic matter in wetland RFA in the Neotropics comes from studies of pre-Columbian RFA in French Guiana coastal savannas. McKey *et al.* (2010) found evidence for the wholescale removal of topsoil from an area about 100 m from a complex of RFs and postulated that it was transported and added to the RFs. In the same site, Iriarte *et al.* (2012) found evidence of low fire frequency during a 300-year period of RFA, suggesting slash-and-mulch rather than slash-and-burn management of fallow vegetation.

Are RFs always perennial structures?

Archaeologists have assumed that in pre-Columbian South America individual RFs, like *chinampa* ridges, were long-lived structures. Is this true of present-day RFA in Africa?

In the cuvette region of the Congo Republic, according to the farmers we interviewed, RFs are indeed perennial structures. Once built, they are maintained and added to over time. After each cropping-and-fallow cycle, vegetation around the field is cut or uprooted and added, along with the topsoil clinging to the roots, to the top of the mound before the new crop is planted (see Fig. 10.2a, b). In the Bangweulu basin, however, individual RFs are short-lived structures. We have seen that in the apparent historical precursors of wetland RFA in that region, i.e. *fundikila* mounds in non-flooded savannas, mounds are regularly levelled and their humic soils spread, sometimes even before crops are planted. As RFA moves progressively into poorly drained environments, mounds are left in place and crops are planted on them. However, even in these seasonally flooded environments, mounds are flattened from one cultivation cycle to the next. At the end of the fallow period (usually 1–3 years after cultivation), vegetation is cleared and put in piles in the spaces between the mounds. Then the old mounds are dug up and the earth is transported on to the piles of dead vegetation. In effect, the whole grid of RFs is moved one space diagonally on the checkerboard with each new cycle of cultivation.

Even the largest RFs we observed in Bangweulu are levelled after a single cropping cycle. On Lunga Bank, a deeply flooded area with relatively rich soil in the central part of the basin, we observed round mounds up to 1.7 m tall and over 3 m broad (McKey *et al.*, 2014; see Fig. 10.3a). Cassava stakes are planted in these large mounds, in all parts except for the lowermost 70 cm, which is virtually certain to be flooded in all years. As the water begins to rise during the high-water season, harvesting starts at the bottom and progresses upward. The construction of these very large mounds must require much more labour than elsewhere. Nevertheless, they are flattened after a single year of cassava cultivation. The following year, the flattened area once covered by RFs is planted in rice. When asked about the reason for this rotation, farmers immediately provided a demonstrative explanation. Digging into the soil with a machete to show us the roots, with each stroke they unearthed one or more large white grubs feeding on the roots, the rhizophagous larvae of an unidentified scarabaeid beetle. Larvae of several *Phyllophaga* spp. are known cassava pests throughout the tropics (Bellotti and van Schoonhoven, 1977). According to the farmers, if they planted cassava every year, the crop would be destroyed by these pests. After 1 year in rice, mounds are rebuilt and the RFs once again are planted in cassava.

The striking difference in longevity of individual RFs between the cuvette region of the Congo and the Bangweulu basin is remarkable. This difference appears to be rooted in the historical origin of wetland RFA in the Bangweulu basin from *chitimene* and *fundikila* systems in forest and upland savannas, respectively. Are there also functional reasons that help to explain why mounds throughout Bangweulu – not just those in Lunga Bank that are flattened to make way for rice in alternate years – are regularly levelled? This question is so far unanswered.

What do our observations of the frequent destruction and reconstruction of RFs in the Bangweulu basin say about the assumption of archaeologists that pre-Columbian RFs were perennial structures? Were some South American RFs periodically destroyed and rebuilt, as in present-day Bangweulu, or were they all permanent structures, as in the cuvette region of the Congo? This question appears to be intractable with currently available data. One important factor must be taken into account: with only wooden tools at their disposal, pre-Columbian farmers may have been reluctant to destroy and frequently reconstruct RFs. Present-day RF farmers have access to metal tools and presumably can move larger volumes of earth with a given expenditure of time and energy.

Is cultivation continuous, or interrupted by fallow periods?

Although soil fertility varies considerably in areas of South America where pre-European RFs are found, even among sites in a single region (compare Rodrigues *et al.*, 2015, 2016a, b), the soils are often quite infertile. For example, in parts of the Beni llanos of Bolivia, soils are so acid and leached that problems of aluminium toxicity appear (Rodrigues *et al.*, 2016a). Furthermore, pests and pathogens in lowland environments likely built up rapidly to levels that would have required fallow periods.

In the two regions where we have begun to study wetland RFA in Africa, only exceptionally rich soils, like those rich in organic matter in the Lunga Bank in the Bangweulu basin, can be cultivated year after year. And, as described above, even there pests have led to the adoption of crop rotation in alternate years. In all other sites in these regions, soils are acid, weathered and of low mineral fertility, so that farmers must manage organic matter deftly. Fallow periods are everywhere integrated into cropping systems and agricultural production appears to depend on fallow periods that are long enough to allow vegetation to grow and accumulate nutrients in plant biomass. In general, in both the cuvette region of the Congo and the Bangweulu basin, farmers practising wetland RFA leave fields fallow for at least 2–3 years between cropping cycles. Mound cultivation and wetland RFA in the Bangweulu basin are certainly much more intensive, in terms of both labour and productivity per unit area, than the kind of agriculture that preceded them. In the 1980s, the *fundikila* grassland-mound system supported population densities of 12 persons/km^2, compared with a carrying capacity of 2–6 persons/km^2 for large-circle *chitimene* in miombo woodland

(Stromgaard, 1989). The diffusion of cassava in these systems in the second part of the 20th century enabled both greater production and shortening of fallow periods, making possible population densities 10–15 times higher than in *chitimene* (Holden, 2001). However, *fundikila* and wetland RFA are still 'extensive' in the sense that they are not permanently cultivated; fallow periods, albeit short, are required for sustainability.

RFA and Resource-Concentration Mechanisms in Natural Ecosystems of Seasonally Flooded Savannas

Spatial self-organization and resource concentration in natural ecosystems

Humans are not the only organisms that build elevated structures in seasonally flooded savannas. Long before humans built RFs, various soil animals, principally earthworms and social insects, built mounds in these environments, thereby escaping flooding. In doing so, they concentrated soil in well-drained islands. Earth-mound landscapes made by soil animals are widespread in seasonally flooded savannas (McKey *et al.*, 2014). Interestingly, these landscapes show striking regularity, as do RF landscapes built by humans. In the case of soil organisms, however, the regularity is not planned, but instead emerges from natural processes that incorporate feedback loops, most often the combined actions of individual soil engineers on the abiotic environment. Processes acting at local spatial scales generate patterns seen at very large scales. The mechanisms that produce such emergent regularity are termed self-organizing.

Spatial self-organization of ecosystems characterizes a diversity of environments, ranging from seasonally flooded savannas to semi-arid shrublands, that all have one thing in common: a critical resource – water in semi-arid shrublands, well-drained soil in seasonally flooded savannas – is in short supply, and soil organisms modify the distribution of this resource, concentrating it in some patches and depleting it in areas between these patches (Rietkerk *et al.*, 2004; Rietkerk and van de Koppel, 2008). In the mosaic landscape thus created, resource-rich patches can shelter organisms that could not persist if resources were spread evenly over the landscape.

The mechanisms leading to spatial self-organization have been well studied in semi-arid shrublands (e.g. Meron, 2012), but work on this theme in seasonally flooded savannas is in its infancy. In the *surales* landscapes of the Orinoco Llanos of Colombia, large earthworms build towers in seasonally flooded savannas that allow them to respire while feeding in the waterlogged soil. As the worm has a permanent gallery system and continually returns to the same spot to breathe, carrying a load of excrement each time, the tower becomes a mound. As the soil carried to the mound comes from the basin surrounding the mound, the basin becomes deeper, preventing the formation of a new mound nearby. This threshold minimum distance between mounds, combined with high density of the worms, leads to a highly regular earth-mound landscape (Zangerlé *et al.*, 2016). 'Termite savannas' are also frequent in seasonally flooded savannas, as termite colonies build large mounds that enable them to escape flooding. In the case of termites, inter-colony competition is thought to drive spacing (Bonachela *et al.*, 2015).

Well-drained soil is not the only resource concentrated in mounds by these 'soil engineers'. In the *surales* mounds, earthworms during the high-water season continually transport large quantities of soil from the waterlogged basin to the aerobic environment of the tops of mounds, enhancing mineralization of the contained organic matter (Cunha *et al.*, 2016). Termite mounds, in particular, are well known to be 'hotspots' of nutrients that facilitate plant growth (Bonachela *et al.*, 2015). It is intriguing that a key element of environmental management by RF farmers – the concentration of scarce resources for plants in elevated structures that comprise a small fraction of the total land area – is also a key feature in the functioning of natural ecosystems in seasonally flooded savanna environments. This observation leads to an interesting question: to what extent do RF farmers take advantage of natural resource-concentration mechanisms in these environments? How do natural and cultural mechanisms interact?

RF farmers take advantage of natural mechanisms of resource concentration

Self-organized structures of natural origin are incorporated into biocultural landscapes. In Zambia, termite-mound islands in the *dambos* and floodplains are favoured sites for constructing fields, as they are already elevated and well drained, and are islands of fertility (Fig. 10.4a). In fact, the co-opting of termite mounds as farm plots may be widespread in the infertile soils of the miombo woodland region. *Macrotermes* mounds are favoured placements for large-circle *chitimene* slash-and-burn plots in miombo woodlands in Tanzania (Mielke and Mielke, 1982). Pre-Columbian wetland RF farmers in the Beni Llanos of Bolivia may have similarly incorporated termite mounds into RFs (Fig. 10.4b).

Other vestiges of pre-Columbian RFA in South America indicate that RF farmers and soil-engineering animals may also have interacted in other ways. McKey *et al.* (2010) showed that soil engineers preferentially colonize the vestiges of raised fields in French Guiana coastal savannas, and that their soil-transporting and soil-stabilizing activities have contributed to maintaining mounds against erosion. They also suggested that these activities could have been important during the period when RFA was being conducted, helping to maintain resource concentration in fallow RFs. In landscapes in Bolivia's Beni Llanos today, the distribution of soil engineers – termites, earthworms and bunchgrasses – similarly shows clear relationships to the distribution of RFs (see Fig. 1 in Cunha *et al.*, 2016) and one wonders how non-human soil engineers may have interacted with RF farmers when these fields were active.

Conclusions

Exploring the indigenous knowledge and practices associated with RFA today suggests a more nuanced vision than those that have emerged from work in the Neotropics. Wetland RFA is neither a technique that in some magical way can bring about a 'Green Revolution' in acid, infertile hydromorphic soils, nor is it a failed system on which research should be abandoned. Rather, it emerges as a system that has adapted to resource-poor environments by spatially concentrating resources for crops on a small proportion of the total surface. The resources thus concentrated include well-drained soil, but scarce nutrients are also concentrated. Nutrient management is a key function of building RFs in the Old World, even in seasonally flooded savannas,

Fig. 10.4. Farmers convert termite mounds into raised fields. (a) Raised fields constructed on top of a large *Odontotermes* termite mound, Bangweulu basin, Zambia. (Photograph by the first author.) (b) Landscape in the western Beni Llanos, Bolivia. The round structures are mounds of an unidentified termite species. The smallest mounds are about 2 m in diameter. In some places, approximately two to five mounds appear to have been 'bridged' to make short ridge-shaped raised fields. (Mosaic of several photographs taken using the Pixy© drone by Delphine Renard.)

and the same is likely to have been true in the Neotropics. However, the increases in nutrient concentration obtained by management are ephemeral, requiring constant renewal of organic matter, and thus may leave no traces detectable by archaeologists, in strong contrast to *terra preta* anthroposols.

Resource concentration by RF farmers mirrors certain traits of natural ecosystems in seasonally flooded savannas and other highly constraining environments, whose functioning is dominated by resource-concentration mechanisms driven by soil engineers. RF farmers also take advantage of, and act synergistically with, these natural mechanisms of resource concentration.

The only failing of wetland RFA is its failure to live up to unrealistic expectations that are based on a fundamental misunderstanding of its nature. In areas of South America where it once flourished, wetland RFA may not work today. It is labour intensive, and where human population density is low it is both unnecessary and infeasible. However, where wetland RFA exists today in Africa it should be given more attention by all those interested in food security and its relation to biodiversity. Although its potential for agricultural production may be less than that of more intensive alternatives (e.g. industrial-scale rice production) in the short term, its potential for combining food production and the maintenance of ecosystem services of wetland environments may be unmatched. In working with wetlands, rather than against them, RFA has the potential to achieve food production without destroying wetlands and the biodiversity and ecosystem services they provide. Farmers in wetland RFA appear to use the biodiversity of wetlands to concentrate and to enhance the availability of scarce resources. They also appear to have fashioned other resource-concentration mechanisms that function in ways analogous to the natural mechanisms. However, the indigenous knowledge of RF farmers has barely been plumbed. Exploring this knowledge will help us better understand the nature of this intriguing way of farming wetlands.

Acknowledgments

The research reported on here was funded by grants from the Institut Universitaire de France, the National Institute of Ecology and Environment (INEE) of the CNRS (Centre National de la Recherche Scientifique), the TOSCA committee of the Centre National d'Études Spatiales, and the GDR Mosaïque (GDR 3353, CNRS/INEE). The ideas presented here profited from discussions and fieldwork with Axelle Solibiéda (Congo), Mélisse Durécu (Zambia) and Anne Zangerlé (Bolivia). Martine Hossaert-McKey (CNRS, Montpellier) and Leonor Rodrigues (University of Bern, Switzerland) provided helpful comments on the manuscript. We are grateful for the assistance in logistical matters provided by Professor Jean-Joël Loumeto and Dr Joseph Yoka (Université Marien Ngouabi, Brazzaville, Congo Republic), by Mashuta Kalebe, National Agricultural Research Institute, Chilanga, Zambia and by Dr José Luis Aramayo (Museo Noël Kempff Mercado, Santa Cruz, Bolivia).

References

Armillas, P. (1971) Gardens on swamps. Archaeological research verifies historical data on Aztec land reclamation in the valley of Mexico. *Science* 174, 653–661. doi: 10.1126/science.174.4010.

Bai, Y. Sun, X., Tian, M. and Fuller, A.M. (2014) Typical water-land utilization GIAHS in low-lying areas: the Xinghua Duotian agrosystem example in China. *Journal of Resources and Ecology* 5, 320–327. doi: 10.5814/j.issn.1674-764x.2014.04.006.

Balée, W. and Erickson, C.L. (2006) Time, complexity and historical ecology. In: Balée, W. and Erickson, C.L. (eds) *Time and Complexity in Historical Ecology. Studies in the Neotropical Lowlands.* Columbia University Press, New York, pp. 1–17.

Barba, J. (2003) Campos elevados. In: *Moxos: una Limnocultura. Cultura y Medio Natural en la Amazonía Boliviana.* Centre d'Estudis Amazònics, Barcelona, pp. 89–92.

Baveye, P.C. (2013) Comment on 'Ecological engineers ahead of their time: the functioning of pre-Columbian raised-field agriculture and its potential contributions to sustainability today' by Delphine Renard *et al.* *Ecological Engineering* 52, 224–227. doi: 10.1016/j.ecoleng.2012.11.011.

Bellotti, A.C. and van Schoonhoven, A. (1977) World distribution, identification, and control of cassava pests. In: Cock, J., MacIntyre, R. and Graham, M. (eds) *Proceedings of the Fourth Symposium of the International Society for Tropical Root Crops, Cali (Colombia), 1 Aug 1976*. International Development Research Centre, Ottawa, pp. 188–193.

Bonachela, J.A., Pringle, R.M., Sheffer, E., Coverdale, T.C., Guyton, J.A., Caylor, K.K., Levin, S.A. and Tarnita, C.E. (2015) Termite mounds can increase the robustness of dryland ecosystems to climatic change. *Science* 347, 651–655. doi: 10.1126/science.1261487.

Chapin, M. (1988) The seduction of models: chinampa agriculture in Mexico. *Grassroots Development* 12, 8–17.

Coe, M.D. (1964) The chinampas of Mexico. *Scientific American* 211, 90–98.

Cunha, L., Brown, G.G., Stanton, D.W.G., Da Silva, E., Hansel, F.A. *et al.* (2016) Soil animals and pedogenesis: the role of earthworms in anthropogenic soils. *Soil Science* 181, 110–125. doi: 10.1097/SS.0000000000000144.

Denevan, W.M. (1992) The pristine myth: the landscape of the Americas in 1492. *Annals of the Association of American Geographers* 82, 369–385. doi: 10.1111/j.1467-8306.1992.tb01965.x.

Denevan, W.M. (2001) *Cultivated Landscapes of Native Amazonia and the Andes*. Oxford University Press, New York.

Denevan, W.M. and Turner, B.L. (1974) Forms, functions and associations of raised fields in the Old World tropics. *Journal of Tropical Geography* 39, 24–33.

Erickson, C.L. (1994) Raised fields as a sustainable agricultural system from Amazonia. Paper presented in the Symposium Recovery of indigenous technology and resources in Bolivia at the 18th international congress of the Latin American Studies Association, Atlanta, 10–12 March. Available at: http://repository.upenn.edu/anthro_papers/14 (accessed 31 October 2016).

Erickson, C.L. (1995) Archaeological methods for the study of ancient landscapes of the Llanos de Mojos in the Bolivian Amazon. In: Stahl, P. (ed.) *Archaeology in the American Tropics: Current Analytical Methods and Applications.* Cambridge University Press, Cambridge, UK, pp. 65–95.

Erickson, C.L. (2008) Amazonia. The historical ecology of a domesticated landscape. In: Silverman, H. and Isbell, W.H. (eds) *Handbook of South American Archaeology*. Springer, Berlin, pp. 157–183.

Frederick, C.D. (2007) Chinampa cultivation in the Basin of Mexico. In: Thurston, T. and Fisher, C. (eds) *Seeking a Richer Harvest: The Archaeology of Subsistence, Intensification, Innovation, and Change.* Springer, New York, pp. 107–124.

Glaser, B. and Woods, W.I. (eds) (2004) *Amazonian Dark Earths: Explorations in Space and Time*. Springer, Berlin.

Hagmann, J. (2009) The 'Chinampas' of the Valley of Mexico. In: Hülsebusch, C. (ed.) *Mexico 2009 – Interdisciplinary Study Tour. Sustainable Rural Development in the Light of Globalisation. A Study Tour to Mexico*. DITSL GmbH, Witzenhausen, Germany, pp. 125–135. Available at: http://www.uni-kassel.de/fb11agrar/fileadmin/datas/fb11/OPATS/Dokumente/mexico_2009_report.pdf#page=124 (accessed 3 November 2016).

Holden, S. (2001) A century of technological change and deforestation in the miombo woodlands of northern Zambia. In: Angelsen, A. and Kaimowitz, D. (eds) *Agricultural Technologies and Tropical Deforestation*. CAB International, Wallingford, UK, pp. 251–269.

Iriarte, J., Power, M.J., Rostain, S., Mayle, F.E., Jones, H. *et al.* (2012) Fire-free land use in pre-1492 Amazonian savannas. *Proceedings of the National Academy of Sciences of the USA* 109, 6473–6478. doi: 10.1073/pnas.1201461109.

Kirch, P.V. (1978) Indigenous agriculture on Uvea (western Polynesia). *Economic Botany* 32, 157–181. Available at: http://www.jstor.org/stable/4253923 (accessed 8 April 2017).

Landarte Puertas, D.-G. Steenbergen, F.V., Mehari Haile, A., Kool, M. and Gebreegziabher Embaye, T.-A. (2016) *Flood Based Farming Systems in Africa*. Available at: http://spate-irrigation.org/wp-content/uploads/2015/03/OP5_Flood-based-farming-in-Africa_SF.pdf (accessed 3 November 2016).

Lee, K. (1997) Apuntes sobre las obras hidrauclicas prehispánicas de las llanuras de Moxos. *Paititi* 11, 24–26.

Lehmann, J., Kern, D.C., Glaser, B. and Woods, W.I. (eds) (2003) *Amazonian Dark Earths: Origin, Properties, Management*. Kluwer, Boston, Massachusetts.

Lombardo, U. (2013) *The diffusion of unsuccessful innovations: the myth of raised field agriculture*. Available at: http://umba-moxos.blogspot.fr/2013/01/the-diffusion-of-unsuccessful.html (accessed 31 October 2016).

Lombardo, U., Canal-Beeby, E., Fehr, S. and Veit, H. (2011) Raised fields in the Bolivian Amazonia: a prehistoric green revolution or a flood risk mitigation strategy? *Journal of Archaeological Science* 38, 502–512. doi: 10.1016/j.jas.2010.09.022.

Mann, C.C. (2005) *1491: New Revelations of the Americas before Columbus*. Alfred A. Knopf, New York.
Martín, J.G., Mendizábal, T., Schreg, R., Cooke, R.G. and Piperno, D. (2015) Pre-Columbian raised fields in Panama: first evidence. *Journal of Archaeological Science: Reports* 3, 558–564. doi: 10.1016/j.jasrep.2015.08.018.
McKey, D., Rostain, S., Iriarte, J., Glaser, B., Birk, J.J., Holst, I. and Renard, D. (2010) Pre-Columbian agricultural landscapes, ecosystem engineers, and self-organized patchiness in Amazonia. *Proceedings of the National Academy of Sciences of the USA* 107, 7823–7828. doi: 10.1073/pnas.0908925107.
McKey, D., Renard, D., Zangerlé, A., Iriarte, J., Adame Montoya, K.L. *et al.* (2014) New approaches to pre-Columbian raised-field agriculture: ecology of seasonally flooded savannas, and living raised fields in Africa, as windows on the past and the future. In: Rostain, S. (ed.) *Amazonia. Memorias de las Conferencias Magistrales del 3er Encuentro Internacional de Arqueología Amazónica*, pp. 91–136. Ministerio Coordinador de Conocimiento y Talento Humano e IKIAM /Secretaría Nacional de Educación Superior, Ciencia, Tecnología e Innovación / Tercer Encuentro Internacional de Arqueología Amazónica, Quito, Ecuador. doi: 10.13140/RG.2.1.4539.0808. Available at: https://www.researchgate.net/publication/283296249_New_approaches_to_pre-Columbian_raised-field_agriculture_the_ecology_of_seasonally_flooded_savannas_and_living_raised_fields_in_Africa_as_windows_on_the_past_and_future (accessed 3 November 2016).
McKey, D., Durécu, M., Pouilly, M., Béarez, P., Ovando, A., Kalebe, M. and Huchzermeyer, C.F. (2016) Present-day African analogue of a pre-European Amazonian floodplain fishery shows convergence in cultural niche construction. *Proceedings of the National Academy of Sciences of the USA* 113, 14938–14943. doi: 10.1073/pnas.1613169114.
Meggers, B.J. (1971) *Amazonia: Man and Culture in a Counterfeit Paradise*. Aldine, Chicago, Illinois.
Meron, E. (2012) Pattern-formation approach to modelling spatially extended ecosystems. *Ecological Modelling* 234, 70–82. doi: 10.1016/j.ecolmodel.2011.05.035.
Mielke H.W. and Mielke Jr, P.W. (1982) Termite mounds and chitemene agriculture: a statistical analysis of their association in outhwestern Tanzania. *Journal of Biogeography* 8, 499–504. doi: 10.2307/2844616.
Mollard, E. (1999) Les hortillonnages en Thaïlande et dans le monde. *Journal d'Agriculture Traditionnelle et de Botanique Appliquée* 41, 165–184. doi: 10.3406/jatba.1999.3717.
Renard, D., Iriarte, J., Birk, J. J., Rostain, S., Glaser, B. and McKey, D. (2012) Ecological engineers ahead of their time: the functioning of pre-Columbian raised-field agriculture and its potential contributions to sustainability today. *Ecological Engineering* 45, 30–44. doi: 10.1016/j.ecoleng.2011.03.007.
Rietkerk, M. and van de Koppel, J. (2008) Regular pattern formation in real ecosystems. *Trends in Ecology and Evolution* 23, 169–175. doi: 10.1016/j.tree.2007.10.013.
Rietkerk, M., Dekker, S.C., de Ruiter, P.C. and van de Koppel, J. (2004) Self-organized patchiness and catastrophic shifts in ecosystems. *Science* 305, 1926–1929. doi: 10.1126/science.1101867.
Rodrigues, L., Lombardo, U., Fehr, S., Preusser, F. and Veit, H. (2015) Pre-Columbian agriculture in the Bolivian lowlands: construction history and management of raised fields in Bermeo. *Catena* 132, 126–138. doi: 10.1016/j.catena.2014.08.021.
Rodrigues, L., Lombardo, U., Canal-Beeby, E. and Veit, H. (2016a) Linking soil properties and pre-Columbian agricultural strategies in the Bolivian lowlands. The case of raised fields in Exaltación. *Quaternary International* (in press). doi: 10.1016/j.quaint.2015.11.091.
Rodrigues, L., Lombardo, U., Trauerstein, M., Huber, P., Mohr, S. and Veit, H. (2016b) An insight into pre-Columbian raised fields. The case of San Borja, Bolivian lowlands. *Soil* 2, 367–389. doi: 10.5194/soil-2016-27.
Rostain, S. (2012) *Islands in the Rainforest: Landscape Management in Precolumbian Amazonia*. Left Coast Press, Walnut Creek, California.
Saavedra, O. (2006) El sistema agrícola prehispánico de camellones en la Amazonía Boliviana. In: Valdez, F. (ed.) *Agricultura Ancestral. Camellones y Albarradas. Contexto Social, Usos y Retos del Pasado y del Presente*. Abya Yala, Quito, Ecuador, pp. 295–311.
Saavedra, O. (2009) *Culturas Hidraulicas de la Amazonia Boliviana*. OXFAM, La Paz, Bolivia.
Siame, J.A. (2006) The Mambwe mound cultivation system. *LEISA Magazine* 22(4), 14–15.
Sillitoe, P. (1996) *A Place Against Time. Land and Environment in the Papua New Guinea Highlands*. Harwood Academic, Amsterdam.
Stromgaard, P. (1989) Adaptive strategies in the breakdown of shifting cultivation: the case of Mambwe, Lamba, and Lala of northern Zambia. *Human Ecology* 17, 427–444. doi: 10.1007/BF00889499.
Stromgaard, P. (1990) Effects of mound-cultivation on concentration of nutrients in a Zambian miombo woodland soil. *Agriculture, Ecosystems and Environment* 32, 295–313. doi: 10.1016/0167-8809(90)90167-C.

Swartley, L. (2002) *Inventing Indigenous Knowledge: Archaeology, Rural Development, and the Raised Field Rehabilitation Project in Bolivia*. Routledge, London.

Torres-Lima, P., Canabal-Cristiani, B. and Burela-Rueda, G. (1994) Urban sustainable agriculture: the paradox of the chinampa system in Mexico City. *Agriculture and Human Values* 11, 37–46. doi:10.1007/BF01534447.

Vasey, D.E., Harris, D.R., Olson, G.W., Spriggs, M.J.T. and Turner II, B.L. (1984) The role of standing water and water-logged soils in raised-field, drained-field and island-bed agriculture. *Singapore Journal of Tropical Geography* 5, 63–72. doi: 10.1111/j.1467-9493.1984.tb00146.x.

Whitney, B.S., Dickau, R., Mayle, F.E., Walker, J.H., Soto, J.D. and Iriarte, J. (2014) Pre-Columbian raised-field agriculture and land use in the Bolivian Amazon. *Holocene* 24, 231–241. doi: 10.1177/0959683613517401.

Woods, W.I., Teixeira, W., Lehmann, J., Steiner, C., WinklerPrins, A.M.G.A. and Rebellato, L. (eds) (2009) *Amazonian Dark Earths: Wim Sombroek's Vision*. Springer, New York.

Zangerlé, A., Renard, D., Iriarte, J., Jimenez, L.E.S., Montoya, K.L.A., Juilleret, J. and McKey, D. (2016) The surales, self-organized earth-mound landscapes made by earthworms in a seasonal tropical wetland. *PloS One* 11(5), e0154269. doi: 10.1371/journal.pone.0154269.

11 Andean Cultural Affirmation and Cultural Integration in Context: Reflections on Indigenous Knowledge for the *In Situ* Conservation of Agrobiodiversity

Christopher Shepherd*

Understanding and mobilizing indigenous, traditional or local knowledge in the Third World has become an increasingly important instrument for the management of natural resources. This is no less the case for initiatives concerning the management of agricultural biodiversity in Peru's highlands, where an astonishing variety of potatoes and associated tubers (*mashua*, *oca* and *ulluco*) have been domesticated over millennia as part of a uniquely Andean indigenous cultural and agricultural system. Yet it is only relatively recently that Peruvian state and non-governmental organizations (NGOs) have come to appreciate the intrinsic value of agrobiodiversity and related knowledge, technology and cultural practices, and recognize the need for planned *in situ* conservation to complement what might be thought of as an endogenous conservation practice. Thus, a new type of institutionalized *in situ* conservation has emerged which draws on, mobilizes and revives millennia-old 'conservation' in strategic ways (Brush, 2000).

Deploying indigenous or local knowledge and cultural practices tends to entail complex negotiations with local communities rather than simply identifying relevant traditional knowledge and putting it to use in any straightforward manner (Pottier *et al.*, 2003). This applies equally to the *in situ* conservation of agrobiodiversity in the Peruvian Andes. There, negotiations are complex for two principal reasons. Firstly, intervening organizations have different understandings of why *in situ* conservation matters and what should be done to conserve agrobiodiversity. In Peru, given a historical tradition of defending indigenous logics and a deep tension between conventional and alternative approaches to development, these understandings are generally polarized between two approaches. One approach follows a more exclusive 'culturalist' logic by affirming precisely those traditions that are rooted in indigenous Andean technology, practice and cosmology and are known to be favourable to the regeneration of agrobiodiversity. This approach is commonly called cultural affirmation. Another approach follows a more integrated, developmentalist strategy whereby aspects of Andean culture and technology are incorporated into, and modified to suit, broader logics and external priorities, including those linked to Western science as well as to the growth of regional, national and international markets. I have termed this cultural integration (Shepherd, 2010).

Secondly, negotiations are complex due to the considerable heterogeneity among the Andean peasant communities that might be targeted for

* E-mail: chris.shepherd@anu.edu.au

in situ conservation. This heterogeneity concerns the variable history and current distribution of existing stocks of diverse crops and landraces – the agrobiodiversity – and the manifold practices that have sustained and continue to sustain that agrobiodiversity, which is referred to simply as agrodiversity (Brookfield, 2001). As a rule of thumb, the uneven spread of agrobiodiversity and agrodiversity relates to the degree to which communities are recognizably 'traditional' or 'modern'. Greater distance from urban centres, higher altitudes and limited transportation (i.e. relative isolation) still provide the principal measure by which Andean communities retain traditional agrodiverse forms; conversely, relative proximity to urban centres, lower altitudes (i.e. valleys) and good transportation are indicative of greater departure from Andean diversity traditions and a corresponding insertion of modern forms.[1]

Planned *in situ* conservation in the Peruvian Andes therefore takes place at the intersection of differently motivated intervening organizations and the spatial and technological heterogeneity of peasant communities in which agrobiodiversity itself has assumed a remarkably uneven profile, not only across regions but also across communities within a given area and even between families in the same community. As Brush (2000) noted, regenerating and conserving biogenetic resources by deploying local (or indigenous) knowledge may involve recourse to knowledge that is in fact not local, given the partial or complete departure from the social, technological and biological components of agrodiversity. In the Peruvian case, the complexities of restoring agrobiodiversity are not only dramatized by the different ideological positionings of organizations, but also obfuscated by them, since a certain legitimizing rhetoric on 'Andean culture' infuses most environmentalist interventions into indigenous spaces. It is, in other words, politically incorrect to diminish 'culture'.

Optimizing indigenous knowledge therefore requires more than an understanding of the knowledge and practices themselves that are useful for *in situ* conservation. It is also necessary to understand the current distribution and availability of these practices in any given place as well as the non-local or non-indigenous technologies that are, have been or could be introduced to assist. In addition, optimizing indigenous knowledge means critically examining the institutions, their strategies and incentives as well as the broader institutional networks that have been enrolled for the purpose of *in situ* conservation of agrobiodiversity. The general thrust of this chapter is, then, that the mobilization of knowledge for the *in situ* conservation of agrobiodiversity must be understood not only in the context of indigenous knowledge, culture and livelihoods, but also in the context of indigenous heterogeneity, institutions, markets and science networks.

The chapter is divided into four more sections. The second section briefly covers the emergence of Peruvian understanding of local and indigenous knowledge on agrobiodiversity and how Peruvian institutions have come to recognize and address the problem of loss of agrobiodiversity. The third section explores the cultural affirmationist approach through a number of NGOs who work in Peru's southern and central highlands. The fourth section delivers a contrasting case study in cultural integration, The Potato Park, which has been established in Cusco's province of Calca. The chapter ends with a discussion.

Approaching Agrobiodiversity, Indigenous Knowledge and Culture

In a land where nearly half of its 30 million inhabitants are indigenous, Peru's largest two indigenous minorities are the highland Quechua and Aymara, spread over approximately 6000 peasant (*campesino*)[2] communities. Their habitat falls within one of the most significant areas of agricultural biodiversity in the world and is represented as one of Vavilov's famous 'centres of origin' with a strikingly high level of diversity in tubers, most of which spans eight species of potato and an estimated 3000 landraces (Zimmerer, 1996). The wide dissemination of high-yielding varieties produced by national and international research centres in the latter decades of the 20th century, however, incurred substantial agrogenetic loss in many parts of Peru (Dueñas *et al.*, 1992). This was particularly the case from the time of the agrarian reform (1969–1974). While that reform brought an end to centuries of feudal domination of indigenous people, it also marked the beginning of a clumsy government attempt to modernize indigenous highlanders through integrated rural development programmes (Cleaves and Scurrah, 1980). These programmes led to a substantial loss of agrobiodiversity and with that went the erosion of indigenous knowledge, such as phenotypical knowledge of landraces and numerous associated agroecological practices (Ploeg, 1993; Apffel-Marglin, 1998). Yet as Zimmerer (1996) reminded us, media images of a catastrophic agrobiogenetic resource 'wipe-out' have been exaggerated.

During the time of science's extraction of germplasm for improving potato and other tuber varieties, the proposition that Andean indigenous knowledge could optimize anything apart from poverty and deprivation was unthinkable. It was no coincidence that 1972 was the year in which the International Potato Centre (*Centro Internacional de la Papa*, CIP) in coastal Lima (at La Molina) was inaugurated, reflecting both the huge store of potato germplasm in the proximate Andes as well as the desire to improve that germplasm not only to modernize 'backward' Andean agriculture but also to ensure a global supply of high-yielding varieties. That indigenous peoples had long been *in situ* stewards of those resources was as much taken for granted as the emphasis of international and state institutions was directed to *ex situ* conservation (see Shepherd, 2005b).

In respect to all crops, not just the potato, the emergent political ecologies of food production in highland Peru took shape at the negotiated interface between development agents and indigenous farmers who had long been drawn to market possibilities yet with an enduring interest in meeting their subsistence needs (Gelles, 2000; Shepherd, 2004). The response of the peasantry to development programmes was therefore mixed and ultimately pragmatic, consisting of ongoing assessments of markets, returns on labour and produce, the ecological viability of particular crops and so on (Mayer and Glave, 1992). Particularly in places where market agriculture was less viable, peasants continued to value diversity as the primary means for eking out a livelihood.

While Peru's highly bureaucratic development apparatus systematically dismissed 'traditional Andean agriculture' for its inefficiencies and low productivity, a vibrant academic research tradition in anthropology, economics and geography looked to the rationality and logics behind Andean agriculture and culture (e.g. Golte, 1980), documenting its manifold technical, social and cosmological intricacies (e.g. Morlon *et al.*, 1992) and exploring the 'co-determination' of subsistence and market systems (Mayer and Glave, 1992). If one Peruvian organization, however, can be credited with converting theory into practice and openly challenging the historically entrenched perspective of a deficient indigenous Andean agriculture, it was the Lima-based NGO PRATEC – the *Proyecto Andino de Tecnologías Campesinas*, or Andean Project for Campesino Technologies (see Apffel-Marglin, 1998). In the 1980s, PRATEC began to document the content of *campesino* or indigenous peasant technology in many parts of the highlands, and with the support of other NGOs set about reviving or affirming these practices. Among those organizations was a sizeable group of NACAs (*Nucleos Andinos de Afirmación Cultural*, or Andean Nuclei for Cultural Affirmation) as well as a handful of NGOs that sought alternatives to standard development models. Agrobiodiversity became the cornerstone of cultural affirmation because it was indeed central to traditional Andean agro-practice and all manner of Andean knowledge, technology and cosmology could be holistically shown to be part of diversity-oriented food production. For its opposition to the underlying tenets comprising the monocultural and commercial direction of the Green Revolution, agrobiodiversity also fuelled the anti-modernist, pro-Andean, indigenist demand for cultural preservation (Rengifo and Ishizawa, 1997).

The PRATEC approach nevertheless remained marginal until the end of the 20th century, at which point *in situ* conservation of agrobiodiversity became an international priority and the stock of indigenous traditional knowledge came to seen as a key resource. Suddenly, PRATEC came into its own, and even state institutions, which had always delegitimized all that was 'Andean' agriculture and culture, looked towards PRATEC for inspiration. The state, however, could not easily shed old habits, and *in situ* conservation was converted into a means to an end, namely, conserving agrobiodiversity as a resource for genetic improvement, market development and as a necessary adjunct to *ex situ* storage. As a consequence, state institutions remained weak in promoting comprehensive *in situ* agrobiodiversity conservation, for they fell back into selecting 'the best' native varieties for agroindustrial purposes. For cultural affirmationists, *in situ* conservation was an end in itself, corresponding to the ambition to preserve the integrity of indigenous agrocentric and cosmological life and arrest the influence of nefarious modernity. Some actors steered a middle path, treating agrobiodiversity as both a cultural phenomenon and as a broader resource for commerce and science.

Cultural Affirmation

The Centre for Agricultural and Livestock Services (*Centro de Servicíos Agropecuarios*, CESA) set itself up in Cusco's north-eastern province of Paucartambo in the mid-1980s. After toying with conventional development models based on high-yielding seed, irrigation and inputs, it instituted a new programme in 1994, 'The Culture of Biodiversity'. CESA was among the first NGO agents to implement *in situ* agrobiodiversity conservation in the southern Peruvian Andes following the PRATEC model of cultural affirmation. The programme was brought to a number of districts of Paucartambo, including Colquepata, which consisted of more than 30 communities spread along both sides of the precipitous Q'enqomayo Valley. I first visited some of these communities in 1997 with CESA's director, Luís Revilla. One of these communities, Miscahuara, was made up of 40 families living in small adobe huts scattered over a mountainside. Covering more than 1000 ha, the terrain extended from the relatively warm valley at 3000 m above sea level to the grassy upper slopes at over 4600 m, where the winds were frigid and the peaks occasionally snowed under. About one-third of the total terrain was dedicated to rainfed cultivation, another third to grazing, and much of the lower area had been given over to eucalyptus forest under a government programme. Raising cows, sheep, camelids, chickens, pigs and guinea pigs was an important but secondary component to the community's primary means of subsistence, namely, the cultivation of tubers, maize, wheat, *tarwi* (*Lupinus mutabilis*), oats, beans, quinoa and barley; of these, potato was the staple. The 'vertical ecology' of multiple ecological levels had been critical to the development of crop diversity and, in particular, the considerable number of potato landraces. Miscahuara was a typical peasant community of Colquepta. Yet across the area there were small differences in the crop configuration and diversity. With regard to the potato, some had more diversity, others had less, and some varieties were scarce, confined to particular ecological niches. Overall, farmers' stocks of native seed had dwindled over the years and some varieties had disappeared altogether (Zimmerer, 1996). Miscahuara was considerably wealthy, for it had retained as many as 40 varieties.

If agrobiodiversity in tubers was customarily attributed to the geography and climate, CESA placed emphasis on the role of local people (Shepherd, 2005a). Given the historical importance of seed exchange in generating and maintaining diversity, CESA's primary focus was placed on the circulation of seed. In the local understanding, seed distribution was said to follow seed paths – *los caminos de las semillas*. Indeed, seed 'walked', reflecting Andean cosmology in which all entities from rivers and rocks to mountains and trees are imbued with life; the Quechua people consider seed to be alive and agentive in a world where nothing is inanimate (Allen, 1988). To devise a programme of intervention, CESA learned about the existing and past dynamics of seed exchange, seeking information from *campesinos* as to which type of seed was circulated, whence and to where. For instance, considerable exchange took place between Miscahuara and two communities on the other side of the Q'enqomayo Valley: Sipascancha Alta and Soncco. Each community revealed a unique seed distribution 'map' that CESA derived from a combination of *campesino* knowledge of the phenotypic traits and names of various seed as well as the physical action and experience (or memory) of carrying seed from place to place. As such, agrodiversity as it was experienced by *campesinos* was converted into institutional knowledge. Knowledge of seed circulation was, in turn, linked to other, more specific intra-community aspects of *campesino* knowledge and seed movements. Seed was found to move internally across cultivated plots and among families; occasionally seed was bought and sold, but more commonly it was exchanged, often in tandem with *ayni* (reciprocal labour relations among kin or the *ayllu*). Seed was also found to move from the middle altitudes (*q'ata*) to the higher altitudes (*loma*) where it would be cultivated according to the *chuki* tillage method and be 'refreshed' (*refrescar la semilla*).[3]

CESA attributed the loss of agrobiodiversity to 'the diminishment in the trajectories and exchange of seed' that had resulted from standard development (Shepherd, 2005a). The identification of conventional development as the primary cause was a simplification of a more complex process, but it nevertheless served to legitimize an intervention that in spirit ran contrary to the demonized Green Revolution and the widespread push, shared by the state and most NGOs, for the market articulation of peasant communities. CESA's aim was to revitalize the fluidity of seed exchange in order to recuperate stocks in native varieties of potato (as well as other tubers, beans and medicinal plants). In doing so, the organization dropped the term 'promotion' (of outside technologies), replacing it with 'accompaniment'; CESA employees accompanied communities in accessing and restoring their own knowledge, technologies, and cultural traditions. Accompaniment suggested that indigenous or local knowledge was innate and local, if partially eroded, requiring little outside expertise. In each of the targeted communities, CESA tended to work with groups of 10–20 of the most enthusiastic 'conservationist' households, to whom it encouraged recuperation of seed by facilitating the exchange of seed and plants along routes that were favourable to diversity.

CESA accompanied *campesino* conservationists in all aspects of the agricultural cycle, from tillage, sowing and pruning through to harvest and post-harvest phases, including storage (Figs 11.1 and 11.2). Each of these implied a rich store of local knowledge and practice, which CESA glossed as 'nurturing the *chacra* (field)'. For example, tillage involved knowledge of the many possible types of tillage configurations (conducive to rain-fed irrigation and erosion prevention), soil types, climatic conditions and plot rotation, and sowing drew on knowledge for the selection

Fig. 11.1. A *campesino* family of the Q'enqomayo Valley in their potato *chacra*. (Photograph courtesy of CESA (*Centro de Servicios Agropecuarios*), Cusco.)

Fig. 11.2. The director of CESA, Luís Revilla, emphasizes the importance of agrobiodiversity conservation in an information session with prospective 'conservationists'. (Photograph courtesy of CESA (*Centro de Servicios Agropecuarios*), Cusco.)

of seed, weather prediction, the lunar cycle, where exactly (depth and spacing) the seed should be planted and how locally procured organic fertilizer should be applied (Pérez Baca, 1996). Throughout the cycle, farm tools (such as the emblematic tillage tool, the *chaquitaclla*) had to be deployed in particular ways (Morlon *et al.*, 1992). For the labour-intensive activities such as tillage and harvest, labour also had to be organized, often in the form of non-simultaneous reciprocal work exchanges (*ayni*) among the extended family (*ayllu*) but sometimes also for community-wide working bees (*faena*). Given a cultural association with fertility, women were found to occupy specialized roles in agrodiversity, such as in seed selection, planting and harvesting.

The fields (*chacras*) were construed as the fertile 'home' where seed, the *Pacha* (the living Earth) and people came together in dialogue. Maintaining communication with the *Pachamama* (Mother Earth) and the *Apus* (the mountain deities) through ritual was deemed as critical to agrodiversity as were the more technical aspects of farming. Ritual offerings of coca preceded key moments in the agricultural cycle. In this way, the permission of the *Pachamama* and mountain deities was sought to begin work, bless the seed, protect the plants and deliver a good harvest. Ritual enhanced the affective dispositions of the farmers: work was performed joyfully with joking, laughter and singing, emoting a sense of empathy and love for the *chacra* and for Nature (Revilla, 2014).

In a given community or *ayllu*, any of the above material, technical, ritual, technological or affective dimensions of agrodiversity might be weak or non-existent. The presence of seed or particular varieties was of course highly irregular, given the intrusion of market agriculture and the impact of increasing reliance on wage labour outside the communities. Some facets of traditional knowledge had fallen by the wayside, *ayni* reciprocities had eroded given substantial out-migration, and invocations to the *Pachamama* were in decline, especially on account of the new evangelical churches (e.g. the *Maranatas* who detected sinful idolatry in Andean rituals). In this context of partial erosion, CESA's job was to 're-invigorate' the local or once-local practices by convening discussions and exchanges. In general, enough traditional practices remained around the area to reinvigorate. To remember agrodiversity and know-how, *campesinos* had recourse to their own memories or they could consult with community elders. Inter-communal reunions were also convened to acquire lost seed or knowledge and to share experiences. In a later permutation of the project, the *campesinos* of Colquepta and Pacuartambo districts travelled to Puno to procure seed there, share techniques and reinforce the importance of ritual (Revilla, 2014). Throughout, CESA implored *campesinos* to value their knowledge and revive their customs. It also contributed with some altogether new 'appropriate technologies', such as the making and applying of organic liquid fertilizer and pesticides resourced entirely from native plants; only in these cases did promotion assume its familiar form of 'technicians' imparting their expert knowledge to *campesinos* (see Shepherd, 2004).

The success of CESA's intervention was of course judged on diversity counts. When I revisited the area in 2005 and again in 2014, substantial varietal loss had been recuperated. In Miscahuara, for example, levels of agrobiodiversity had doubled. This diversity was displayed at various times at agricultural fairs in district and departmental (or state) capitals. Clearly, the *campesinos* had become proud of their recuperated diversity and were pleased that the government and public had begun to take an interest. All along, CESA had offered small incentives (such as tools) to the targeted communities to win *campesinos*' attention and to keep them participating as scheduled in the organized activities. Fundamentally, however, the project succeeded because the aims of the 'Culture of Biodiversity' resonated with local values, preferred diets, and living in ritualized harmony with nature; in Quechua, this is known as *sumaq kausay*, or living well (or 'fit livelihood') (see Zimmerer, 1996; Shepherd, 2010). CESA had also tapped into a genuine feeling of nostalgia for the *papitas* ('little' potatoes, expressing affection) that had been lost, and an enthusiasm for welcoming the *papitas* back home again.

In 2000, funding from FAO (Food and Agriculture Organization of the United Nations) gave rise to a national *in situ* conservation programme that enrolled a gamut of governmental and non-governmental actors across the Andes. About 30 of PRATEC's sister NGOs, the NACAs, joined CESA in agrobiodiversity conservation. They took their agendas to a range of areas that differed in terms of traditions and extant agrodiversity. I came to research the NACAs in 2007, under a grant from the Wenner-Gren Foundation. In Puno, I followed one NACA, *Chuyma Aru*, to the Aymara community of Patacancha (province of Moho) on the north-eastern side of Lake Titicaca. Here, there was a large number of landraces in key staples, including potatoes. *Chuyma Aru*'s appreciation of the connection between ritual life and seed exchange was more explicit than I had seen in the case of CESA in Paucartambo. Apart from the routine rituals that accompanied sowing and harvest, seed exchange was shown to be embedded in musical rituals and dance festivals. For example, one festival was the annually celebrated *Kullarani* where a troupe of couples danced their way from one community to the next, picking up seed along the way. As one *campesino* put it, 'the fresh seed wakes the *chacras* that have been frightened by the frosts of the foregoing season' (Shepherd, 2010).

To promote these – what Zimmerer (2002) described as the multi-scaled networks for seed provisioning – *Chuyma Aru* did what CESA did: it picked out the farming and ritual activities that were conducive to diversity, and supported them by speaking them up, attending them, documenting them, providing small incentives and being generally interested in a topic that, in the past, had failed to excite any outsider interest at all. Like CESA, *Chuyma Aru* also facilitated exchanges with other communities and hosted events, bringing *campesinos* together from different parts to talk about agrodiversity, share experiences and, of course, exchange seed. In Patacancha, the intervention worked for the same reasons that it worked in Paucartambo: cultural affirmation and agrodiversity accompaniment linked in with existing notions of what constituted a good diet and a good life as well as with practices that were affectively and ritualistically woven into the fabric of everyday life. In Ayacucho, another part of Peru, I visited a community called Warki in the district of Chuschi (province of Cangallo) where another NACA, *Asociación Pacha Uyway* (APU), was active. Again, the same mode of accompaniment was evident; similar kinds of practices and rituals were singled out as relevant to diversity; and a similar level of success ensued.

When NACAs endeavoured to implement *in situ* conservation in communities closer to 'home' – that is, closer to the departmental (or state) capitals where they had their offices and from where they ran their programmes – they encountered more difficulty. The experience of *Asociación Wari* (AWAY) illustrates the point. AWAY was active in the province of Socos, little more than half an hour's drive from the capital of Ayacucho. In social, technological and biological terms, there was little agrodiversity here. Standard agricultural development had left the region well integrated into the urban market of Ayacucho, and agrobiodiversity had taken a dive. Because there was barely any local stock of expertise, practice or seed that AWAY could draw on, accompaniment would not be the right word to describe the relationship between AWAY staff and local *campesinos*; it was rather a kind of inverse promotion of exactly those forms that had long been derided by conventional organizations, organizations that found drawbacks rather than potentials in 'Andean tradition' and who had effectively contributed to the disappearance of diversity-based agriculture.

Now, AWAY was sending the opposite message to *campesinos* in Socos, urging them to take up the agricultural practices and the attendant social organization and rituals that had long been abandoned. In Socos, pro-Andean inverse promoters used many strategies to make their point. They provided seed and incentives. But what was more notable was their emphasis on the importance and relevance of the truly 'Andean' over what was an inauthentic, 'non-Andean' import. They urged *campesinos* to respect what was 'theirs' over what came in 'from the outside'. AWAY promoters questioned the anti-Andean religious messages of the new evangelical churches and they ran down what they saw as the infiltration of urban values, including competition, individualism, consumerism and environmental exploitation. AWAY vouched for the Andean values of beauty, love, respect, harmony, reciprocity and autonomy that went hand in hand with Andean cosmology. In such places, it was harder for organizations to make headway; agrobiodiversity did not resonate strongly with local views of what it meant to get ahead – pursuing market agriculture, setting up small businesses, accessing the cash economy, securing higher education for children, owning private transport, modernizing their dwellings, and availing of communication technologies. Moreover, the purchasing of food (rice, bread, noodles) had become common and the traditional subsistence diet had long been in decline. Agrodiversity revival clearly struggled to take root in places such as this.

From Cultural Affirmation to Cultural Integration

Should one take the serpentine, unpaved road from Colquepata (CESA's territory), up and over the high pass (4500 m above sea level) and down towards the township of Pisac (en route to Cusco), one will notice that the roadside communities on the Pisac side appear increasingly less traditional: the communities are more densely populated, the dwellings are larger with tin roofs, and there are more shops, hostels, road traffic, irrigation and market production. However, if one might have suspected that more development equated to less agrobiodiversity, as in Socos (where AWAY was operational), this was not so. Indeed, this had once been a hotspot of agrobiodiversity, with over 750 local varieties of tuber. Cultural affirmation would have done well here, but there were clearly other options for conserving agrodiversity, and The Potato Park (*El Parque de la Papa*) was one.[4]

The Potato Park is essentially an amalgamation of five communities (Sacara, Chawaytire, Pampallaqta, Paru Paru and Amaru) originally pulled together under a number of themes: protecting territorial rights, knowledge, resources, and advancing a kind of endogenous development, including agrobiodiversity conservation. A local NGO by the name of ANDES (*Asociación para la Naturaleza y Desarrollo Sostenible*, or Organization for Nature and Sustainable

Development) helped to organize The Potato Park in 2000 when it proposed to the five aforementioned communities as well as a sixth, Cuyo Grande, that the land disputes which had long divided them should be transcended and that they should cooperate with each other for the common good. Following this new commitment to cooperation and the pronounced interest in agrobiodiversity, an agreement emerged between The Potato Park, ANDES and CIP to repatriate hundreds of potato landraces and other tuber varieties that CIP had collected in the vicinity in the 1970s and had kept in cold storage ever since. With the support of CIP and ANDES, *campesinos* at The Potato Park worked to 'repopulate' almost 440 additional tubers, and 160 more came in through exchanges with other local communities, to be cultivated and 'held in trust'. By 2015, approximately 1400 types of locally cultivated potato and associated tubers made The Potato Park the most dense site of this type of agrobiodiversity in the world. To repatriate so many, a team of 'local knowledge experts' was assembled to draw on and mobilize the stock of local potato knowledge that had been retained by many of the communities' elder folk.

ANDES understood this local knowledge as embedded in what it called the 'Andean ecological logic'. The Andean ecological logic was a sort of 'discursive package' that spoke of a complementarity within 'a community' or *ayllu*. The ANDES interpretation of *ayllu* was broader than that commonly deployed by development institutions, where the term simply denotes a community of human kin. For ANDES, the *ayllu* referred to a reciprocity consisting of three sub-categories expressed in Quechua: the human community *ayllu* was the *runa* (the 'people'); the *ayllu* of *salq'a* encompassed all that was wild and mostly beyond human control (such as rain, rivers, rocks, winds, lightning, and wild plants and animals); and the *ayllu* of the *aukis* referred to the sacred realm of the *Pachamama* and the mountain deities (*Apus*). All three sub-categories had to be in communication with each other to maintain cosmological harmony and balance. The Andean ecological logic was in fact a conceptual offshoot of cultural affirmation.

The Andean ecological logic and the *ayllu* elaborated by ANDES, however, did not form a package that was naturally reflected in the communities in a holistic, ideological sense; institutional discourses always operate as a particular interpretative overlay in order to make sense of local phenomena in a way that lends an instrumental structure to planning and intervention. Nevertheless, the elements of *ayllu* persisted in the extant traditional agricultural and religious practices. After a decade-and-a-half of exposure to the language of ANDES, however, particular members of the communities spoke the *ayllu* package with some fluidity, as if it were their own. Those most adept at the new *ayllu* talk were of course the local 'traditional knowledge experts'; these experts needed to articulate the *ayllu* as a quasi-Andean cultural discourse as much as know which local knowledge and practices were required to plant out the varieties trucked up from the coast. A core of eight such local knowledge experts were chosen and each one received a salary from ANDES. According to the director of ANDES, Alejandro Argumedo, these indigenous Quechua-speaking professionals were employed to 'support the communities from a participatory and cultural perspective'. In other words, they articulated the Andean cultural essence of The Potato Park to everybody who lived within its bounds and to those who came to pay a visit. To community insiders and visitors alike, the local knowledge experts had to become publicly adept at defining what was culturally 'theirs' and what was 'not theirs', what was Andean and what was not. This ontological division was critical to making sense of the very mixed cultural and agricultural terrain and encouraging not only the recuperation of traditional practices from a much broader constituency of local diversity-savvy farmers (and arguably the real experts in local knowledge), but also their presentation to outsiders. The *ayllu* in fact functioned as a dual instrument for internally oriented promotion as well as externally oriented public relations.

If the *ayllu* package as a holistic discourse was more or less alien to the locals, the knowledge and practices that the package 'contained' were more or less familiar to them, particularly to the aforementioned elders, who had kept up their agrodiverse practices and could be called on to teach them to others. This was the promotional task of the local knowledge experts; they had to mobilize people and knowledge to incorporate the varieties coming in from CIP to the substantial existing farmer stocks. If the exhibition of Andean idealism as well as the practical recourse to specific traditional knowledge resembled cultural affirmation, the resemblance stopped there. The Potato Park incorporated market and scientific logics that went well beyond the traditionally 'Andean'. The Potato Park relied heavily on incentives, which came in the form of salaries to the local knowledge experts, training and associated benefits, and various other economic benefits that derived from an ever expanding range of market- and science-oriented activities.

For instance, at the higher altitudes, experimentation with a selection of promising native varieties was conducted to see which ones were most resistant to pests and viruses as well as adaptable to a hotter climate, so that these could be improved. CIP, INIA (National Institute for Agrarian Research) and some university agronomy faculties were behind this experimentation. ANDES underlined the fact that over the past 30 years, climatic warming meant that conditions apt for potato cultivation had jumped in altitude by 350 m, to a record height of over 4500 m above sea level. At these higher altitudes, science divided the *loma* (the highest ecological level) into transections of 100 m, measuring different variables (e.g. climatic and hydrological) and observing outcomes (e.g. growth rate and pest resistance) at each transection, with and without the addition of certain inputs (e.g. lime and carbonate). Because these activities involved local *campesinos*, they were referred to as 'participatory improvement' experiments.

Other participatory improvement experiments sought to identify varieties that were high in antioxidant content: 'There's a huge potential market,' exclaimed the director of ANDES excitedly, 'and if these varieties only grow above 4000 metres then this can benefit exclusively the poorest populations in our country who live at high altitudes.' ANDES foresaw the creation of niche markets with 'soft intellectual property rights' (e.g. collective trademarks and community registers) secured for the indigenous communities responsible for primary production. In these experiments, CIP crunched the data, ANDES acted as a kind of intermediary (and translator), and a group of local *campesinos* performed the experimental labour as well as contributed with their own knowledge and practice: where to plant, how to till and sow, which tools to use, when and how to tend the crops, when to harvest and so on. The varieties, however, were not planted together (*en mezcla*) as tradition would dictate, but rather separately. Nevertheless offerings were made to the *Apus* and the *Pachamama* at various stages of the experimental period to guarantee beneficent tutelary mountains and deities. For those who analysed the results, these offerings were inconsequential, that is, not a rational variable requiring measurement. Through this 'live laboratory in climate change and adaptation', stated ANDES, 'we can determine the optimum altitudinal level for specific varieties and for the purpose of improvement'. It should be noted that The Potato Park and ANDES were vociferously opposed to genetically modified organisms (GMOs) and hybrids.

Andean communities customarily reserve a portion of the crop to use as seed in what is known as vegetative reproduction. Segregating the best seed (generally the smallest potatoes with a healthy appearance and with more nodes) and meticulously storing it (often in high-altitude holes in the ground covered by layers of straw) are skills that all Andean potato farmers must learn; the women frequently pick out the best seed immediately following the harvest, to be planted the following year. At The Potato Park, this method continued to be practised. However, The Potato Park also superimposed another, outside method, which entailed pollen-based multiplication to produce disease-free tuberlets (Figs 11.3 and 11.4). For the imported propagation method, local knowledge experts received training at CIP, and back in their communities they were able to implement the learned techniques with the aid of *in situ* greenhouses and propagation facilities. Greenhouse tuberlets gave rise to pre-basic and certified basic seed, which was part of a broader, prospective commercial venture which would see virus-free seeds sold to other farmers in the area. Virus-free seed was also available for farmers within The Potato Park itself, and if I suggested to ANDES that these techniques were replacing indigenous ones of seed selection and removing the pathogens by planting the seed out at high altitudes (*refrescar la semilla*), I stood to be corrected: I was assured that it was rather a question of complementarity between indigenous and non-indigenous techniques. Either way, it appeared that the labour of

Fig. 11.3. Growing out tuberlets in the greenhouse. (Photograph courtesy of ANDES (*Asociación para la Naturaleza y Desarrollo Sostenible*), Cusco.)

seed selection and the *caminos de las semillas* (seed paths) were being modernized in a way that Peru's cultural affirmationists would be reluctant to endorse.

Addressing another complementarity, that between *in situ* and *ex situ* conservation (also known as 'dynamic conservation'), altogether new *caminos* took seed even further afield. Botanical (sexual) seed derived from the plants' mature berries had been taken to Europe for storage (Fig. 11.5). At the time of research, two *campesinos* had been flown to Norway with 750 varieties of botanical seed to deposit at the Svalbard Global Seed Vault. Images of Peruvian *campesinos* in ponchos 'at the North Pole' appeared in national newspapers and in-flight magazines (Fig. 11.6). *Campesinos* at The Potato Park insisted that only they had access to the

Fig. 11.4. The farmers extract pollen from the flowers to produce new offspring for seed multiplication. (Photograph courtesy of ANDES (*Asociación para la Naturaleza y Desarrollo Sostenible*), Cusco.)

Fig. 11.5. A peasant examines the berries (fruit) from which sexual seed (or botanic seed, also known as True Potato Seed) will be derived. (Photograph courtesy of ANDES (*Asociación para la Naturaleza y Desarrollo Sostenible*), Cusco.)

ex situ supply in the event that their *in situ* seed should be lost through a catastrophic event. ANDES defended the action, stating that 'traditional knowledge need[ed] to accommodate science'.

Equally anathematic to the cultural affirmationists would be The Potato Park's adoption of entrepreneurship. Ever since the advent of structural adjustment and neoliberalism in the late 1980s and 1990s, small business development had become the favoured model. The Potato Park responded to this market logic, forging links with travel agencies in Cusco and pushing itself as a tourist attraction. ANDES, meanwhile, helped to incorporate The Potato Park as an Indigenous Biocultural Heritage Territory and assisted the communities in the logistical organization of conducting tours, not only for tourists but also for students on educational exchanges. Local knowledge experts and other locals were trained as 'tour guides' for this novel form of agroecotourism and education. In a rhetorical effort to 'Andeanize' the dominant market mindset in which the initiative was grounded, the tours were said to facilitate *sumaq kausay* (good living or 'fit livelihood'). The tour guides would shed their modern attire, don their traditional woven hats (*chullo*) and slip into their ponchos before taking visitors to the high-altitude experimental sites, the greenhouses, the seed propagation facilities and the restaurant, where a delectable platter of guinea pig and native potato was served. Visitors were also led to the fields where they could pick up a *chaquitaclla* and turn over a piece of terrain, sow a few seeds, dig out some potatoes (depending on the time of year), savour some coca or take part in an offering of leaves to one of the resident *Apus*. At the time of research over a thousand visitors had been chauffeured through the

Fig. 11.6. *Campesinos* at the Svalbard Global Seed Vault, Norway. (Photograph courtesy of ANDES (*Asociación para la Naturaleza y Desarrollo Sostenible*), Cusco.)

five communities, generating a considerable income, which was then distributed among the experts, guides and cooks, while a portion was designated to the communities. To manage the business, some community members had received scholarships to be trained in administration and accounting so that, said the director of ANDES, 'they can do it themselves and we can withdraw little by little'. Judging by the renovated houses and private cars of some of the local knowledge experts as well as their trips to Norway and to various indigenous summits in Latin America, The Potato Park has become as much about upward mobility, travel and indigenous politics as about the restoration of agrodiversity; indeed, all of these are now intertwined. (In a comment to a draft of this chapter, ANDES remarked that revenue derived through the activities of The Potato Park was not the only driver of economic betterment.)

The Potato Park has tended to omit references to what might be called its 'modern politics', unless pressed; academics also participate in this oversight in their essentialistic analyses of culture and agriculture (e.g. Garrett Graddy, 2013), as if The Potato Park were a prototype of tradition or *ayllu*. One might leave The Potato Park having learned all about how many varieties there are, how they are planted, and how integral is the relationship to the *Pachamama*, but nothing about the political fractures that run between the communities and inside them as certain actors vie for power and access to what has become a lucrative operation. One community, Cuyo Grande, was withdrawn from The Potato Park because, according to the director of ANDES, 'certain powerful people sought to control the flow of tourism and channel incoming visitors through their own hotels'. In his view, those people belonged to that category of indigenous people who leave their communities, work elsewhere (in mining or construction, for example), and later return to their home communities having accumulated resources – money, vehicles, knowledge – to take over the community leadership. In the case of Cuyo Grande, such people had been able to access international funding to construct accommodation under the rubric of *turismo vivencial* or homestays. These people were allegedly governed by an 'individualistic ethic', which ran counter to The Potato Park's egalitarian and democratic principles. Other inequities extended beyond The Potato Park's boundaries; several neighbouring communities wanted to become members when they saw The Potato Park making money. (It was surely not just agrobiodiversity per se that motivated them.) For obvious reasons, The Potato Park's member communities resisted newcomers entering the fold.

Discussion

Both cultural affirmation and integrationist modes of *in situ* conservation (such as The Potato Park) avail of or reconstitute a core body of indigenous, traditional or local knowledge and practices. The institutional mechanisms for cultural affirmation are lightly interventionist with limited incentives, relying more on the established motivation of farmers compelled to eke out a subsistence livelihood. Cultural affirmation is aimed at restoring agrobiodiversity by promoting that which is traditionally Andean, and resists importing new technologies (unless shown to be easily absorbed without disruption to 'culture') and resorting to 'non-Andean' opportunities; squabbles over equity (who benefits and how much) are rare because participation is open. Integrationism offers more incentives, is highly interventionist, and is grounded not only in subsistence needs but also in market logics and scientific networks and methods; in some instances, science and tradition work side by side, while in other instances scientific methods may overlay or replace traditional ones; equity squabbling is inevitably accentuated when more external resources are accessed and a politics of exclusion arises. As we saw with The Potato Park, Andean culture (that integrationist approaches claim to be protecting or regenerating) is compromised, since there is liberal recourse to non-traditional knowledge, technology, practices and market possibilities.

In those cases where NGOs and government institutions have found ways to incorporate some native varieties into commercial enterprises, these are dubitable *in situ* conservation measures given the focus on select varieties. (These approaches arguably perpetuate the erosion of the agrobiodiversity.) Similarly, the market-oriented experimentation that takes place in The Potato Park might appear to run counter to the objectives of *in situ* conservation, since it is based in varietal selection. However, the extensive conservation of existing diversity plus the repatriation that was The Potato Park's *raison d'être* from the outset counts as a legitimate mode of *in situ* conservation. There is no doubt that The Potato Park sought to preserve aspects of the contents that fell within the category of the 'Andean ecological logic', but it also sold that logic in various ways to maintain a supply of diverse germplasm for the commercialization of viable native varieties as well as to perform an 'Andeanism' for tourists seeking an authentic experience of indigeneity.

With considerable incentives, The Potato Park style of *in situ* conservation could work almost anywhere in the Andes (cf. Brookfield, 2001); the integrationist approach needs outside incentives, whether in the form of market-based profits, salaries or training from NGOs or other organizations (training may even include the 'relearning' of traditional farming). If The Potato Park represents a concentrated site of *in situ* conservation and experimentation aimed at supplying certified native seed and complementing *ex situ* conservation, it is difficult to imagine that there would be any demand for more than one potato park in Peru's southern Andean region. Yet the biocultural heritage territory model with a central interest in *in situ* conservation writ large is currently being replicated in other parts of the Andes, as it is in other parts of the world where agrobiodiversity is abundant.

Approaches based in cultural affirmation can be more widely applied than parks and reserves of various kinds, in relation not only to *in situ* conservation of tubers in the Andes but also to a broader range of *in situ* conservation strategies. Cultural affirmation seems to work best in those places where agrodiversity is already well represented in the form of seed, knowledge and practice. Yet it can have a positive impact even where there are only remnants or memories of traditional agrodiverse cultivation. The main requisite is to have farmers sufficiently interested in agrodiversity to welcome the new seed back into their fields if and when the seed is made available. Communities as a whole do not have to be involved; a few 'conservationist farmers' will suffice. Local farmer support, furthermore, may appear in the most unlikely places, since a degree of disillusionment with market-based approaches to agriculture is common (often due to poor economic returns and the marked increase in pests and pathogens that have accompanied the Green Revolution) as well as a certain nostalgia for the farming practice 'of our grandfathers' (*de los abuelos*).

Peasant communities are forever adapting to new situations and responding to new opportunities. Improved roads and transportation, the growth in communications technology, expanding labour markets, encroaching mining operations, the flourishing 'small-business' mentality of *campesinos*, the growth of remittances and consumerism, the intrusion of private land markets, the rise of the new 'anti-*Pachamama*' evangelical churches and the increased physical and social mobility across highland regions, are giving new meaning to what it means to be Andean and indigenous. Some communities close to urban centres or around mining areas are becoming more densely populated, while other, more remote communities are emptying out, leaving only the elders behind to tend the crops and animals. This strains traditional indigenous forms; *faenas* (communal working bees), for instance, are becoming rarer, *ayni* (reciprocal labour) has succumbed to contract labour, non-monetary exchange (*treque*) loses ever more ground to monetary exchange, Andean ritual is being neglected and previously cultivated land lies dormant. *In situ* agrodiversity conservation will have to work out new ways to accommodate these changes, ways which will not always fit neatly with the ideology of cultural affirmation. Indeed, The Potato Park serves as one example of creative thinking to a changing Andean peasantry that is ever more connected to phenomena outside the communities and ever more embedded in global networks.

Forces emanating from beyond indigenous farming communities, however, do not exclude traditional factors such as the preference for a diverse subsistence diet, love for the *chacra*, and a deeply affective and religious relationship with seed, plants and Nature. If for some, this sounds like romantic indigenism, it is not; it remains the reality for many peasant communities, despite all the aforementioned changes. Andeanism and tradition continue to provide a rich repertoire of cultural imagery, meaning and identity as well as a highly developed store of practical knowledge by which to pursue *in situ* agrobiodiversity conservation. The feeling of diversity – farmer pride in diverse landraces, connectedness to land and community, love of plants, taste of produce – will always remain key to the *in situ* enterprise (whether that conservation is planned by institutions or is a mode of autochthonous conservation) (see Brush, 2000). Local knowledge is certainly as much about feeling and connection as it is about cognition and technique, and the delineation of 'cultural or Andean essences' is all part of a negotiated process where Andeanism corresponds to a locally valid system of values and ideals (Shepherd, 2010). Regrettably, cultural affirmation has fallen into decline as an institutional strategy, due principally to lack of funding and the abiding concern for business. More attention therefore needs to be paid to cultural affirmation and the type of accompaniment described in this chapter, which genuinely seek to restore the full gamut of agrobiodiversity to the fields while respecting, if not helping to recreate, the integrity of indigenous Andean culture. Creating dense hotspots of agrobiodiversity, such as The Potato Park, is an interesting and innovative proposal, but should not be taken as a replacement for widely distributed, culturally embedded, conservation.

Acknowledgements

Fieldwork for this article was supported by two Wenner-Gren Foundation post-PhD fieldwork grants for research on agrobiodiversity conservation (in 2007) and mining (in 2015). I also thank Luís Revilla of CESA and Alejandro Argumedo and Tammy Stenner of ANDES for their comments on a draft of this chapter, and PRATEC, APU, AWAY and Chuyma Aru for their support of earlier fieldwork.

Notes

[1] The binary that divides 'the traditional' from 'the modern' is always problematic, but less so in this case since there is a notable correspondence between the presence of agrobiodiversity and what is commonly understood to be traditional Andean agriculture (see Morlon *et al*., 1992).

[2] *Campesino* is generally translated as peasant. In the context of this chapter, the term refers to small-scale highland farmers who live in indigenous communities and whose mixed agricultural and livestock practices are oriented primarily towards subsistence, with varying degrees of market articulation.

[3] While *loma* and *qhata* (or *q'ata*) equate to altitudinal figures, representations of metres above sea level are rarely meaningful to peasants. As Ploeg (1993) observes, the peasant experience of altitude is flexibly interpretive: the terms up (*arriba*) and down (*abajo*) are linked to a nexus of 'rules of thumb' that specify agricultural conditions, yet which appear inaccurate when taken solely as indications of relative altitude in the quantifiable (scientific) sense. In Spanish *refrescar la semilla* means to cultivate a generation of seed at high altitude in order to revitalize it and free it of disease. The Quechua *chuki* (*labranza cero* in Spanish) is the simplest of numerous forms of tillage.

[4] The figure of 750 is that given by the managing NGO, ANDES. Informants outside the potato park have questioned the figure. It is, infact, highly unlikely that there ever existed an original stock of 750 local varieties within what are now the potato park confines. If there are indeed 750 such 'local varieties' there today, these would have been sourced from far and wide.

References

Allen, C.J. (1988) The Hold Life Has: Coca and Cultural Identity in an Andean Community. Macmillan, London.

Apffel-Marglin, F. (1998) Introduction: knowledge and life revisited. In: Apffel-Marglin, F. (ed.) The Spirit of Regeneration: Andean Culture Confronts Western Notions of Development. Zed Books, London, pp. 1–50.

Brookfield, H. (2001) Exploring Agrodiversity. Columbia University Press, New York.

Brush, S.B. (2000) The issues of in situ conservation of crop genetic resources. In: Brush, S.B. (ed.) Genes in the Field: On-Farm Conservation of Crop Diversity. International Plant Genetic Resources Institute, Rome, pp. 3–26.

Cleaves, P.S. and Scurrah, M.J. (1980) Agriculture, Bureaucracy, and Military Government in Peru. Cornell University Press, Ithaca, New York.

Dueñas, A., Mendivil, R., Lovaton, G. and Loaiza, A. (1992) Campesinos y Papas: A propósito de la variabilidad y erosión genética en comunidades campesinas del Cusco. In: Degregori, C.I., Escobal, J. and Marticorena, B. (eds) Péru: el Problema Agraria en Debate SEPIA IV, Seminario Permanente de Investigación Agraria (SEPIA). Universidad Nacional de la Amazonía Peruana (UNAP), Iquitos, Peru, pp. 287–309.

Garrett Graddy, T. (2013) Regarding biocultural heritage: in situ political ecology of agricultural biodiversity in the Peruvian Andes. Agriculture and Human Values 8. doi: 10.1007/s10460-013-9428.

Gelles, P. (2000) Water and Power in Highland Peru: The Cultural Politics of Irrigation and Development. Rutgers University Press, New Brunswick, New Jersey.

Golte, J. (1980) La Racionalidad de la Organización Andina. IEP, Lima.

Mayer, E. and Glave, M. (1992) Papas Regaladas y Papas Regalo: Rentabilidad, Costos e Inversión. In: Degregori, C.I., Escobal, J. and Marticorena, B. (eds) Péru: el Problema Agraria en Debate SEPIA IV, Seminario Permanente de Investigación Agraria (SEPIA). Universidad Nacional de la Amazonía Peruana (UNAP), Iquitos, Peru, pp. 87–120.

Morlon, P., Orlove, B. and Hibon, A. (eds) (1992) Tecnologías Agrícolas Tradicionales en los Andes Centrales: Perspectivas para el desarrollo. COFIDE – Proyecto Regional de Patrimonio, Lima.

Pérez Baca, L. (1996) Crianza de Papas en Paucartambo–Qosqo. Centro de Servicios Agropecuarios, Cusco, Peru.

Ploeg, J.D. van der (1993) Potatoes and knowledge. In: Hobart, M. (ed.) An Anthropological Critique of Development: The Growth of Ignorance. Routledge, London, pp. 209–228.

Pottier, J., Bicker, A. and Sillitoe, P. (eds) (2003) Negotiating Local Knowledge: Power and Identity in Development. Columbia University Press, New York.

Rengifo, G. and Ishizawa, J. (1997) Los Caminos Andinos de las Semillas. In: Rengifo, G. and Ishizawa, J. (eds) Los Caminos Andinos de las Semillas. Proyecto Andino de Tecnologías Campesinas, Lima, pp. 1–43.

Revilla Sta. Cruz, L. (2014) Costumbres de las Papas Nativas. CESA, Cusco, Peru.

Shepherd, C.J. (2004) Agricultural hybridity and the 'pathology' of traditional ways: the translation of desire and need in postcolonial development. Journal of Latin American Anthropology 9(2), 235–266.

Shepherd, C.J. (2005a) Agricultural development NGOs, anthropology, and the encounter with cultural knowledge. Culture and Agriculture 27(1), 35–44.

Shepherd, C.J. (2005b) Imperial science: the Rockefeller Foundation and agriculture in Peru. Science as Culture: Special Issue on Postcolonial Technoscience 13(14), 113–137.

Shepherd, C.J. (2010) Mobilizing local knowledge and asserting culture: the cultural politics of in situ conservation of agricultural biodiversity. Current Anthropology 51(5), 629–654.

Zimmerer, K.S. (1996) Changing Fortunes: Biodiversity and Peasant Livelihood in the Peruvian Andes. University of California Press, Berkeley, California.

Zimmerer, K.S. (2002) Social and agroenvironmental variability of seed production and the potential for collaborative breeding of potatoes in the Andean countries. In: Cleveland, D.A. and Soleri, D. (eds) Farmers, Scientists, and Plant Breeding: Integrating Knowledge and Practice. CAB International, Wallingford, UK, pp. 83–107.

12 The Indigenous Knowledge of Crop Diversity and Evolution

Stephen B. Brush*

The numerous colours, shapes, sizes and tastes within crop species are a window into cultures and knowledge systems. This chapter relates indigenous knowledge to crop diversity and evolution by focusing on two regions of crop domestication: Mesoamerica and the Andes. While indigenous knowledge systems encompass many domains relevant to agriculture (e.g. hydrology, soils, climate, pests and pathogens), the knowledge of crop diversity is better studied than other domains. One goal here is to show how studying indigenous knowledge of crop diversity is useful to understanding the broader topic of crop evolution. Ultimately, that understanding must depend on fuller understanding and integration of the many domains of indigenous knowledge employed in producing food.

Potatoes (*Solanum tuberosum*) originated in the Andes and maize (*Zea mays*) in Mesoamerica. Greater diversity in these crops exists there than elsewhere; the crops have extraordinary cultural salience; and the relationship between crop diversity, evolution and indigenous knowledge is readily observable. Factors that affect this are the length of crop evolution, the presence of wild crop ancestors, and the relative cultural and nutritional importance of local domesticates.

Introduction: Crop Evolution

Biological evolution, including crop evolution, is defined as a change in gene frequencies among populations, observable in population structure that may suggest eventual speciation. Crop evolution traces its roots to gradual practices that established agriculture as the predominant human mode of existence. Plant gatherers and eventually early cultivators induced domestication through selection for favourable traits (e.g. non-shattering seed heads) and moving plants between micro-habitats (Harlan, 1992). Two key processes describe crop evolution: domestication; and the diversification at the sub-species and varietal levels. While indigenous knowledge associated with original domestication has been lost in the millennia since the Neolithic, we can comprehend some aspects of the knowledge of Neolithic people in contemporary approaches to non-domesticated plants (Casas *et al.*, 2007).

Domestication and diversification, the fundamental processes of crop evolution, push diversity in opposing directions. Domestication reduced diversity by taking wild species through the 'bottleneck' of human selection. Dubcovsky and Dvorak (2007) noted that crops have lost 40–70% of the genetic diversity of their wild

* E-mail: sbbrush@ucdavis.edu

ancestors. After passing the domestication bottleneck, crops have generally shown a trend towards morphological differentiation – a flowering of varieties below the species level (Brush, 2004). At first, this was probably a function of the diffusion of crops into new areas and habitats, followed by the accumulation of diversity derived from farmers' selection to meet changing production conditions and satisfy aesthetic and other cultural incentives. This differentiation has not raised diversity of crops to that of wild ancestors, but it has resulted in speciation in some crops. The expansion of morphological diversity appears to be halted and reversed by forces in contemporary agriculture: scientific crop breeding, commercial and state-sponsored seed distribution, and industrial agriculture.

Darwin (1896) provided the fundamental model of crop evolution that still frames crop science. Like its natural counterpart, variation and selection characterize crop evolution, but it is by distinguishing two types of selection: natural and 'artificial'. Artificial selection refers to human-mediated selection and, in turn, subdivides into two distinct types: conscious (intentional) and unconscious (unintentional). Although the term artificial is now rarely used, Darwin's framework still directs us to the role of human variation and agency in promoting variation in crop species.

The establishment of scientific crop breeding in the early 20th century triggered the systematic collection and cataloguing of crop diversity. The resulting inventories reveal the accomplishment of generations of farmers in amassing diversity, apparent in vast numbers of varieties with distinct morphological characteristics (morphotypes) – grain colour, taste or processing qualities, length of growing period, cold or drought tolerance, and so on. Gene banks have accumulated 327,392 accessions of maize and 98,285 for potato (FAO, 2010). The most common source is local varieties collected from farmers, especially in centres of origin, but they also represent the accumulated diversity of farming around the world. The accepted term for these accessions is landraces (Zeven, 1998), a term applied to traditionally derived varieties that are recognized as distinct by farmers. Above all, the numbers reflect the plasticity of crop species and farmers' ability to find and maintain variation.

Investigating crop diversity

The effort to identify the sites of the origins of crops culminated in the early 20th century with the work of Vavilov (1926), who brought to fruition research begun in the 19th century. Two of Vavilov's insights were that diversity within crop species is unevenly distributed and that geographical concentration of crop diversity was evidence of domestication. The convergence of three forms of information forms the contemporary approach to understanding domestication: (i) crop diversity; (ii) the presence of wild ancestors; and (iii) evidence from archaeology and archaeobotany (Brush, 2004). Crop diversity was originally recognized by the presence of numerous landraces known to farmers and measured in plant characteristics such as seed size or tuber shape. Contemporary recognition includes traits not directly visible in the plant, such as enzyme systems (isozymes) and, most recently, molecular variation such as that observed in short sequence repeats (SSRs or micro-satellites) (e.g. Vigouroux *et al.*, 2008). These newer methods have sharpened our understanding of the locations of domestication and the ancestry of contemporary crops, illustrated by the case of maize. Doebley (1990) located maize's origins in the Rio Balsas drainage in southern Mexico. Robust genetic evidence (Vigouroux *et al.*, 2008) documented its subsequent spread and diversification. This genetic work was buttressed by extensive research in archaeology, biogeography and linguistics (Staller *et al.*, 2006).

Botanists and crop scientists struggled with classification of large collections of crop varieties. The initial effort to classify crop species relied on variation in morphology and agronomic characteristics – the same traits that are visible to farmers and under selection. Crop scientists first organized diversity by constructing lists of key descriptors of crop species, subspecies and varieties. The size of collections, variation among landraces grown under different conditions and uncertainty about the significance of different traits were daunting obstacles. The discovery of methods to measure neutral traits (e.g. isozymes and SSRs) contributed significantly to overcoming them.

Potato diversity

Potatoes, native to the central Andes, illustrate crops' potential for morphological differentiation and how the tools used to assess diversity have changed and affected classification. Like crop classification in general, potato taxonomy is constantly under revision. Among the first and most widely used approaches to classifying the tuber was that of Hawkes (1947), who relied on morphological variation in potato botany. This approach framed the development of morphological descriptors used to classify the world collection (Huamán *et al.*, 1977). In his initial collecting trip to the Andes in 1939, Hawkes (1951) collected 1210 samples, primarily of cultivated types – 30% from Bolivia and 40% from Peru. One of his first publications about these collections dealt with Indian potato names: 223 Aymará and Quechua terms that are used monomially or binomially to label potato varieties (Hawkes, 1947). He noted that 'Indian peasants grow 50–100 separate and distinct kinds in one field alone' (Hawkes, 1947, p. 2).

Hawkes (1990) eventually recognized 12 potato species with four polyploid groups: diploids ($2n = 24$, five species), triploids (two species), tetraploids (four species) and pentaploids (one species). Species were determined using 75 morphological traits: stem and habit characteristics (5), leaf characteristics (33), floral and fruit characteristics (25) and tuber characteristics (12). For tuber characteristics alone, there are 12 different traits, each with multiple possible expressions. The most recent revision of cultivated potato classification, based on molecular data (Spooner *et al.*, 2014), radically simplifies its taxonomy, reducing the number of species to four and combining polyploid groups that earlier defined species.

Maize diversity

Our understanding of maize diversity in Mesoamerica follows a similar path to that of potatoes. Like the Andean tuber, the Mesoamerican grain was first systematically collected and classified in the mid-20th century. Vavilov played a role in making initial collections, but American and Mexican researchers were responsible for the first species-wide classification. As with potatoes, the initial classification rested on morphological characters. The descriptors for maize were developed by an international team in the 1940s (Wellhausen *et al.*, 1952) that measured 47 traits: geographical distribution (1), vegetative characteristics (7), tassel characteristics (8), external ear characteristics (11), internal ear characteristics (12) and physiological, genetic and cytological characteristics (8). The product of the research by Wellhausen *et al.* was to classify 2000 Mexican varieties into 25 races, based on similarities and contrasts among the characters.

By 2000, the list of significant maize descriptors focused on fewer data points and added isozymes to the data used for classification. The definitive study of diversity across Mexican maize landraces (Sanchez-Gonzáles *et al.*, 2000) studied 35 morphological traits and 21 isozymes in a sample of 209 accessions. Their analysis dramatically increased the number of races to 59. While morphological traits remain important in understanding maize diversity, research has increasingly shifted to molecular data. This data has allowed researchers to connect domesticated maize to a regionally specific wild ancestor (*Z. mays* L. subsp. *parviglumis*) and to estimate the diffusion and diversification of maize throughout the Western Hemisphere (Vigouroux *et al.*, 2008). This work, based on allelic variation at SSR loci, demonstrates the direct link between diffusion away from the Rio Balsas drainage and diversification observed in morphologically distinct races and regional clusters of races.

The use of morphological characters requires that plants be grown in controlled environments, ideally for at least 2 years, to detect the interaction of genetic background and environment in determining characters. Although molecular analysis requires elaborate laboratory apparatus, it is more efficient compared with the controlled comparison of plant materials. Yet, while crop science has shifted to greater emphasis on measuring molecular variation, the association between genetic and morphological profiles of crop populations is still prized. This association is essential because of the significance of farmer selection in crop evolution and the fact that conscious selection is ultimately based on morphology. This association is especially evident in the work on maize, where the original unit of classification (race) – based on morphology – is still useful to classification based on molecular markers.

Nevertheless, the link between morphological and molecular approaches is limited by two factors. Firstly, morphology is continually variable and is most visible over relatively large geographical distances. For instance, at the local level, mixtures of maize races are the norm. While these can be sorted into 59 races (Sanchez-Gonzáles *et al.*, 2000), molecular methods developed so far permit the distinction of only four regional clusters. Secondly, the evidence of population structure based on morphology may not always be matched by genetic structure (Pressoir and Berthaud, 2004a, b; Perales *et al.*, 2005) (see below).

Anthropology's Engagement with Crop Diversity

Botanists generally eschew the abundance of diversity below the species level, preferring to focus on nature's more permanent building blocks. This aversion is also common among ethnobotanists, whose work on classification tends to neglect the variety level that is essential to crop classification. Berlin *et al.* (1973, p. 216) observed: 'Varietal taxa (i.e. further divisions of specific taxa) are rare in most folk biological taxonomies.' In contrast, indigenous knowledge systems in agriculture focus on varietal diversity, providing exceptions to this observation. Examples are easily found, such as Aymará potato farmers in the Andes (La Barre, 1947), Makushi cassava cultivators in Guyana (Elias *et al.*, 2000), or taro growers in Vêtuboso, northern Vanuatu (Caillon *et al.*, 2006). In each of these cases, folk varietals are numerous, rising above 200 in the case of Aymará potato varieties.

Although some early ethnographers studying indigenous knowledge of cultivators (e.g. La Barre, 1947) noted elaboration at the variety level, anthropology's engagement with crop diversity began in the 1980s when human ecologists began to look to diversity as a tool to better understand the nature of traditional agricultural systems (e.g. Brush *et al.*, 1981). This coincided with the growing prominence of biological diversity in scientific research and policy, especially the awareness of the loss of biological diversity accompanying tropical deforestation.

Farmer-mediated selection

Focusing on connections between ethnobotanical knowledge and crop evolution opens new perspectives and questions about this knowledge. Most importantly, it shifts the perspective from anthropological linguistics and cognitive anthropology to human ecology, from a focus on the structure of folk taxonomies and variation between lexical sets to behaviour and the dynamics of crop populations. The new perspective's goal is to inquire how ethnobotanical knowledge influences artificial selection and thus crop evolution.

Contrasting crop science and indigenous knowledge

As with comparisons between formal botanical science and ethnobotanical knowledge of plants, it is pertinent to understand the differences and similarities between the complementary knowledge systems about crop diversity found in crop science and ethnobiology. Indigenous knowledge about crop diversity is rarely as thoroughly articulated as its plant science counterpart. Indigenous knowledge is bound by geography, language and culture and its classificatory effort does not strive to overcome the effect of genotype by environment interaction. Nor is it based on replication of measurement and statistical analysis. While issues of over- and under-classification are significant for crop science, farmers do not seem particularly concerned over synonyms. Farmers are familiar with many of the same morphological traits as breeders, but farmers' classification usually rests on fewer traits. Ultimately, the plant science of crop diversity depends on statistical analysis of data from measuring numerous traits studied under controlled comparison. Measurement and controlled comparison are not part of indigenous knowledge of crop diversity.

Ethnobotanical knowledge is taxonomic but less extensive than or comprehensive as plant science taxonomies. Indigenous knowledge is most evident in the practice of naming varieties with names implying differences and unique combinations of traits. These may not be formally enumerated or even articulated. An especially knowledgeable farmer might explain the differences between two varieties, but names are usually used without specific reference to a trait list. Beyond the variety level, some indigenous crop taxonomies, e.g. Andean potato classification, group varieties into populations that are adapted to different ecological zones. Such grouping is rarely relevant in crop science.

Another difference between crop science and ethnobiology is indigenous farmers' focus on plant parts intended for primary use, e.g. tubers and seed-bearing organs. These parts always constitute a significant portion of descriptor lists, undoubtedly owing to the fact that they have been the objects of conscious selection throughout crop evolution. Farmers may be cognizant of pre-harvest characteristics but are primarily reliant on post-harvest selection. Contemporary plant science uses many more morphological traits than those from edible organs and relies on traits that are only observable in pre-harvested fields or in laboratories. Finally, indigenous knowledge systems for crops may focus on perceptual distinctiveness (Boster, 1985) – on aspects such as colour or leaf form that are readily observable but not necessarily functional in terms of linkage to higher yields, better storage qualities, etc. Boster (1985) found this focus among Aguaruna cassava cultivators, but it should be noted that other researchers studying indigenous cassava cultivators elsewhere in South America (e.g. Elias *et al.*, 2000) find perception of functional aspects connected to productivity.

Variety names, crop type and indigenous knowledge

Studying varietal selection and folk knowledge associated with varieties is the starting point for understanding farmer-mediated selection. Variety naming is the most evident and abundant source of information about indigenous knowledge of crop diversity. Names imply recognition of subspecies differences that are important to farmers, such as yield, resistance to pests and pathogens, drought and frost tolerance, storage properties, nutritional value and culinary qualities.

Farmer-mediated selection is logically understood as a rational choice among varieties, but seeing selection as a purely rational choice exercise is complicated by several factors. Firstly, as Boster (1985) warned, it cannot be assumed that perception of varietal distinctiveness rests on perception of functional traits (e.g. yield). Conceivably, selection for perceptual distinctiveness might run counter to natural selection based on functional traits. Secondly, population-based selection, rather than variety-based selection, may indicate selection for sets of varieties that are adapted to specific environments within a farming system, i.e. ecotype selection. Finally, farmers may not be explicit about a variety's attributes beyond those suggested by the name (e.g. colour) but this does not mean that implicit understanding of other attributes is lacking. Implicit knowledge might only be evident in how different varieties are managed, such as where they should be planted.

Differences among crop species are relevant to understanding the nature of indigenous knowledge of crop diversity. Of particular importance is the method of propagation, whether by sexually derived seed or asexually (vegetatively) derived tubers, bulbs, rhizomes or corms. Reproductive differences are clearly reflected in indigenous classification, as illustrated in the cases of potatoes and maize. Vegetative propagation of potatoes permits individual handling of 'seed', allowing farmers to observe and select among different types or varieties. The result is an elaboration of morphotypes that can be easily managed as individual varieties. In contrast, sexual reproduction and planting by seed lead to bulk management of populations that share morphological similarity (e.g. grain colour) but which may be genetically heterogeneous. The amount of mixture and genetic diversity in seed-based populations depends on: (i) whether the crop is self-pollinated, such as wheat, or out-crossing, such as maize; and (ii) the amount of rigour that farmers bring to selection, for instance whether the seed is selected before harvest or afterwards.

The potato knowledge of Andean farmers reflects the complex agroecological space they manage and adaptation of potato species to different environments (Zimmerer, 1996). Four folk species are recognized (Table 12.1) and differentiated according to tuber characteristics, altitude range (ecology), use and diversity. This folk system overlaps with plant science that recognizes distinct species occupying different environments in Andean agriculture. The greatest diversity exists within potatoes cultivated in the mid-altitudes. Individual households commonly maintain 35 or more morphotypes (Brush *et al.*, 1981), and hundreds of varieties can be found in small regions. CIP (2006) reports 500–600 varieties in Huancavelica Department of Peru and 150 in the small Chugay District of La Libertad (CIP, 2015). Names emphasize tuber

Table 12.1. Folk species of Andean potatoes (from Brush, 1992).

Folk Species	Ecology	Uses	Phenotype	Polytype
Mikhuna papa	Broad adaptability; mid-altitudes: 2500–3700 m	Boiling; soups; frying	Non-bitter tubers; highly variable	Very high
Haya papa	Frost resistant; high altitude: 3700–4100 m	Processing by freeze-drying into *chuño*	Bitter tubers	Low
Araq papa	Weedy species; low–medium altitude: 2500–3200 m	(Rarely used); boiling; soups	Non-bitter tubers	Low
Atoq papa	Wild species; all altitudes	Not used	Small tubers	None

characteristics (colour, form, texture) but Andean farmers also rely on ecology and plant characteristics in classifying potatoes. Huancavelican farmers in Peru reference 22 morphological characteristics in addition to tuber characteristics (CIP, 2006). Individually named potato morphotypes appear to be genetically similar within villages but this similarity erodes with distance from the village level (Quiros *et al.*, 1990).

The contrast between indigenous knowledge of Mesoamerican maize and Andean potatoes is useful. Individual maize farmers in Mesoamerica manage many fewer named varieties that Andean potato farmers: one to three maize varieties versus 35+ potato varieties per household (Brush *et al.*, 1981; Perales *et al.*, 2005). Maize varieties are best understood as populations of out-crossing individuals. These populations are distinguished by grain characteristics (colour, size, type), ear characteristics (ear size and shape) and plant characteristics (short versus long growing cycle; tassel length). Grain colour dominates the naming system of maize and different types (e.g. short or long season) are sometimes named only by colour, though they are separated for planting in different cycles.

As mentioned above, studies of maize populations using morphological versus genetic measures show that morphological segregation does not automatically indicate genetically distinct populations (Pressoir and Berthaud, 2004a, b; Perales *et al.*, 2005). Lack of genetic separation may also characterize potato varieties grown over large regions, but asexual propagation appears to promote greater local genetic convergence among potatoes varieties (Quiros *et al.*, 1990). Appropriately for a sexually reproduced and out-crossing species, maize populations are genetically very dynamic over small geographical ranges.

While different propagation for potatoes and maize results in variety-based versus population-based selection, Andean and Mesoamerican farmers are similar in distinguishing their respective crops according to management strategies. Environmental heterogeneity (e.g. altitudes, soils) and socio-economic heterogeneity (subsistence versus commercial production) are clearly reflected in indigenous knowledge, selection and crop population distribution.

The sum of selection-related knowledge can be characterized as 'ideotype' (Donald, 1968), a concept used by crop breeders to organize the art and craft of creating new varieties. For indigenous farmers, ideotype is based on morphology and performance in specific areas in their production systems. Ideotypes might not be accompanied by explicit descriptions but they are culturally salient (Benz *et al.*, 2007). Cultural salience, moreover, commonly has a clear gender component evidenced in the relatively greater extent of, and sometimes quite different, knowledge among women than men (Chambers and Momsen, 2007). Gender differences may derive from women's exposure to diversity during food storage and preparation, though crop diversity should be understood as a product of households.

Indigenous crop ideotypes are likely to play roles in seed exchange beyond specific localities. The conventional view of crop diversity (e.g. Harlan, 1992) rests on the assumption that varieties are local and maintained because of superior adaptation to local environments. Ethnographic research has challenged this view by finding widely dispersed seed flow for potatoes in the Andes (Brush *et al.*, 1981) and maize landraces in Mexico (Louette *et al.*, 1997). Louette *et al.*'s contribution was to show that seed flow was so ubiquitous and significant as to make maize landraces genetically 'open' systems. Similar findings have now been widely reported for many crops in centres of traditional agriculture (Zeven, 1999). A major outcome of this finding was to reorient the concept of landraces from being local populations to metapopulations that are spatially dispersed and interconnected (Alvarez *et al.*, 2005). Since variety names may not be widely used spatially (Quiros *et al.*, 1990), farmers are likely to seek varieties that conform to types that they know and trust, i.e. ideotypes.

Unconscious Selection

Although Darwin directed attention to unconscious selection, he never clearly defined it and this counterpart to conscious selection has been little researched. The movement of crops into new habitats accompanying migration following domestication is the conventional example of unconscious selection, but such movement is not solely prehistoric, as shown in the 'Columbian exchange' between Eurasia and the New World. Zohary (2004) provided examples of agricultural practices that make unconscious selection an active force in contemporary crop evolution: maintenance by seed or vegetative propagation, the plant part for which the crop is grown, crop production practices (tilling, sowing, and reaping). Nevertheless, distinguishing between conscious and unconscious selection in response to these factors is a delicate, perhaps impossible, task.

A role for unconscious selection is bolstered by the observation by Boster (1985) of selection for perceptual distinctiveness, since it may not be based on rational choice for crop adaptation and utility. A role for unconscious selection is particularly evident in indigenous maize cultivation in Mesoamerica, where the limits of rational choice relating to diversity are apparent. In particular, ubiquitous seed flows and relatively uniform uses of maize imply a force other than rational choice in maintaining diversity. Seed flow suggests that maize races can be widely adaptable to different environments. Sanchez-Gonzáles and Goodman (1992) divided Mexican maize races into six biogeographical groups and areas of adaptation. Each of these six biogeographical areas contains seven to ten morphologically distinct races. While maize seed exchange might be confined to races within a specific biogeographical area, post-harvest selection and cross-pollination mitigate against conscious selection or rational choice in maintaining this number of races in a particular biogeographical area. With few exceptions, the use of different races is similar, based on processing to remove the pericarp and grinding into meal to make tortillas and tamales. A rational choice model should predict concentrating on races with higher yields resulting in the eventual loss of races except the highest-performing ones. Yet numerous races continue to coexist in single maize environments.

Maize researchers began to address this puzzle by looking at the role of social origin in structuring maize populations and by inquiring whether population structure within specific regions was connected to specific communities or social groups. Initial work in the Central Valleys of Oaxaca demonstrated that maize populations were morphologically but not genetically differentiated by community (Pressoir and Berthaud, 2004a, b). Perales *et al.* (2005) then showed that maize populations in adjacent municipalities dominated by Tzeltal and Tzotzil ethnolinguistic (Mayan) groups in highland Chiapas belonged to two maize races (i.e. were morphologically distinct) but were not genetically differentiated. Tzeltal farmers' Comiteco maize is superior to Tzotzil farmers' Olotón maize, yet Comiteco has not replaced Olotón in Tzotzil villages. The projected reason for a disconnect between social origin and genetic structure is that sufficient seed flow and cross-pollination among maize populations prevents structure. Farmer selection successfully segregates populations according to morphology, thus masking genetic mixing.

Two other Mesoamerican case studies did, however, find genetic structure of maize populations related to social origin. Van Etten *et al.* (2008) found that maize populations in the highlands of western Guatemala were both morphologically and genetically differentiated by distance between relatively close communities, though regional differentiation was muted by seed flow and the diffusion of improved maize varieties. The effect of social origin on crop diversity was confirmed by Orozco-Ramírez *et al.* (2016). They found that social origin (Mixtec and Chatino ethnolinguistic affiliation) in their Oaxaca study region shapes both morphological and genetic diversity of maize more than environmental (altitudinal) variation. This suggests that social factors constraining seed flow between Mixtec and Chatino farmers in the adjacent municipalities affect maize's genetic diversity.

Ethnobiologists working on sorghum in Africa have replicated these findings. Analysing data from across Niger, Deu *et al.* (2008, p. 910) found that 'the association of individual ethnic groups with specific geographical regions causes strong geographical × ethno-cultural interactions in the structure of crop genetic diversity'. To disentangle the contribution of ethnolinguistic diversity from that of environmental differentiation, Labeyrie *et al.* (2014) studied sorghum populations grown by three ethnolinguistic groups (Chuka, Mbeere and Tharaka) in eastern Kenya. They concluded that in this uniform agroecological environment, ethnolinguistic diversity has a discernible impact on the distribution of sorghum varieties and their genetic spatial patterns.

Leclerc and Coppens d'Eeckenbrugge (2012) synthesized the research on sorghum and maize and noted that morphological and genetic differentiation of crop populations was strongly related to gene flow and thus influenced by seed exchange. If seed exchange is influenced by social factors, such as ethnolinguistic identity, then these factors act in an indirect, or unconscious, way to affect crop populations. Leclerc and Coppens d'Eeckenbrugge (2012) expanded the classical genotype by environment interaction formula (G × E), which heretofore has defined crop evolution, to G × E × S by adding social component (S). The social component of crop evolution functions in both conscious and unconscious selection and can be seen as the product of factors such as perception, classification and seed management that are the purview of ethnobiology.

Conclusion

Modelling indigenous knowledge, crop diversity and crop evolution

The preceding discussion of potatoes and maize illustrates the need for multiple analytical perspectives and types of information to link indigenous knowledge, crop diversity and crop evolution. Diversity is measured in different ways, and farmers' distinctions based of crop morphology may not be reflected in an underlying genetic structure of crop populations. Knowledge is apparent in folk taxonomies but covert knowledge is also at play. Both intentional and unintentional selection operate to order variation.

While general ethnobiological knowledge has been modelled (e.g. Berlin *et al.*, 1973), no comparable model exists for crop diversity. One challenge is to link indigenous knowledge to observed patterns of diversity. Leclerc and Coppens d'Eeckenbrugge (2012) provided a useful model for understanding anthropological analysis of indigenous knowledge of crop diversity. It showed four approaches for analysing behaviour located at the societal or individual levels and directed at determining meaning and function (Fig. 12.1). Figure 12.2 builds on this model but focuses on the connections between indigenous knowledge, crop diversity and crop evolution that have been the subject of this chapter. This model shifts the vertical axis of Leclerc and Coppens d'Eeckenbrugge (2012) from the level of analysis to variation versus selection, and the horizontal axis from focus on meaning versus function to knowledge versus social organization. Specific areas of data and research in the internal space of both figures largely overlap. Juxtaposing the two figures shows the scope and organization of anthropological approaches to indigenous knowledge of crop diversity and their usefulness to understanding the structure and dynamics of crop populations, i.e. to crop evolution.

Guidelines for understanding farmer-mediated selection

1. Farmer knowledge of crops is taxonomic, recognizing generic, variety and sub-variety levels. This is especially true for asexually propagated plants whose seed is individually handled and when different ecotypes, feral plants and wild relatives are present.
2. Seed handling affects selection. Small grains are selected in bulk. Large fruits and vegetatively

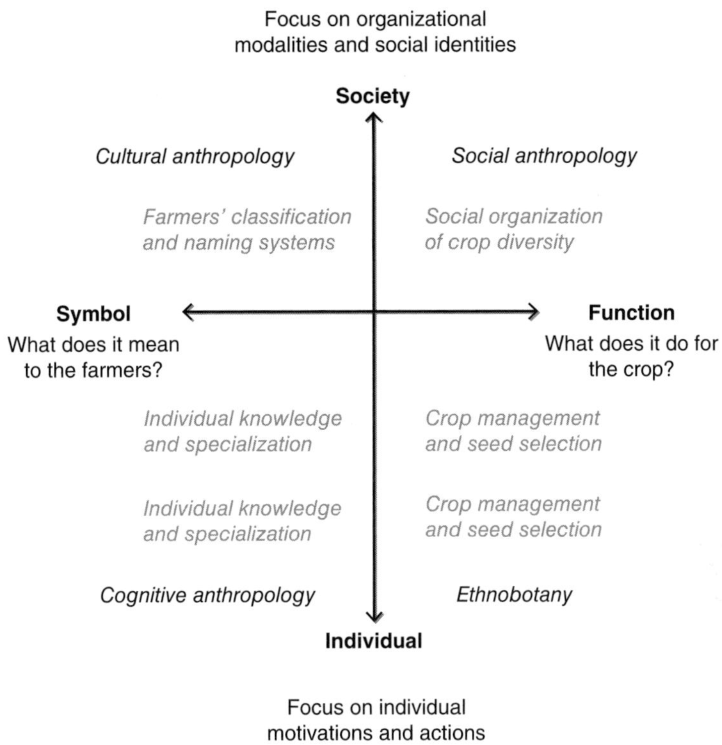

Fig. 12.1. Anthropological study of crop diversity (adopted from Leclerc and Coppens d'Eeckenbrugge, 2012).

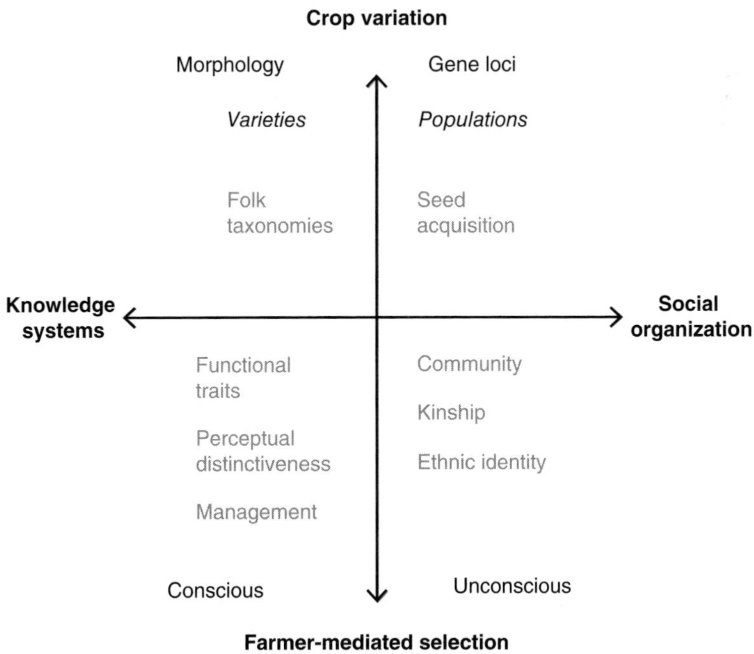

Fig. 12.2. Human components of crop evolution: indigenous knowledge and crop diversity.

propagated crops are selected as individuals, providing for greater morphological diversity.

3. Selection for perceptual distinctiveness is frequent, often emphasizing colour. Perceptual distinctiveness may be both for non-functional traits (e.g. not linked to yield) and functional traits (e.g. bitterness in potatoes linked to frost resistance).

4. Selection for ecotype is frequent – by seed characteristics (e.g. bitter potatoes) or population sampling where better-adapted plants do not necessarily exhibit different seed characteristics (e.g. long-season or short-season maize).

5. Knowledge and selection emphasize the primary part of the crop for use (e.g. seed head or tuber).

6. Farmer selection maintains distinct morphotypes but may not maintain genetically distinct populations.

7. Farmer knowledge and selection is usually associated with only a few plant traits but implicit knowledge is embedded in variety nomenclature.

8. Ethnobotanical knowledge of crop diversity rests on 'ideotypes' – idealized types that are culturally salient.

9. Women and men often have different understandings of crop diversity and women are usually more knowledgeable.

10. Unconscious selection linked to social origin and seed flow can be significant in structuring diversity.

References

Alvarez, N., Garine, E. Khasah, C., Dounias, E., Hossaert-McKey, M. and McKey, D. (2005) Farmers' practices, metapopulation dynamics, and conservation of agricultural biodiversity on-farm: a case study of sorghum among the Duupa in sub-Sahelian Cameroon. *Biological Conservation* 121, 533–543.

Benz, B., Perales H. and Brush, S.B. (2007) Tzeltal and Tzotzil farmer knowledge and maize diversity in Chiapas, Mexico. *Current Anthropology* 48, 289–300.

Berlin, B.O., Breedlove, D.E. and Raven, P.H. (1973) General principles of classification and nomenclature in folk biology. *American Anthropologist* 75, 214–242.

Boster, J.S. (1985) Selection for perceptual distinctiveness: evidence from Aguaruna cultivars. *Economic Botany* 39, 310–325.

Brush, S.B. (1992) Ecology, biodiversity and modernization in Andean potato agriculture. *Journal of Ethnobiology* 12, 161–185.

Brush, S.B. (2004) *Farmers' Bounty: Locating Crop Diversity in the Contemporary World.* Yale University Press, New Haven, Connecticut.

Brush, S.B., Carney, H.J. and Huamán, Z. (1981) Dynamics of Andean potato agriculture. *Economic Botany* 35, 70–85.

Caillon, S., Quero-Garcia, J. and Lescure, J.-P. (2006) Nature of taro (*Colocasia esculenta* (L.) Schott) genetic diversity prevalent in a Pacific Ocean island, Vanua Lava, Vanuatu. *Genetic Resources and Crop Evolution* 53, 1273–1289.

Casas, A., Otero-Arnaiz, A., Pérez-Negrón, E. and Valiente-Banuet, A. (2007) *In situ* management and domestication of plants in Mesoamerica. *Annals of Botany* 100, 1101–1115.

Chambers, K.J. and Momsen, J.H. (2007) From the kitchen and the field: gender and maize diversity in the Bajío region of Mexico. *Singapore Journal of Tropical Geography* 28, 39–56.

CIP (2006) *Catálogo de Variedades de Papa Nativa de Huancavelica – Perú.* Centro International de la Papa (CIP), Lima.

CIP (2015) *Catálogo de Variedades de Papa Nativa de Chugay – La Libertad – Perú.* Centro Internacional de la Papa (CIP), Asociación Pataz, Instituto Nacional de Innovación Agraria (INIA), Lima.

Darwin, C. (1896) *The Variation of Animals and Plants Under Domestication* [Original 1868]. D. Appleton and Company, New York.

Deu, M., Sagnard, F., Chantereau, J. Calatayud, C.D., Hérault, D. *et al.* (2008) Niger-wide assessment of *in situ* sorghum genetic diversity with microsatellite markers. *Theoretical and Applied Genetics* 103, 903–913.

Doebley, J.F. (1990) Molecular evidence and the evolution of maize. *Economic Botany* 44, 6–27.

Donald, D.C. (1968) The breeding of crop ideotypes. *Euphytica* 17, 385–403.

Dubcovsky, J. and Dvorak, J. (2007) Genome plasticity a key factor in the success of polyploid wheat under domestication *Science* 316, 1862–1866.

Elias, M., Rival, L. and McKey, D. (2000) Perception and management of cassava (*Manihot esculenta* Crantz) diversity among Makushi Amerindians of Guyana (South America). *Journal of Ethnobiology* 20, 239–265.

FAO (2010) *The Second Report on the State of the World's Plant Genetic Resources for Food and Agriculture*. Food and Agriculture Organization of the United Nations, Rome.

Harlan, J.R. (1992) *Crops and Man*, 2nd edn. American Society of Agronomy and Crop Science Society of America, Madison, Wisconsin.

Hawkes, J.G. (1947) On the origin and meaning of South American Indian potato names. *Journal of the Linnean Society (Botany)* 53, 205–250.

Hawkes, J.G. (1951) The Commonwealth potato collection. *American Potato Journal* 28, 465–471.

Hawkes, J.G. (1990) *The Potato: Evolution, Biodiversity, and Genetic Resources*. Belhaven, London.

Huamán, Z., Williams, J.T., Salhuana, W. and Vincent, L. (1977) *Descriptors of the Cultivated Potato*. International Board for Plant Genetic Resources, Rome.

La Barre, W. (1947) Potato taxonomy among the Aymara Indians of Bolivia. *Acta Americana* 5, 83–103.

Labeyrie, V., Deu, M., Barnaud, A., Calatayud, C., Buiron, M. *et al.* (2014) Influence of ethnolinguistic diversity on the sorghum genetic patterns in subsistence farming systems in eastern Kenya. *Plos ONE* 9, e92178.

Leclerc, C. and Coppens d'Eeckenbrugge, G. (2012) Social organization of crop genetic diversity: the G×E×S interaction model. *Diversity 2012*, 1–32.

Louette, D., Charrier, A. and Berthaud, J. (1997) *In situ* conservation of maize in Mexico, genetic diversity and maize seed management in a traditional community. *Economic Botany* 51, 20–39.

Orozco-Ramírez, Q., Ross Ibarra, J. and Brush S. (2016) Maize diversity associated to social origin and environmental variation in southern Mexico. *Heredity* 116(5), 477–484.

Perales, H., Benz, B. and Brush, S. (2005) Maize diversity and ethnolinguistic diversity in Chiapas, Mexico. *Proceedings of the National Academy of Science of the United States of America* 102, 949–954.

Pressoir, G. and Berthaud, J. (2004a) Population structure and strong divergent selection shape phenotypic diversification in maize landraces. *Heredity* 92, 95–101.

Pressoir, G. and Berthaud, J. (2004b) Patterns of population structure in maize landraces from the Central Valleys of Oaxaca in Mexico. *Heredity* 92, 88–94.

Quiros, C., Brush, S.B., Douches, D.S., Zimmerer, K.S. and Huestis, G. (1990) Biochemical and folk assessment of variability of Andean cultivated potatoes. *Economic Botany* 44, 254–266.

Sanchez-Gonzáles, J.J. and Goodman, M.M. (1992) Relationships between Mexican races of maize. *Economic Botany* 46, 72–85.

Sanchez-Gonzáles, J.J., Goodman, M.M. and Stuber, C.W. (2000) Isozymatic and morphological diversity in the races of maize of Mexico. *Economic Botany* 54, 43–59.

Spooner, D.M., Ghislain, M., Simon, R., Jansky, S. and Gavrilenko, T. (2014) Systematics, diversity, genetics, and evolution of wild and cultivated potatoes. *Botanical Review* 80, 283–383.

Staller, J., Tykot, R. and Benz B. (2006) *Histories of Maize: Multidisciplinary Approaches to the Prehistory, Linguistics, Biogeography, Domestication, and Evolution of Maize.* Elsevier, Amsterdam.

van Etten, J., Fuentes López, M., Molina Monterroso, L. and Ponciano-Samayoa, K.M. (2008) Genetic diversity of maize (*Zea mays* L. ssp. *mays*) in communities of the western highlands of Guatemala: geographical patterns and processes. *Genetic Resources and Crop Evolution* 55, 303–317.

Vavilov, N. (1926) *Studies on the Origin of Cultivated Plants*. Institute of Applied Botany and Plant Improvement, Leningrad.

Vigouroux, Y., Glaubitz, J., Matsuoka, Y., Goodman, M.M., Sánchez, G. and Doubly, J. (2008) Population structure and genetic diversity of New World maize races assessed by DNA microsatellites. *American Journal of Botany* 95, 1240–1253.

Wellhausen, E.J., Roberts, L.M., Hernandez X.E. and Mangelsdorf, P.C. (1952) *Races of Maize in Mexico, Their Origin, Characteristics, and Distribution*. The Bussey Institution, Harvard University, Cambridge, Massachusetts.

Zeven, A.C. (1998) Landraces: a review of definitions and classifications. *Euphytica* 104, 127–139.

Zeven, A.C. (1999) The traditional inexplicable replacement of seed and seed ware of landraces and cultivars: a review. *Euphytica* 110, 181–191.

Zimmerer, K.S. (1996) *Changing Fortunes: Biodiversity and Peasant Livelihood in the Peruvian Andes*. University of California Press, Berkeley.

Zohary, D. (2004) Unconscious selection and the evolution of domesticated plants. *Economic Botany* 58, 5–10.

13 Investigating Farmers' Knowledge and Practice Regarding Crop Seeds: Beware Your Assumptions!

Daniela Soleri* and David A. Cleveland

Supporting small-scale traditionally based agriculture is increasingly seen as a viable and desirable alternative to the large-scale industrialized food production system (IAASTD, 2009; De Schutter, 2010). We and others believe that local farmers' knowledge (FK) and practice is an essential element for successful, mutually respectful partnerships between farmers and formally trained scientists to support and improve traditionally based agriculture in ways that are socially equitable and environmentally sustainable. Much of our research conducted with colleagues has focused on understanding FK and management of crop genetic diversity. We have emphasized observation and quantitative analysis of farmers' practices and their results to understand their biological and genetic consequences, and elicited farmers' knowledge through posing questions based in the contexts of their experiences (Soleri and Cleveland, 2009).

This work grew out of testing the general hypothesis, based on our informal experiences, that the cognitive abilities of resource-poor and often illiterate farmers are no different than those of other humans, including scientists, and that farmers' practices make sense in their local contexts. In the process of conducting this research we have also identified the untested assumptions of researchers that contrast with our general hypothesis. Testing those assumptions has helped us to formulate specific alternative hypotheses that speak directly to strategies for agricultural research and change. In the process, we have tested hypotheses about many aspects of FK and practice and have been most surprised, and learned the most, when our own assumptions have not been supported. In this chapter we describe three examples of such surprises and how they pushed us to a more profound and useful understanding of our own knowledge and assumptions, as well as FK.

Common Assumptions about Farmer Knowledge

Much of the discussion of and research on FK is based on its assumed degree of similarity with formal Western scientific knowledge (SK) in terms of its reflection of empirical reality and its capacity to effectively address farmers' needs. Here (Box 13.1) we outline three common essentializing views of farmers and associated assumptions about FK and practice that require testing and add our own fourth perspective.[1]

* Corresponding author; e-mail: soleri@geog.ucsb.edu

Box 13.1. Views and assumptions

(a) The ignorant farmer
FK is different from and inferior to SK. The professionalization of science contributed to the divide between farmers and the formally trained researchers who were often part of organizations that defined science as a unilineal progression from informal inferior knowledge to superior ways of understanding the world. 'Development' was premised on the assumption that FK is different from and inferior to formal Western knowledge, even among scientists who empathized with farmers.
(b) The barefoot scientist farmer
FK is similar but inferior to SK. Documentation of some positive empirical outcomes of applying FK that were consonant with the outcomes of applying Western SK, and with the goals and logic of Western society, produced another view of FK. In this view FK is similar to SK, but hampered by lack of tools and methods. From this perspective farmers were seen as diminutive versions of formally trained scientists.
(c) The wise farmer
FK is different from and superior to SK. The view that FK is superior to SK was based in part on evidence that FK often seemed to have positive and sometimes superior efficacy in, and sometimes beyond, local contexts, and based on social and environmental failures of formal science in agricultural development. This frequently led to a value-based argument for the superiority of FK and belief in a mythical wise farmer who was ecologically and ethically better than scientists. It became easy to conflate challenging negative stereotypes through empirical research with an uncritical reification of FK.
(d) The complex farmer
Our overarching hypothesis is based on a holistic, dynamic view of knowledge – depending on the context, FK can be both different from and similar to SK, and both superior and inferior to SK in terms of its efficacy in advancing social and environmental sustainability. For example, both farmers and scientists are often able to describe accurately or predict empirically verifiable outcomes of crop genotype × environment interactions under conditions within their own experiences, but not able to do so under conditions beyond their experiences (see below).

By taking the complex farmer approach, we thought we could avoid the unfounded, essentializing assumptions of other common perspectives on FK. However, it was unavoidable that we also made untested assumptions about FK, including farmer goals and motivations. These were sometimes based on what we thought they *should* be, influenced by our own values and our understanding of Western science. But because we were also aware that our assumptions could misrepresent farmers' values and experiences in the absence of actual research, we tested our assumptions as hypotheses, sometimes with surprising results. This chapter describes three examples of such surprises. We also suggest how others doing similar research can minimize chances of undermining their understanding with their own untested assumptions. But first we describe our basic methodology.

Our Methodology

Our overarching methodological goal was to investigate farmers' knowledge about agriculture in farmers' own terms, that is, using elements and contexts they were familiar with, or could easily imagine. A key component of this methodology is what we refer to as an ontological comparator, which allows us to avoid, as much as possible, using SK as the standard for evaluating FK when testing hypotheses (Soleri and Cleveland, 2005, 2009). We define ontological comparators as basic models of reality that can be used as relatively neutral common referents to evaluate both the farmer's and the scientist's understanding of empirical reality, and expectations based on that understanding. As Western scientists, we defined these comparators using the most basic knowledge as described by Western science, about which there is no disagreement, although scientists often disagree about the interpretation and use of this basic knowledge at higher levels. For example, we would consider the statements that 'water moves from higher to lower levels' and that 'plants need water to produce a harvest' to be ontological comparators. Scientists and others recognize these simple statements, yet can disagree about their application due to their different experiences and assumptions and the different contexts in which they are applied; for example, the amount of water that should be applied to a field to optimize yield. The fact that scientists disagree about how the basic knowledge of the ontological comparators is applied supports their use for investigating FK and SK.

Another key component of our methodology is scenarios, which are often based on ontological comparators and which create hypothetical situations within which farmers can apply their knowledge (Soleri and Cleveland, 2005). Scenarios depict genotypes, environments and situations with which farmers are familiar, such as the variable resource-limited environments of their fields. But scenarios can also present novel situations that farmers can imagine and extrapolate to, based on their experience, such as uniform growing environments without significant resource limitations. We often use props to make it easier for farmers to participate, such as photographs of plants, seeds from their own harvests, or stones of different sizes to represent different amounts of annual rainfall. We have found that farmers from many different countries, growing different crops in different environments, all participate enthusiastically in responding to these scenarios.

For example, we documented farmers' expectations for phenotypic variation across familiar and novel growing environments, and their distinction between high and low heritability phenotypic traits (Soleri and Cleveland, 2001; Soleri *et al.*, 2002). While we found that both farmers and plant breeders are aware of the components of phenotypic variation (Box 13.2, eq. 1), and use that knowledge to achieve their goals, both FK and SK were also affected by the limited experiences of farmers and scientists. We found that most farmers may not see or have access to genetic variation for low heritability traits such as yield, even though such variation exists, because it is hidden by the high level of environmental variation in their fields (Fig. 13.1). This means they may not carry out seed selection for traits like yield, even though it is important to them.

On the other hand, plant breeders may not be aware of qualitative genotype-by-environment interactions among different crop populations because they have not experienced the range of variation in growing environments that farmers work with (Ceccarelli, 1996). This means that their assumption about phenotypic traits like yield can be biased by their relative lack of experience, for example in predicting the performance of varieties across environments (Cleveland

Box 13.2. Equations describing basic biological models used

In our research we have primarily used ontological comparators about the relationship between plant phenotypes, plant genotypes and plant environments, and about the determinants and results of plant selection, using as ontological comparators the basic biological models described in the following equations:

- **the source of phenotypic variation:**

$$V_P = V_G + V_E + V_{GXE} \quad \text{(eq. 1)}$$

which states that phenotypic variation (V_P) is the result of variation in genotype (V_G), environment (V_E) and genotype-by-environment interaction (V_{GXE});

- **the heritability of that phenotypic variation:**

$$H^2 = V_G/V_P \quad \text{(eq. 2)}$$

which states that broad-sense heritability (H^2) is the proportion of phenotypic variation accounted for by genotypic variation; and

- **the determinants of the response to selection:**

$$R = h^2 S \quad \text{(eq. 3)}$$

which states that the phenotypically expressed genetic gain due to selection (R) is the product of narrow-sense heritability (h^2), and the selection differential (S), where S is the phenotypic difference between the selected individuals and the population from which they are selected.

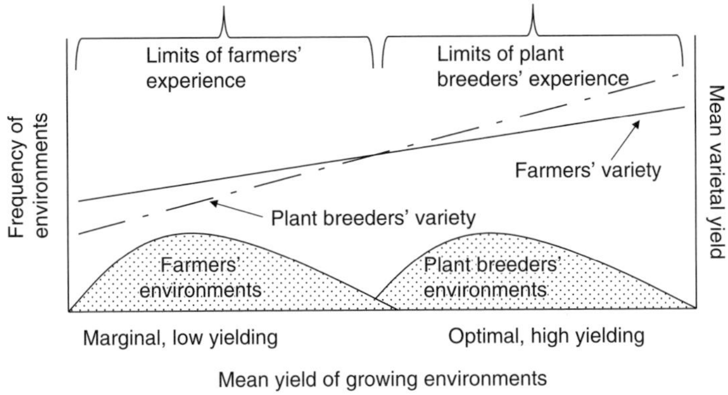

Fig. 13.1. Different experiences affect knowledge and practice. (From Soleri *et al.*, 2002, used with permission, partially based on Ceccarelli (1989, 1996)).

and Soleri, 2002). This example of farmers and scientists agreeing on some aspects of empirical reality, but not on others, supports our use of ontological comparators. Different contexts, and different experiences, values and goals, can result in differences in interpretations and expectations between farmers and scientists, but also among farmers, and among scientists.

In addition to questions based on an ontological comparator, we also elicited farmers' own opinions of a technology, based on their personal experiences and values. This was one part of the first research example described below.

Example 1: Farmers' Perceptions of Risk and Transgenic Maize

Discussions about transgenic crop varieties became particularly heated in 2001 when Quist and Chapela (2001) reported transgenes present in a small sample of traditional maize plants in Oaxaca, Mexico. The polarized debate included declarations about what this meant for small-scale traditional farmers, and advocates with conflicting views claiming to represent farmers' best interests. For example, the flow of transgenes to farmers' traditional maize varieties was described as 'crime against all the indigenous peoples and farmers who have for millennia protected [maize], for humanity to be able to enjoy' (Melina Hernández Sosa, of UNOSJO, an Oaxacan non-governmental organization (NGO), cited in Vélez Ascencio, 2003), or alternatively as a welcome addition which augments Mexican farmers' varieties and agriculture and which they are lucky to get for free (AgBio View, 2002). However, there was no systematic attempt to document and understand what farmers themselves thought about transgenesis or transgenic varieties (TGVs) and the risk those might pose; that is, there was no attempt to engage farmers beyond discussing their preferences for a final product in the form of a maize variety.

Eliciting preferences among a limited number of predefined finished technologies has been a common way to use FK in agricultural development. For example, researchers in Kenya asked maize farmers if they had trouble with stem borers (a major pest in the area) and interpreted their affirmative answer as implying that they wanted transgenic stem borer-resistant maize (KARI and CIMMYT, 2007). To address the lack of direct comments by farmers, we conducted research on perceptions related to transgenic maize with 334 farmers reliant on maize for food and livelihood in six communities, two each in Mexico, Guatemala and Cuba (Soleri *et al.*, 2005, 2008).

What did we hypothesize?

Our goal was to ask farmers' opinions regarding fundamental aspects of transgenic maize in order to facilitate their evaluation of the complex issues involved. We wanted to document their responses so that they could be included in discussions about policy and practice. We hypothesized that farmers would reject the new technology, based on our assumption that they were committed to their traditional farming and maize varieties and were distrustful of unfamiliar technologies introduced from the outside, and their associated risks.

How did we test our hypotheses?

To elicit FK about transgenic varieties in a manner that did not over-simplify the varieties or eliminate farmers' perceptions of risk or opinions on key aspects of them, we deconstructed transgenic maize (Fig. 13.2) into the transgenic technology itself, the variety it was present in, and its possible consequences. We first asked farmers about the technology of transgenesis per se, by describing a process in which a property from another kind of plant or animal could be inserted into maize seed in a laboratory by scientists (Fig. 13.2a). We stated that when planted the seed would grow normally, but the plant would have that inserted property, using the example of resistance to damage by caterpillars.[2] After this description we asked farmers if they thought the process was good, bad, or depended on the outcome. We then asked farmers about varieties that were different combinations of technology (transgenic or not) and with different genetic backgrounds (farmer and modern varieties) (Fig. 13.2b). This is because transgenesis could in theory be used in farmers' varieties, not only in the proprietary hybrids where it has been applied, and evaluations of transgenesis and genetic backgrounds are often confounded. We did this by asking farmers to rank their preferences among four maize varieties: their own local farmers' variety (FV); a familiar modern variety (MV) in the form of a hybrid sold in local seed stores; their FV with 'properties' (e.g. transgenic, TGFV); and the MV with 'properties' (e.g. transgenic, TGMV). Finally, we asked farmers about some of the potential consequences of using a transgenic maize variety in their fields (Fig. 13.2c), including the need to acquire seed of a new variety from the formal seed system after 6–7 years when yields of currently used varieties declined (as might occur due to the evolution of resistance in the caterpillar that had been controlled by that variety), and the cost of that seed (Table 13.1).

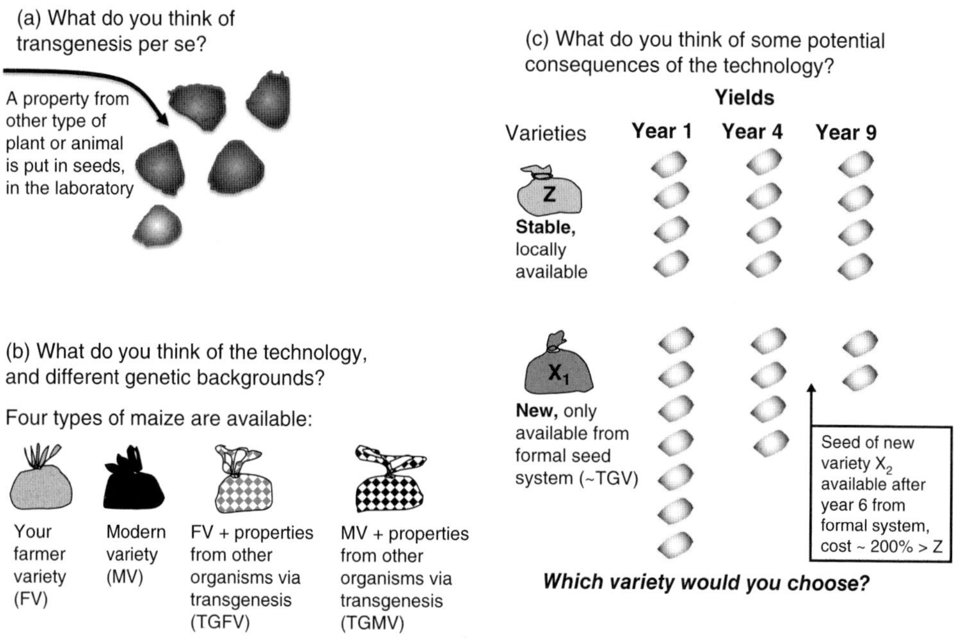

Fig. 13.2. Deconstructing transgenic maize for farmer assessment.

Table 13.1. Farmers' opinions and risk perceptions regarding transgenic maize: questions and responses (data from Solani *et al.*, 2005).

			Country		Community	
Question	Our assumption	Actual finding (summary)	Name (n)	%	Name (n)	%
Is transgenesis per se good, bad, or depends only on the outcome?	Farmers would be unified in rejecting the technology per se, saying it is bad / unacceptable because it is unknown, may be considered a violation of culturally significant maize, and from an outside system that some mistrust	Our assumption, tested as a hypothesis, was not supported by a majority of farmers; responses varied substantially by country and by community	Responses that transgenesis is unacceptable			
			Cuba (114)	42%	La Palma, Pinar del Rio (56)	28%
					Mayorquín, Holguin (58)	55%
			Guatemala (110)	17%	El Rejón, Sacatepequez (55)	33%
					La Máquina, Suchitepequez (55)	2%
			Mexico (110)	43%	Santa Inéz Yatzeche, Oaxaca (55)	51%
					Comitancillo, Oaxaca (55)	35%
			Total (334)	34%		
Which genetic background is preferable for maize for your family to eat?	Farmers would be pro-FV,[(a)] they would prefer local FV-based varieties above all others because of preferred culinary and gastronomic qualities; their top two most favoured varieties would be FV and TGFV[(b)]	Our assumption, tested as a hypothesis, was not supported. The most common ranking pattern was anti-transgenic, ranking any non-transgenic variety as better than any transgenic ones; their top two most favoured varieties were FV and MV[(c)]	Anti-transgenic responses			
			Cuba (114)	64%	La Palma, Pinar del Rio (56)	50%
					Mayorquín, Hollguin (58)	78%
			Guatemala (110)	76%	El Rejón, Sacatepequez (55)	80%
					La Máquina, Suchitepequez (55)	73%
			Mexico (110)	74%	Santa Inéz Yatzeche, Oaxaca (55)	71%
					Comitancillo, Oaxaca (55)	76%
			Total (334)	71%		
Which variety would you choose? (See Fig. 13.2)	Farmers would reject variety representing TGV[(d)] because of consequences	Our assumption, tested as a hypothesis, was supported for the large majority of farmers	Farmers choosing stable variety, not one with potential local consequences of TGV[(d)]			
			Cuba (114)	NA	La Palma, Pinar del Rio (56)	NA
					Mayorquín, Holguin (58)	NA
			Guatemala (110)	83%	El Rejón, Sacatepequez (55)	93%
					La Máquina, Suchitepequez (55)	73%
			Mexico (110)	90%	Santa Inéz Yatzeche, Oaxaca (55)	86%
					Comitancillo, Oaxaca (55)	94%
			Total (232)	86%		

[(a)]FV, farmers' variety; [(b)]TGFV, transgenic farmers' variety; [(c)]MV, modern variety; [(d)]TGV, transgenic variety

These questions were designed to elicit farmers' opinions based on their values and experiences, based in part on the ontological comparator that variation in a plant's phenotype is determined in part by what it consists of, i.e. by V_G as well as V_E (Box 13.2, eq. 2). In other words, farmers would understand that changing the makeup of plants, by inserting a property of another plant or animal, would change the plants' phenotypes.

What did we find?

We were surprised by the results, with more than 60% of farmers responding that if the technology has potential to increase their maize yields, they were open to considering it. Still, when asked to rank the four varieties for sowing, most farmers gave ranking patterns that placed any variety without transgenes (FV, MV) higher than transgenic varieties (TGFV, TGMV); that is, the technology itself may not be inherently negative, but that did not mean they would automatically accept it. However, farmers in one Guatemalan community, with high levels of agricultural commercialization and industrialization and already using MVs, ranked all varieties very closely, but with transgenic varieties favoured over non-transgenic ones. For eating, everyone preferred FVs, but overall ranking patterns showed 53% and 70% of farmers strongly disfavoured transgenic maize for both sowing and eating, respectively (Soleri *et al.*, 2008). Finally, a large majority of farmers did not accept the likely consequences of the technology, including dependence on the formal seed system for seed acquisition, higher seed costs and the need to replace varieties periodically due to declining yields. For the farmers we interviewed, increased dependence on formal seed systems, and the associated costs and risk, were substantial disincentives.

How did this change our understanding?

Based on our analysis of farmers' responses, and informal comments they made while discussing the questions, it became clear that our assumption regarding farmers' opinion of transgenesis per se was simplistic and ignored their lived reality. Instead of a simple rejection of technology and resistance to change, many of these farmers are looking for ways to improve farming and their lives in general. Their answers showed that they were open to change that could help them to do this, including consideration of novel technologies. However, farmers did not accept novel technology without question; they were not naïve and did not automatically trust new technology and its effects on their health and farming. Their responses indicated caution regarding both growing and eating the produce of a novel, unknown technology like the one we described. And that technology is unacceptable if it means increased reliance on systems beyond their control, like markets or the government for seed acquisition, or forms of exploitation which they have experienced in the past and which continue today. Our questions focused on some of the most basic issues surrounding maize TGVs; of course there are other aspects of TGVs to be investigated that we did not address in our research. But even with these limited questions we learned that to portray farmers as simply rejecting transgenic maize oversimplifies them as anti-technologists whose rigid culture excludes them from assessment of new technologies for potential adaptation to their circumstances and needs. It is equally true that a majority of these farmers would not willingly accept this technology based on the concerns expressed in response to questions (b) and (c) in Fig. 13.2. Farmers may not want TGVs, but many are interested in technologies and methods that can support and improve their farming and community in ways consistent with their values, their resources, and their social and biophysical environments.

Example 2: What Maize Farmers Expect and Accomplish with Seed Selection

On-farm seed selection and conservation is a characteristic of traditionally based farming systems, is critical for *in situ* conservation of crop genetic diversity and is increasingly considered part of a strategy for adaptive, resilient agriculture under changing conditions (e.g. Murphy *et al.*, 2005; Vernooy *et al.*, 2015). Research on seed networks in traditional farming systems has shown that they are often a combination of self-provisioned seed as well as frequent local exchanges, and less frequent experimentation with novel material, including MVs (e.g. Louette *et al.*, 1997; Soleri *et al.*, 2005). These practices are usually characterized by less stringent genetic criteria in selection and choice compared with those of scientists, allowing gene flow that generates and maintains crop genetic diversity, as has been documented in a number of studies, for example wild-crop gene flow in sorghum (Mutegi *et al.*, 2012) and clonal–sexual propagule gene flow in cassava (Elias *et al.*, 2001; Duputie *et al.*, 2009; McKey *et al.*, 2010).

On the other hand, precision by some farmers in selecting seed and choosing seed lots for diverse criteria can be greater than that of scientists, which can also function to maintain diversity. Indeed, at least some farmers are capable of extremely fine-grained classification and discrimination when they believe that this will be agronomically helpful (e.g. Worthington *et al.*, 2012) (see below).

What did we hypothesize?

We undertook this research to quantify and document farmers' seed selection and its outcomes. Using the response to selection as our ontological comparator (Box 13.2, eq. 3), our hypothesis was that when selection differentials are substantial – that is, when the difference between the mean of the selected plants is significantly different than the mean of the entire population they are a part of – then farmers, like plant breeders, are using seed selection to genetically change and improve their varieties. This was based on our assumption that, like formally trained scientific plant breeders, farmers believe this cumulative change will create crop populations that better serve their needs; that is, they are selecting for heritable change (Table 13.2). This widespread assumption is part of the legacy of the way in which selection has been conceptualized since Darwin (Cleveland and Soleri, 2007b). It was also based in part on our previous research with these farmers showing that they understood the difference between high and low heritability traits in their crops (Box 13.2, eq. 2) (Soleri and Cleveland, 2001). In retrospect, we fell into the trap of assuming that farmers were in a sense barefoot scientists, applying astute insights to overcome the methodological limitations they were working with, but having the same goals as plant breeders: a heritable, cumulative response to selection.

How did we test our hypothesis?

We designed a number of experiments to understand what a small sample of farmers ($n = 13$) expected to accomplish, and did accomplish, when they selected their own maize seeds on farm. These experiments quantified components of selection practice (Fig. 13.3), allowing us to investigate the contribution of each component to the response to selection. To do this, these experiments were based on the ontological comparators $H^2 = V_G/V_P$ and $R = h^2 S$ (Box 13.2, eq. 2 and eq. 3). In addition, we asked farmers why they selected seeds, instead of simply using a random sample from the same population, and in further research presented a much larger sample of farmers ($n = 380$) with a scenario (Fig. 13.4) based on the response to selection formula, comparing the outcomes when using selected versus randomly sampled seeds from the same original population.

What did we find?

When farmers' actual selection practices were quantified, we found that despite selection differentials (S) comparable to those achieved by plant breeders and an understanding of heritability (h^2), response (R) to farmers' key selection criteria (seed and ear size) was zero (Soleri *et al.*, 2000). That is, they were not producing a cumulative genetically based change in their maize populations as a result of their seed selection. Yet seed selection for those criteria was virtually universal among farmers, despite being costly in terms of their time and in a number of other ways (Table 13.3).

When asked directly why they selected seeds, instead of using a random sample, most farmers replied that this was their custom. In comparing the outcomes when using selected versus randomly sampled seeds in the scenario

Table 13.2. What maize farmers accomplish and expect to accomplish with seed selection (data from Soleri *et al*., 2000, 2002 and authors' surveys).[a]

			Selection differential (S) (response to selection (R) ≅ 0)				
Question	Our assumption	Actual finding (summary)	Location (n)	Ear diameter	Ear weight	Ear length	100-grain weight
What response to selection do farmers *accomplish* with their maize seed selection?	Significant selection differentials (S); cumulative response (R) in form of directional phenotypic change in selection criteria	Significant selection differentials, but no significant phenotypic change over generations ($R = 0$)	Oaxaca, Mexico (13)	0.95	1.14	0.88	0.85

			Proportion of farmers answering advantages of seed selection only last 1 year			
Question	Our assumption	Actual finding (summary)	Location (n)	%	Community (n)	%
What response to selection do farmers *expect* with their maize seed selection?	Heritable, cumulative, phenotypic change, improvement, in selection criteria	No change, improvement that persists past 1 year	Oaxaca, Mexico (380)	82%	Four Central Valley communities (199)	81%
					Four Sierra Juárez communities (181)	82%

[a]*T-tests* for all S values significantly different from original 100 ear samples, $p \leq 0.05$.

(see Fig. 13.4), 82% of the 380 Oaxacan maize farmers responded that seed selection provides an advantage in the field for the first year after selection, but no cumulative phenotypic change over 10 years of selection (Table 13.2) (Soleri *et al.*, in preparation, p. 1). That is, their expectation for what selection for these quantitative traits would accomplish is very different from plant breeders' expectations, and different from what we had also assumed.

Initially our findings were discouraging and perplexing to us; they did not fit our expectations, based on our assumptions, and did not seem to make sense given the costs associated with using selected seed. However, the results were not so surprising, given the fact that seed and ear size are quantitative traits strongly influenced by the environment. Still, as is always the case, there was variation among farmers' responses. While most farmers stated they selected large seeds for planting because it was their custom, three farmers among the original 13 commented that the selected larger seeds would grow better than unselected ones, though individual plants from these seeds would not yield more than randomly selected seeds. This comment led us to consider other ways in which selected seeds could provide an advantage over randomly selected ones, focusing on seed size.

Compared with many other crops, seed size in maize is relatively plastic (Sadras, 2007); that is, it can change a lot depending on the maternal plant's growing environment. Studies of maize have documented that larger seed size can provide significant advantages in the early stages of plant growth, including more rapid development (Pommel, 1990; Bockstaller and Girardin, 1994; Revilla *et al.*, 1999). However, existing studies used materials, seed sizing parameters and environmental treatments relevant to

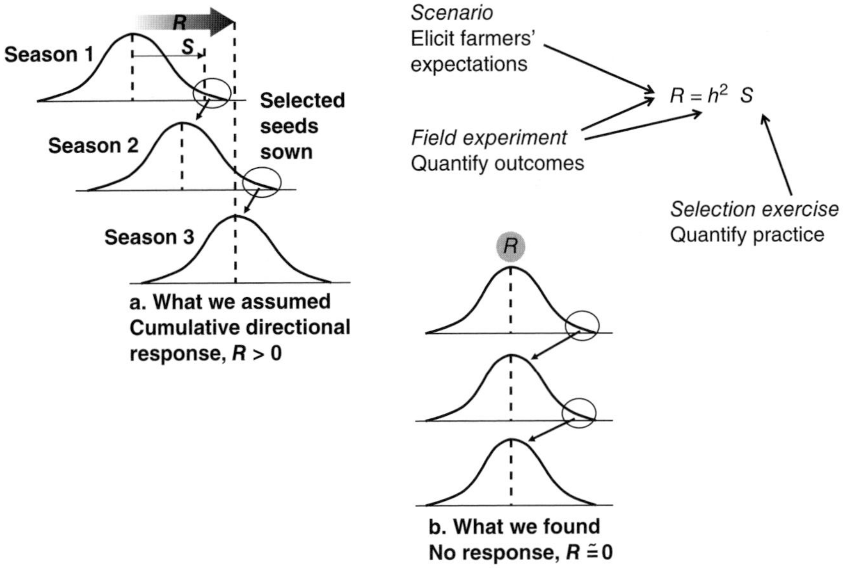

Fig. 13.3. Components of selection practice and how we quantified them.

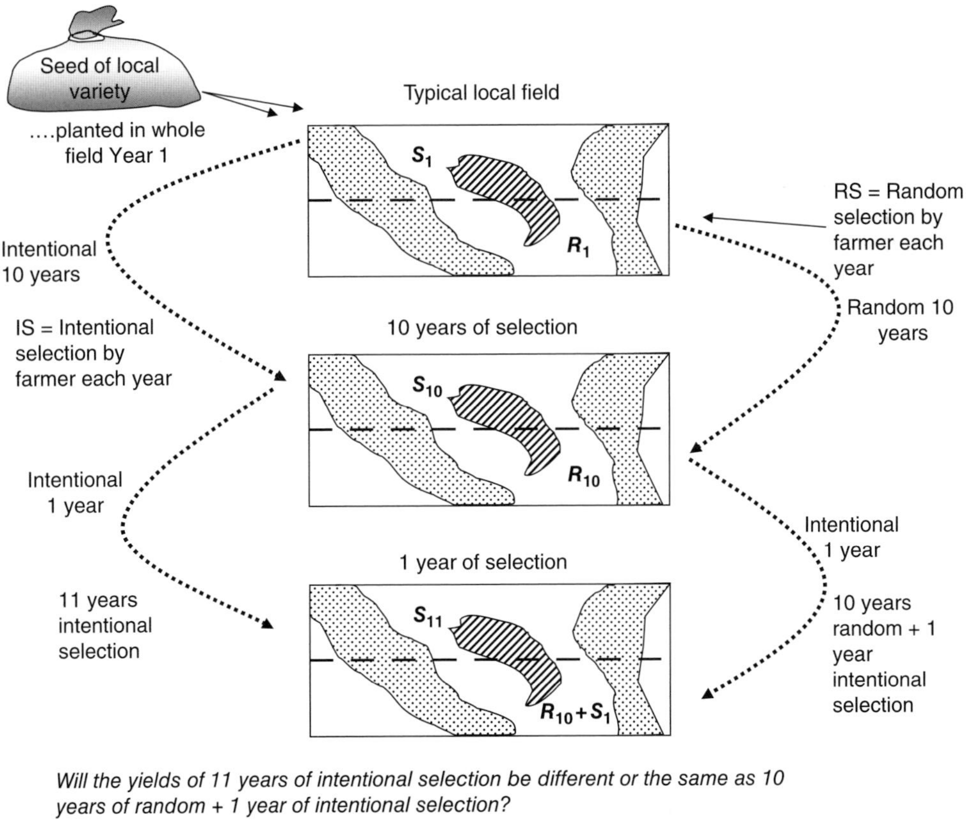

Fig. 13.4. Seed selection scenario presented to farmers.

industrial agriculture in temperate regions. We did not know if benefits would still be present in farmers' varieties or if there would be differences between farmer-selected and random samples of the same maize population. To test this, DS collected seed of a white and yellow maize variety from one farming household in each of two communities in the Central Valley of Oaxaca in December 2009. For each variety the households provided us with seeds they selected for planting and a random sample from the same maize population, all identified at the same time.

In autumn 2010, DS conducted a small greenhouse experiment[3] to see if farmer-selected seed had any advantage over seed that had not been selected (Soleri *et al.*, in preparation, p. 2). We looked at characteristics that are advantageous early in life, including vigour of early growth measured by seedling size and number of leaves. In the majority of comparisons, the selected seed was superior to the unselected seed (Table 13.4), indicating an ecological advantage, even with no genetic response to selection. This finding was consistent with the majority of farmer responses to our selection scenario, indicating that farmer seed selection provides a benefit in the first year, but no heritable, cumulative advantage.

How did this change our understanding?

This research is ongoing, but even with the results to date, it is clear that preconceptions hindered our understanding of the role of seed selection. Despite ascribing to a 'complex farmer' view of FK, we unconsciously made oversimplified assumptions about seed selection influenced by the perspectives of Western plant science,

Table 13.3. Some differences between seed selected for sowing and a random sample from the same population.

	Random sample of seed	Farmer-selected seed	F value	*P*
Seed weight (g)[a]	0.3324a	0.4810b	428.41	0.0001
Seeds/kg	3008	2079	~ 50% > number seeds/kg in random sample	
Market cost (MXP)/kg (December 2009)	7.5	10		

MXP = Mexican pesos

[a]N=140 seeds of each; Tukey's means separation, means followed by different letters are significantly different, $p \leq 0.05$.

Table 13.4. Early growth of farmer-selected and randomly sampled seeds from the same populations of four maize varieties. (In paired comparisons, variety means followed by different letters are significantly different, $P \leq 0.05$.)

		Variety			
Variable	Seed	A	B	C	D
Leaf count A	Selected	2.876a	2.721a	2.607a	2.564a
	Random	2.628b	2.565b	2.579a	2.526a
Plant height A (cm)	Selected	13.702a	13.284a	12.321a	11.736a
	Random	11.879b	11.724b	11.756a	10.524b
Leaf area A (cm²)	Selected	5.694a	6.712a	5.476a	5.726a
	Random	5.084a	5.591b	4.645b	4.478b
Leaf count B	Selected	3.540a	3.435a	3.352a	3.252a
	Random	3.343b	3.065b	3.230a	2.956b
Plant height B (cm)	Selected	25.891a	25.094a	24.139a	21.686a
	Random	22.485b	22.408b	22.251b	19.584b
Leaf area B (cm²)	Selected	13.055a	12.172a	10.947a	9.233a
	Random	9.007b	10.013b	9.559a	7.555b

implying that farmers were barefoot plant breeders. We assumed farmers' seed selection was a practice that identified individual plants with superior phenotypes that were heritable, and would therefore contribute to a cumulative process of microevolution, and that this was their intention. Instead, our research indicates that farmers are well aware of the difficulty of achieving a cumulative response for those traits and do not expect to do so. However, seen in terms of the environments they are working with, the apparent ecological advantages resulting from seed selection make the persistence of that practice easy to understand. While this will need to be investigated further, especially in farmers' growing environments, the data thus far confirm the rationale for farmer seed selection. There may be other advantages as well.

We are now exploring the possibility that, in addition to providing an ecological advantage, this selection may help to conserve the genetic diversity in farmer-managed crop populations. Even though their selection differentials were significant and similar to those sought by plant breeders, the variation in seed size in farmers' seeds is in large part a response to the heterogeneity of the environments where the parent plants grow, for example variations in soil quality or water availability that are common within a field or local area. Plant biologists have observed that in some cases this can 'mask genetic variation' (Sultan, 1987), even protecting genotypes that might otherwise be eliminated (Rice *et al.*, 1993). If this occurs, farmers' selection may serve to protect crop genetic diversity, in addition to other functions.

Thus, what formally trained scientists might see as ineffectual intentional selection by farmers may in fact produce meaningful non-genetic ecological benefits and contribute to maintaining genetic diversity that is useful both locally and globally. A few others have quantified farmer practice with similarly surprising results that have changed our understanding of how traditional agricultural systems function. For example, in cassava propagation in French Guiana (Elias *et al.*, 2001; Pujol *et al.*, 2005), greater vigour and farmer preference for heterozygous seedlings resulted in increased diversity maintenance in this staple crop that is predominantly propagated vegetatively. In contrast to the case with maize, where selection is largely based on non-heritable phenotypic differences, in cassava selection of plants based on phenotypic differences that are heritable led to similar outcomes: the protection of diversity.

Example 3: Consistency and Variation in Farmer-Identified Bean Varieties in One Community

Farmer-named varieties have long been important units of diversity management on farm and widely used basic indicators of diversity for crop genetic resources conservationists. The use of named varieties as diversity indicators has been supported by studies where farmers presented with a number of seed lots have generally agreed on how to classify those lots into varieties (Sadiki *et al.*, 2006). But improving our understanding of what farmer classification means at a community level is challenging, especially when there is variation in both the seeds and the farmers doing the classifying. For example, the extent of community-wide agreement among farmers in sorting unclassified seeds into farmer-named classes is unknown, but affects whether farmer varieties are an over- or under-estimate of the diversity present.

What did we hypothesize?

We undertook this research to investigate bean classifications that were confusing to researchers, by documenting and analysing farmer classifications. As in Example 2, our tendency was to assume that farmers were careful taxonomists and that community membership and farming in similar environments with similar materials would mean high consistency in the way farmers classify their local crop seeds. We thought of named varieties as consistently constructed units, and that within varieties different seed lots were comparable, but that seed lots of different varieties were distinct, and even mutually exclusive. That is, seed A was a member of either variety X or Y, and could not simultaneously be seen as a member of both. The assumption we were testing in our hypothesis was that there was a high level of agreement among farmers about how to classify individual seed phenotypes and the genotypes they represent.

How did we test our hypothesis?

To investigate varietal classification among farmers we quantified practice using a reference sample including seeds from each of the three *Phaseolus* species grown locally (*P. vulgaris*, 87; *P. dumosus*, 73; *P. coccineus*, 75), all collected in the same community in the Sierra Juárez region of Oaxaca, Mexico. By asking each farmer individually to organize the same complete sample into varieties, we were able to quantify farmers' knowledge applied to the same set of seeds. Controlling the seeds themselves allowed a focus on the variation among farmers' classifications, allowing us to document farmer classification and what that means for genetic diversity.

What did we find?

We found that common named varieties represented broad seed morphology types, but that there was little consistency among farmers for how individual seeds were classified, except for one variety (Soleri *et al.*, 2013). The recognition of broad types supported previous findings that when presented with already constituted seed lots, farmers tended to agree on varietal names, but our finding of inconsistency at the level of individual seeds, for example within a species (Table 13.5), indicated that varietal names underestimate diversity. Is this because farmers are not capable of finer levels of discernment and can only form approximations of a shared classification? Our additional research with the same farmers and bean varieties showed this was not the case, but that they were able to discriminate differences within a species down to the sub-racial level and deploy these to different growing environments, as documented by using simple sequence repeat (SSR) DNA markers in farmer-managed *Phaseolus vulgaris* populations (Worthington *et al.*, 2012).

How did this change our understanding?

Our results suggest that inconsistent classifications among farmers may reflect idiosyncratic skills and needs. Variation in farmer knowledge and practice of classification may be another point in the farming system where diversity is preserved, even if it is not what we as researchers may have anticipated, as was the case with maize seed selection in Example 2. It may also be another example of how formal scientific precision could in fact undermine the diversity maintained in community seed systems – diversity that may contribute to farmers' ability to work effectively with the spatial and temporal variability and change present in their bean fields. The type of stringent replicated precision formally trained scientists associate with skilled taxonomies may simply not be necessary, or even desirable, in all seed classification systems.

Table 13.5. Agreement in farmer classification of *Phaseolus* bean species into varieties (data from Soleri *et al.*, 2013) (n = 9 farmer sortings).

Question	Our assumption	Actual finding, summary	Adjusted Random Indicator of mean agreement among farmer classifications (0 = agreement comparable to chance; 1 = complete agreement among classifications)			
			Species (*n* seeds)			
			All species (235)	*P. vulgaris* (87)	*P. dumosus* (73)	*P. coccineus* (73)
Is there agreement among individual farmers' taxonomies of the three *Phaseolus* species they grow?	Farmers are barefoot taxonomists, there will be high agreement among their bean species classifications	Levels of agreement among farmers' classifications are low	0.20	0.31	0.10	0.03

Conclusion

We all seek patterns and meanings that are familiar, or based on the familiar, and sometimes this is all we can see, no matter how empathetic and open we believe ourselves to be. In the case of research on local FK, the assumptions that our pattern-seeking can lead to may be inaccurate. This inaccuracy may risk misunderstanding or misrepresenting farmer practice, and the needs and goals of farmers themselves, or any group or individuals we partner with in research. Positive respectful attitudes towards farmers and their local knowledge are subject to distorting research assumptions, in a similar way that less respectful attitudes are.

Through the examples outlined here, we have found that using methods that seek farmers' knowledge from their perspective are very useful for testing these assumptions and can lead to a deeper and richer understanding of FK and farmer practice. Specifically, it is important to: (i) deconstruct technologies when eliciting FK and opinion to provide a more informative and nuanced understanding than simply asking preferences between finished products; (ii) ask farmers questions, based on their own experiences, that clarify their goals and expectations; and especially (iii) quantify practice so that empirically based assumptions, including our own as researchers, can be tested. These lessons can facilitate successful partnerships between farmers and formally trained scientists to support and improve small-scale agriculture in ways that are socially equitable and environmentally sustainable.

The common assumption that traditionally based farmers successfully conserve crop genetic diversity *in situ* has been supported with evidence of the diversity in terms of species, varieties of a species, varietal heterogeneity and genotypic heterozygosity maintained by farmers. But the different ways in which this diversity is maintained, and what might threaten it, will vary with variation in crop populations, local environments, farmers and other elements of the farming system. In hindsight it seems obvious that we assumed farmers have comparable cognitive abilities as scientists, but failed to consider, or imagine, how the nature of farmer practice and intent could be so different from that of scientists. Anticipating our findings was simply beyond our own capacities as researchers when we started these investigations, but being able to listen to farmers and test what we thought was occurring enabled us to move beyond those limitations.

Progress in understanding farmer knowledge and practice will come from research that compares farmer and scientist knowledge in terms that minimize bias in favour of either, and that emphasizes testing the assumptions all of us have as researchers. The results can be surprising, or even unsettling, but also push us as researchers to explore processes and outcomes we may never have anticipated.

Acknowledgements

We thank the many farmers who have worked with us and whose patience and good humour is an essential ingredient in our understanding described in this chapter; our colleagues on the research described above with assessment of transgenic maize (Flavio Aragón Cuevas, Mario Roberto Fuentes, Humberto Ríos Labrada), maize seed selection (Steven E. Smith) and bean classification (Margaret Worthington, Paul Gepts); and Paul Sillitoe for inviting our contribution to this volume.

Notes

[1] We have discussed these views more extensively elsewhere (Cleveland and Soleri, 2007a), with the four perspectives identified there as: (i) the economically irrational farmer; (ii) the economically rational farmer; (iii) the socio-culturally and/or ecologically rational farmer; and (iv) the complex farmer.

[2] Resistance to caterpillar damage is a common trait of transgenic varieties, based on version of the Bt gene, originally from a soil bacterium, *Bacillus thuringiensis*.

[3] This was a split-split-split plot design using the four farmers' varieties: yellow and white maize each from two families in different communities.

References

AgBio View (2002) *Joint statement in support of scientific discourse in Mexican GM maize scandal*. AgBioWorld Foundation, Auburn, Alabama, in online newsletter *AgBioView*. Available at: www.agbioworld.org/jointstatement.html (accessed 12 April 2017).

Bockstaller, C. and Girardin, P. (1994) Effects of seed size on maize growth from emergence to silking. *Maydica* 39, 213–218.

Ceccarelli, S. (1989) Wide adaptation: how wide? *Euphytica* 40, 197–205.

Ceccarelli, S. (1996) Adaptation to low/high input cultivation. *Euphytica* 92, 203–214.

Cleveland, D.A. and Soleri, D. (2002) Indigenous and scientific knowledge of plant breeding: similarities, differences, and implications for collaboration. In: Sillitoe, P., Bicker, A.J. and Pottier, J. (eds) *'Participating in Development': Approaches to Indigenous Knowledge*. Routledge, London, pp. 206–234.

Cleveland, D.A. and Soleri, D. (2007a) Farmer knowledge and scientist knowledge in sustainable agricultural development. In: Sillitoe, P. (ed.) *Local Science versus Global Science: Approaches to Indigenous Knowledge in International Development*. Berghahn Books, Oxford, UK, pp. 211–229.

Cleveland, D.A. and Soleri, D. (2007b) Extending Darwin's analogy: bridging differences in concepts of selection between farmers, biologists, and plant breeders. *Economic Botany* 61, 121–136.

De Schutter, O. (2010) *Agroecology and the Right to Food*. Report submitted by the Special Rapporteur on the right to food to 16th session of UN Human Rights Council, 20 December 2010 (agenda item 3). Report no. A/HRC/16/49. United Nations General Assembly, New York.

Duputie, A., Massol, F., David, P., Haxaire, C. and McKey, D. (2009) Traditional Amerindian cultivators combine directional and ideotypic selection for sustainable management of cassava genetic diversity. *Journal of Evolutionary Biology* 22, 1317–1325.

Elias, M., Penet, L., Vindry, P., McKey, D., Panaud, O. and Robert, T. (2001) Unmanaged sexual reproduction and the dynamics of genetic diversity of a vegetatively propagated crop plant, cassava (*Manihot esculenta* Crantz), in a traditional farming system. *Molecular Ecology* 10, 1895–1907.

IAASTD (2009) *Synthesis Report with Executive Summary: a Synthesis of the Global and Sub-global IAASTD Reports*. Island Press, Washington, DC.

KARI and CIMMYT (2007) Insect Resistant Maize for Africa (IRMA) Project, Annual Report 2006. Available at: http://www.cimmyt.cgiar.org/english/wpp/gen_res/pdf/irmaAnnRep06.pdf. (accessed 12 December 2007).

Louette, D., Charrier, A. and Berthaud J. (1997) *In situ* conservation of maize in Mexico: genetic diversity and maize seed management in a traditional community. *Economic Botany* 51, 20–38.

McKey, D., Elias, M., Pujol, B. and Duputie, A. (2010) The evolutionary ecology of clonally propagated domesticated plants. *New Phytologist* 186, 318–332.

Murphy, K., Lammer, D., Lyon, S., Carter, B. and Jones, S.S. (2005) Breeding for organic and low-input farming systems: an evolutionary-participatory breeding method for inbred cereal grains. *Renewable Agriculture and Food Systems* 20, 48–55.

Mutegi, E, Sagnard, F., Labuschagne, M., Herselman, L., Semagn, K. *et al.* (2012) Local scale patterns of gene flow and genetic diversity in a crop-wild-weedy complex of sorghum (*Sorghum bicolor* (L.) Moench) under traditional agricultural field conditions in Kenya. *Conservation Genetics* 13, 1059–1071.

Pommel, B. (1990) Influence du poids de la semence et de la profondeur de semis sur la croissance et le 'développement' de la plantule de maïs. *Agronomie* 10, 699–708.

Pujol, B., Muhlen, G., Garwood, N., Horoszowski, Y., Douzery, E.J.P. and McKey, D. (2005) Evolution under domestication: contrasting functional morphology of seedlings in domesticated cassava and its closest wild relatives. *New Phytologist* 166, 305–318.

Quist, D. and Chapela, I.H. (2001) Transgenic DNA introgressed into traditional maize landraces in Oaxaca, Mexico. *Nature* 414, 541–543.

Revilla, P., Butrón, R., Malvar, R.A. and Ordás A. (1999) Relationships among kernel weight, early vigor, and growth in maize. *Crop Science* 39, 654–658.

Rice, K.J., Gordon, D.R., Hardison, J.L. and Welker, J.M. (1993) Phenotypic variation in seedlings of a 'keystone' tree species (*Quercus douglasii*): the interactive effects of acorn source and competitive environment. *Oecologia* 96, 537–547.

Sadiki, M., Jarvis, D., Rijal, D.K., Bajracharya, J., Hue, N.N., Camacho-Villa, T.C. and Burgos-May, L.A. (2006) Variety names: an entry point to crop genetic diversity and distribution in agroecosystems? In: Jarvis, D., Padoch, C. and Cooper, D. (eds) *Managing Biodiversity in Agricultural Ecosystems*. Columbia University Press, New York, pp. 34–76.

Sadras, V.O. (2007) Evolutionary aspects of the trade-off between seed size and number in crops. *Field Crops Research* 100, 125–138.

Soleri, D. and Cleveland, D.A. (2001) Farmers' genetic perceptions regarding their crop populations: an example with maize in the Central Valleys of Oaxaca, Mexico. *Economic Botany* 55, 106–128.

Soleri, D. and Cleveland, D.A. (2005) Scenarios as a tool for eliciting and understanding farmers' biological knowledge. *Field Methods* 17, 283–301.

Soleri, D. and Cleveland, D.A. (2009) Breeding for quantitative variables. Part 1: farmers' and scientists' knowledge and practice in variety choice and plant selection. In: Ceccarelli, S., Weltzien, E. and Guimares, E. (eds) *Participatory Plant Breeding*. FAO (United Nations Food and Agriculture Organization, Rome) in collaboration with ICARDA (International Center for Agricultural Research in the Dry Areas, Aleppo, Syria) and ICRISAT (International Crops Research Institute for the Semi-Arid Tropics, Patancheru, India), pp. 323–366.

Soleri, D., Smith, S.E. and Cleveland, D.A. (2000) Evaluating the potential for farmer and plant breeder collaboration: a case study of farmer maize selection in Oaxaca, Mexico. *Euphytica* 116, 41–57.

Soleri, D., Cleveland, D.A., Smith, S.E., Ceccarelli, S., Grando, S. *et al.* (2002) Understanding farmers' knowledge as the basis for collaboration with plant breeders: methodological development and examples from ongoing research in Mexico, Syria, Cuba, and Nepal. In: Cleveland, D.A. and Soleri, D. (eds) *Farmers, Scientists and Plant Breeding: Integrating Knowledge and Practice*. CAB International, Wallingford, UK, pp. 19–60.

Soleri, D., Cleveland, D.A., Aragón Cuevas, F., Ríos Labrada, H., Fuentes Lopez, M.R. and Sweeney, S.H. (2005) Understanding the potential impact of transgenic crops in traditional agriculture: maize farmers' perspectives in Cuba, Guatemala and Mexico. *Environmental Biosafety Research* 4, 141–166.

Soleri, D., Cleveland, D.A., Glasgow, G.E., Sweeney, S.H., Aragón Cuevas, F., Ríos Labrada, H. and Fuentes Lopez, M.R. (2008) Testing economic assumptions underlying research on transgenic food crops for Third World farmers: evidence from Cuba, Guatemala and Mexico. *Ecological Economics* 67, 667–682.

Soleri, D., Worthington, M., Aragon-Cuevas, F., Smith, S.E. and Gepts, P. (2013) Farmers' varietal identification in a reference sample of local *Phaseolus* species in the Sierra Juarez, Oaxaca, Mexico. *Economic Botany* 67, 283–298.

Soleri, D., Aragón Cuevas, F., Cleveland, D.A., Smith, S.E. and Castro García, H. (in preparation) *In situ* conservation of crop diversity in context: maize farmers' perceptions and practices in Oaxaca, Mexico.

Sultan, S.E. (1987) Evolutionary implications of phenotypic plasticity in plants. *Evolutionary Biology* 21, 127–178.

Vélez Ascencio, O. (2003) Encuentran en la Sierra Juarez maíz transgénico prohibido en EU [Estados Unidos]. *Noticias* (Oaxaca), 9A [newspaper article]

Vernooy, R., Otieno, G., Bessette, G., Fadda, C., Galluzzi, G. *et al.* (2015) *A Novel Strategy to Discover and Use Climate-Adapted Germplasm.* Bioversity International Report, Rome.

Worthington, M., Soleri, D., Aragón-Cuevas, F. and Gepts P. (2012) Genetic composition and spatial distribution of farmer-managed *Phaseolus* bean plantings: an example from a village in Oaxaca, Mexico. *Crop Science* 52, 1721–1735.

14 Traditional Domestic Knowledge and Skills in Post-harvest Processes: A Focus on Food Crop Storage

Patricia L. Howard*

> The snake has seized half the grain, and the hippopotami have eaten the rest. Mice abound in the fields, the locusts descend and the herds devour; the sparrows steal – woe to the farmers! The remains on the threshing floor are for the thieves.
>
> (Inscription on the private tomb of Menna, Royal Scribe Surveyor, in ancient Egypt[1])

> Without storage techniques that can preserve food for long periods of time, production increases would not be feasible and surplus generation would be virtually impossible
>
> (Smyth, 1991, p. 41)

Some two billion people live directly from small-scale agriculture; in sub-Saharan Africa and Asia, 80% of all farmland is managed by smallholders (up to 10 ha) (Altieri and Koohafkhan, 2008). The vast majority of these smallholders produce for subsistence, non-monetary exchange and small-scale marketing using family labour, local technologies, local varieties and local knowledge to manage most tasks involved in agricultural production and in the post-harvest chain. Their crops and other harvested plant materials[2] are often not immediately consumed, and the majority of what is stored is kept on-farm; in the tropics, for example, some 60–80% of all grain produced is stored on the farm using traditional methods (Haines, 1995).

Traditional storage systems are very ancient (see e.g. Panagiotakopulu *et al.*, 1995; Cunningham, 2011). They are the product of long-term co-evolution based on interactions between local cultural, socio-economic and environmental conditions. Structural materials and design are influenced by the availability of local resources, the type of crop or plant materials collected and the forms of stored material (e.g. threshed or shelled grains). Storage conditions and methods are also influenced by and influence crop varietal development and thus have a direct relation with agrobiodiversity. Each type of storage has an associated pattern of management and use. Since storage is well adapted to local conditions, systems are highly varied (Haines, 1995). Given all of these influences, traditional storage systems are subject to continuous innovation as new pests and diseases emerge, new crops and crop varieties are introduced, and social and economic conditions change.

Nevertheless, little is known about traditional on-farm storage systems. Of the information available, most appears to be highly biased. The emphasis in research and policy making is overwhelmingly on crop production, neglecting what happens to crops once they are harvested. Women manage most on-farm post-harvest tasks as part of their domestic work; domestic

* E-mail: patricia.howard@wur.nl

work is largely unpaid, invisible and under-valued. When attention is paid to post-harvest processes, two assertions are often made in tandem: post-harvest food losses are high; and traditional post-harvest systems are the cause, especially traditional storage. Much research therefore seeks to develop and promote use of 'modern' technologies and external inputs.

Here, as the literature permits, evidence is presented that demonstrates that negative assumptions about traditional storage contrast sharply with a cumulative body of research that shows that losses are low and that reducing them further may not be cost-effective. Traditional storage systems are the result of long-term adaptations that entail interaction between local cultural, socio-economic and environmental conditions. Storage involves complex and dynamic sets of traditional knowledge and skills, often referred to as ITK (indigenous technical knowledge). Here, when referring specifically to the knowledge used in the domestic sphere, the term traditional domestic knowledge (TDK) is preferred, as it is often the least recognized of all ITK. Much TDK is corroborated by science, which confirms that traditional systems are generally highly effective, appropriate to smallholder conditions and more sustainable compared with modern chemical and physical storage technologies. Food storage systems that depend upon ethnobotanical and other local technical knowledge are vital to ensuring household food security and also provide both motive and means to conserve agricultural biodiversity.

To Blame, or to Celebrate? Traditional On-farm Storage and Post-harvest Food Losses

It is often argued that, to feed 9.1 billion people by 2050, global food production must increase by 50–90%. But many scientists and policy makers also argue that, by reducing the current rampant waste of food mainly in wealthier countries, as well as what are considered to be high post-harvest food losses in poorer regions, the growing global population can be fed without investing as many resources in agriculture and without generating as many greenhouse gases. The Food and Agriculture Organization (FAO) of the United Nations (UN) estimates that, in Europe and North America, about one-third of the food produced is lost due to waste in the retailing and consumption end of the supply chain (Gustavsson *et al.*, 2011). While less food is produced, wasted and lost per capita in sub-Saharan Africa and South and South-east Asia, losses are concentrated in the immediate post-harvest and storage/transport stages; this is termed post-harvest food loss (PHL).

The current interest in PHL is also related to the food and financial crises of 2006–2008 (World Bank *et al.*, 2011) when food prices soared, leading to increasing food insecurity and political instability. The current focus on PHL in low-income countries is not new: in 1975, the UN General Assembly had declared that PHL was a global concern and should be reduced by 50% by 1985 (World Bank *et al.*, 2011). PHL estimates were generally very high (Greeley, 1982).[3] Reducing PHL presented a 'soft third option' for eliminating poverty and hunger – an alternative or complement to increasing food production or redistributing food. The blame for high PHL was laid squarely on the shoulders of traditional farmers (World Bank *et al.*, 2011, p. 16; see also Hodges, 2012), especially traditional on-farm storage (Boxall, 1989; Haines, 1995):

> Traditional farm-level methods were regarded as inefficient, not because of any specific evidence of high losses, but because they were characterized by 'ancient' often rudimentary practices, uninfluenced by modern food science and technology ... Opinions differed on the root of the problem – peasant farmers' supposed neglect or carelessness, their lack of basic knowledge of food science, their lack of access to improved methods, or their shortage of resources to invest in improved methods.
>
> (Greeley, 1991, p. 8)

Today, PHL estimates range from a low of 10–40% to a high of 50–70%. But, as was the case in the 1970s, these estimates are also highly controversial. Especially in sub-Saharan Africa (SSA), PHL data are 'spotty and scanty,' and 'research is poor' (Affognon *et al.*, 2015, p. 52); the World Bank admits that loss estimates are 'frequently guesstimates' (World Bank *et al.*, 2011, p. 17). Estimates mainly focus on on-farm losses, which are generally calculated for harvesting, drying, processing (e.g. threshing and winnowing), transporting from the field to storage areas, and storage. Traditional storage systems and especially on-farm crop storage are still considered to be the weakest link in the post-harvest chain, demonstrated by Affognon *et al.*'s (2015) meta-analysis of PHL studies in SSA, where some 80% of the loss estimates reviewed are related to crop storage. Modi (2004) argues that attitudes toward on-farm traditional seed storage, which is seen as insignificant to agrobiodiversity maintenance, are also negative as seeds are often stored in rudimentary conditions, such as in baskets buried in the earth with ash, sealed in mud vessels, or packed into raised thatched huts.

These high estimates of storage losses and negative assumptions about traditional practices contrast sharply with a cumulative body of research that employs more rigorous methods to measure on-farm losses among smallholders. This research clearly shows that, in traditional systems, storage losses are usually very low, often under 5% of grain weight in a storage season, and often closer to 2% (Greeley, 1982, 1991; Boxall, 1989; Golob *et al.*, 1999; Modi, 2004, 2007; World Bank *et al.*, 2011; Hodges, 2012).

Greeley (1991) noted that, in the 1970s, high estimates of PHL were often market-driven. Corporations had a vested interest in promoting private or government-sponsored or subsidized programmes for the purchase and use of synthetic pesticides, storage structures and post-harvest processing equipment. Today, high PHL estimates are quoted confidently and are seen to present an opportunity for international private sector investment and profit-taking, as a recent Rockefeller publication attests:

> PHL levels are currently very high across value chains in SSA and action is required as a reduction in post-harvest losses has a direct impact on food security ... the [Rockefeller] model ... focuses on engaging key market makers and market push and pull actors to address loss in their specific food-crop value chains. As an entry point, we would identify the potential partners (e.g. Coca-Cola, WFP, Flour Mills Nigeria, Dangote Group, Cargill, Nestlé, and Unilever) whose food-crop value chains experience significant losses. We propose to work with them to develop appropriate interventions ... [and] identify opportunities to profitably source from smallholder farmers.
>
> (Rockefeller Foundation, 2015)

Whatever the problems are that such corporations may have in sourcing raw materials from African smallholders, experts in PHL argue that 'traditional practices are an unlikely culprit, as farmers have survived difficult conditions over long periods by adapting their practice to prevailing circumstances' (World Bank *et al.*, 2011, p. 16). In general, traditional storage systems are regarded by many scientists as highly sustainable and appropriate. They maximize the use of renewable local resources for storage structures and containers' construction and maintenance; wild and cultivated plant resources are used as repellents and insecticides to prevent storage losses; and family labour is deployed so that storage is both economical and appropriate for smallholder conditions.

Storage losses may be high or increasing among some smallholders, but this is mainly attributed to the intensification and modernization of farming systems (double cropping, introduction of high-yielding varieties (HYVs), the related increase in production and in pest and disease incidence, and consequent changes in crop storage characteristics). This leads to the loss of traditional knowledge and technologies, as discussed below. Many new technologies have been developed or introduced to improve on-farm storage, but have generally failed due to a lack of understanding of local management patterns and their relation with the balance between the storage ecosystem and the external environment (Greeley, 1982, 1991; World Bank *et al.*, 2011). The impact of PHL on livelihoods has been found to be minor compared with other problems that smallholders confront, and reducing such losses is often costlier than the losses themselves.

The Functions of Small-scale Storage

On-farm storage should maintain crop quality and prevent deterioration over the short or long term. Its most essential short-term livelihood function is to provide food between harvests, converting seasonally available wild and cultivated plants into dependable all-season staples, allowing households to meet nutritional requirements year-round. Longer-term storage ensures against future shortages when crops may fail due to weather, storms, or insects and diseases, or distribution networks fail due to warfare, economic and political crises, or natural disasters. It 'provides increased socio-ecological resilience under environmental and climatic stress' (Balbo, 2015, p. 305). Storage also provides access to nutritionally superior traditional (local) foods as an alternative to nutritionally poorer purchased foods (see e.g. Toledo and Burlingame, 2006). It is vital to agriculture, as it ensures seed supplies for future plantings; its role in maintaining crop agrobiodiversity is also crucial. Stored seed ensures that varieties are maintained over time. Stored food is also a form of savings: it can be bartered, gifted, or sold when supplies are otherwise scarce and the value of stored goods is higher. It helps to balance supply and demand, stabilizing market prices.

Storing surpluses fulfils many other social functions with major consequences for social relations and resilience. Anthropologists coined the term 'social storage' to refer to the fact that stored surpluses can be accumulated as wealth. Having stored surpluses means having the ability to redistribute them to gain status, or trade them for other resources and wealth (Cunningham, 2011). Stored food thus allows social stratification and complexity to emerge (Testart, 1982). Storage allows cooperation and exchange (communal sharing), and thus underpins political, social and kinship ties. It allows communities to exploit ecological differences and access different types of food and other resources that are not locally available, as is the case with inter-island exchange networks in the Pacific, or overland exchange networks in 'mainland' Papua New Guinea (Campbell, 2015). These exchange networks permit biologically resource-poor areas to withstand environmental disturbance and stress (Ellen, 2016).

In spite of its significance for food security, sustainability, agrobiodiversity, social differentiation and socio-ecological resilience, very little research has focused on traditional on-farm storage technologies, knowledge, skills and social relations. The vast majority of literature dealing with it is technical (examining efficacy) or archaeological; much in-depth social science literature focuses on prehistoric hunter-gatherer societies.[4] Most contemporary literature concerns grains and pulses and, to a lesser extent, root and tuber crops; less deals with fruit and vegetables. While much research investigates food preparation and consumption ('foodways'), little investigates the specialized local technical knowledge and skills underpinning on-farm storage, intra-community knowledge distribution, or changes in these and their drivers; virtually none exists on knowledge transmission.

Storage ITK and Gender

Post-harvest activities, such as winnowing, threshing, parboiling, drying, storing, distributing, preserving and preparing food, are strongly inter-related in terms of use of space (Smyth, 1991), time and labour (often representing a series of steps carried out sequentially in or near the home), as well as in terms of techniques (the way that plants are harvested influences pre-storage processing, which influences storage, which influences subsequent processing and consumption). These tasks are also often not distinguishable by end-use: products are separated, processed and stored simultaneously for the next crop, for home consumption and for sale. Most often, women are responsible for these tasks and they are considered to be part of the domestic sphere (Symth, 1991; Howard, 2003).

It is widely acknowledged that women are usually the managers of traditional on-farm storage systems (see e.g. World Bank, 2009), though this is also culturally variant. There are barely any studies of intra-cultural knowledge distribution in post-harvest or storage systems. Case studies show that men usually build large storage structures while women are mainly responsible for preparing crops for storage and for storage itself, including seed storage. When men are involved, decisions are frequently shared, though most work still falls to women. Most literature fails to discuss storage divisions of labour or women at all. Affognon *et al.* (2015, p. 60) reported that, of the 213 documents reviewed in their meta-analysis of PHL in SSA, 'only three (or less than 1.5%) explored gender issues, and these appraised participation levels of women and men in postharvest operations and constraints faced by women in adoption of technologies'.

Pre-storage and storage activities usually take place within the home or in the homestead; it is often said that food storage is a 'domestic affair.' In Bangladesh, men consider seed management and storage as women's work – an extension of their 'knowledge of matters in the house'; women say it is similar to house cleaning (Oakley and Momsen, 2007). In Timor, Indonesia (Situmeang, 2013), the 'rule' is that only a woman who has been married in a traditional ceremony has access to crops stored in the attic. No one else can remove food, even if there is nothing to eat. This prerogative belongs to women 'because they cook'; 'women are more capable in managing food ... men are improvident' (Situmeang, 2013, p. 41). In Zimbabwe, informants explained, 'women are more careful, a lot of grain would be lost if winnowing and mixing was done by males.' Men do not have the time: 'A man cannot spend the whole day at home burning manure' (used to smoke stored crops) (Winniefridah and Manuku, 2013, p. 240). Such cultural norms and stereotypes define gender relations and characterize women's work, which may be seen both as skilled and less valuable than men's.

In some cultures, there are more cosmological reasons for women's predominance in storage management. Over much of Asia, only women manage rice storage because of beliefs associated with a rice goddess, the 'mother of the grain' (*Dewi sri* in Thailand and *Bok sri* in Java) (IRRI, 1983; Proctor, 1994; Hamilton, 2004). In the Peruvian Andes, women almost exclusively manage seed and storage; men are forbidden to enter storage areas. In Quechuan cosmology, useful plants are worshipped under the name of 'mother', such as *Mama sara* (maize) and *Mama acxo* (potato). 'Seed' also refers to semen, providing a metaphor between the 'seed' that the male deposits in the womb and that which is sown in the field, collected, and deposited in the home (Tapia and de la Torre, 1993).

When men are directly involved in storage, the division of labour and control over stored products are embedded in gender power relations. Among the Embu, in Kenya, women usually manage storage but, if a crop belongs to a man, men decide the amounts to sell and store and the storage mode. When men cannot purchase storage pesticides, they ask women to store the crops using 'their own' traditional methods (Nathani, 1996). Households may also have multiple stores belonging to different members. In Burkina Faso, sorghum and millet are held in the male household head's granary and men distribute it to meet household needs; they may give part to their wives to do with as they wish. Pulses are women's crops stored in women's granaries (van Liere *et al.*, 1996). In Zimbabwe, women are responsible for stored grain but, when wives and husband have separate stores, women manage their husband's store. If husbands take the active role in storage management, they often give instructions that wives must implement. When men and women have separate stores, 'it was always the women's stores which were exhausted first' (Manda and Mvumi, 2010, citing Douglas *et al.*, 1997). Women must strategize to exercise control over the use of the stores, bargaining with men to enhance their own positions in the process (Manda and Mvumi, 2010).

ITK in a Human-Created Ecosystem

Sinha (1995, p. 9) argued, 'Stored-grain ecosystems in rural agriculture of a developing country ... function like a natural ecosystem.' Farmers design and manage ways to 'manipulate the ecological processes that are invariably involved in the destructive action of undesirable ... pathogens (fungi), and arthropod or vertebrate pests' (Sinha, 1995, p. 7). In the tropics, high temperatures and water activity lead to greater biological activity; biological complexity and diversity are greater and there is a greater range of storage structures and methods to match (Haines, 1995). A good storage system limits high humidity and temperatures and provides protection against insect and rodent attack (Boxall, 1989). Pests are a very significant storage problem; they compete directly with humans for food. Their importance is evidenced in the literature; 81% of the reports reviewed on mitigating PHL in SSA (Affognon *et al.*, 2015) dealt with strategies to combat insect pests in cereals and pulses, including 'variety selection, biological control, improved storage structures, modified atmosphere facilities, and treatment with chemical insecticides' (Affognon *et al.*, 2015, p. 56).

While much ethnobotanical, ethnoentomological and ethnozoological knowledge is required to control storage pests, many other types of knowledge are also required. Managers must know 'how many, how large, and what type' of storage facilities are required (Sinha, 1995), which depends on the ability to forecast storage volumes, forms and consumption requirements, as well as household economic and spatial constraints (Smyth, 1991). They also require strategies for storage design and control, including: the storage potentials and susceptibilities of different crop species and varieties; how to dry plant materials and protect them from rain or excessive moisture; how to construct, maintain and repair drying and storage structures and containers; how to clean everything that comes into contact with plant materials and maintain hygiene throughout the storage period; and how to protect stored plant materials against attack not only by insects, but also by bacteria and fungi, and against theft by rodents, birds and humans. Managers must estimate likely storage losses as well as monitor stored stocks and take measures as problems arise. Knowledge is needed about how much and which stored material must be allocated for planting, consumption, barter, gift-giving and sales, and of how stored products must be distributed, by whom, when and to whom.

Households may need to store many different types of plant materials using different technologies and methods. The required knowledge is not only multiplied by the number of plants, technologies and methods; the interactions between them must also be considered, including the allocation of time, space and other resources entailed; consideration of the quantities and forms in which materials will be stored; the location of different stored materials; and possible pest interactions (Boxall, 1989; Haines, 1995; Stathers *et al.*, 2013). There is a dearth of literature on most of this knowledge. Here, the focus is necessarily on those aspects that are best covered in the literature – the most crucial to avoiding storage losses.

Post-harvest, pre-storage ITK

Drying plant materials prior to storage helps to prevent the formation of bacteria and fungi. Some fungi (e.g. *Fusarium* spp., *Aspergillus* spp.) produce mycotoxins that can be lethal to humans. Drying decreases insect infestation since insects feed on and spread fungal spores, leading to higher moisture content. The heat from sun drying (spreading crops in the open air) eliminates many insects (Haines, 1995; Golob *et al.*, 1999). Special drying structures may be needed to protect against rain while ensuring proper ventilation; some also function as storage structures (Boxall, 1989). In south-western Burkina Faso, sorghum and millet are spread and dried for several weeks on specially built platforms, whereas beans and vegetables are spread on house roofs, and shea nuts are dried over an oven, allowing the nuts to be easily cracked (van Liere *et al.*, 1996).

Research on women's seed management in Bangladesh demonstrated the importance of drying and of the knowledge required to dry different crops. In Tangail District's humid environment, drying is very time consuming and women are taught by their mothers or mothers-in-law. After cleaning, seeds are dried in courtyards in piles of different varieties. According to a respondent, 'because each area of the courtyard receives differing degrees of sunlight and heat, women avoid mixing piles, even if of the same variety, as each pile will dry at varying rates' (Oakley and Momsen, 2007, pp. 100–101). Rice must be stirred all day long, 'to bring the bottom grains to the surface; we make sure it all dries evenly.' Table 14.1 summarizes the tests that women use to determine seed dryness.

Drying is only one of a total of six steps involved in preparing rice for storage. Again in Bangladesh, courtyards are first prepared with a layer of fresh mud mixed with cow dung to keep from contaminating the rice with dirt. Paddy is threshed by bullocks and the straw is dried for fodder. Rice is winnowed and sieved multiple times before being parboiled and soaked; only afterwards can it be dried and stored. Post-harvest labour requirements are very high, 'between 41 per cent to 49 per cent of the total man-days required for its cultivation' (Abdullah, 1985, citing Von Harder, 1975).

After threshing, paddy is soaked for 3–24 h and then placed in vessels lined with straw to keep the rice from burning. Parboiling time depends on how long the rice has been soaked (Abdullah, 1985). 'Women can tell by the smell of

Table 14.1. Women's tests for checking seed dryness in two study villages in Bangladesh (reproduced from Oakley and Momsne, 2007, p. 101, with permission from John Wiley and Sons).

Crop	Test
Rice	Sound of seeds hitting against each other when shuffled in the palm of the hand. When bitten should sound like 'katkata'. Throw in the air with the winnower to feel for weight. Husk comes off readily
Jute	Sound of 'jhan, jhan' when seeds are stirred together. Squeeze to feel for hardness
Wheat	Sound of a gravel stone when seed is bitten. Should break into pieces when bitten; should crumble into powder when chewed
Foxtail millet	Sound of 'machmacha' when seeds are rubbed together
Barley	Seeds feel slippery between the fingers. When bitten should sound like 'tannath'
Proso millet	Husk comes off easily if seeds are rolled against one another
Pulses	Seeds roll easily when stepped on
Mustard	Deep black colour. Shrinks in size. Seeds are slippery and roll when stepped on
Linseeds	Light weight when held in hand
Vegetables/fruits	Break open seed to look for well formed cotyledons. Sound of seeds rubbing against each other in the palm of the hand. Sound of 'madtath' when broken; should break into two pieces. Squeeze seed to check for moisture

paddy, by how much it rises and how it splits, whether the parboiling is complete ... [parboiling] facilitates easy milling and absorption of proteins, vitamins and minerals, reduces the amount of broken rice, stores better and the bran contains approximately 15–30 per cent of oil and is used as food for animals and poultry' (Abdullah, 1985, p. 212). In Sierra Leone, different rice varieties respond differently to parboiling. 'Women ... are often very skilled in determining the amount of heat and time sufficient for a particular quantity' (Kroma, 2002, p. 188).

If rice is not properly dried after parboiling, grains will break when milled or husked and deteriorate in storage. When drying, paddy must be watched continuously so that it is not lost to livestock or birds. In Bangladesh, women dry the rice for 1–3 days (depending on temperature and day length), turning it periodically. If rice is exposed to too much sunlight, it can over-dry and grains will break during milling; insufficient drying leads to storage losses (Abdullah, 1985; see also Kroma, 2002).

Traditional storage structures, and mechanical and cultural controls

Short-term crop storage may be done on the ground or on drying floors, on open wooden platforms, or by suspending plant materials in the air outside (e.g. hung from branches) or indoors (e.g. over an oven). Long-term storage is done in storage baskets, cribs, or specially constructed buildings, especially in very humid environments, since they provide adequate ventilation. In the Yucatan, Mexico, different maize storage structures and forms of storage (shelled, husked, and packed with husks) are related to the length of storage, which in turn is related to the amount of maize removed each day for household consumption (Smyth, 1991).

Gourds, calabashes, or earthenware pots are used to store small quantities of plant materials and may be hermetically sealed using mud, clay, or dung, but this mode is usually too expensive to use with bulkier food grains (Proctor, 1994; Golob *et al.*, 1999). Jars or other large vessels (clay and, increasingly, metal or plastic) are used to store seeds and legumes; especially in dry areas solid-wall bins are used, with a base made of wood, soil, or stone. Silos are usually round or cylindrical shapes, which are less prone to cracking than rectangular shapes, and are made of clay, possibly mixed with straw. Roofs are of thatched grass with an overhang. Underground storage pits are widespread in areas with low water tables (hot, dry climates), such as in Ethiopia, the Sudan and Somalia, but they are also known in more humid areas, such as Fiji. Pits are varied in shape and the entrances may be closed by soil, sand, or a stone sealed with mud (Proctor, 1994; Golob *et al.*, 1999).

Ash and other chemically inert materials such as sand, kaolin, paddy husk ash, wood ash, and clays, lime, salt and diatomaceous earths (Golob, 1997) or even seed are widely used in different types of traditional storage systems. These control insects by closing the spaces between grains, which prohibits movement, and also by damaging the insect's cuticle, leading to dehydration (Golob *et al.*, 1999) or death by suffocation (Wolfson *et al.*, 1991). Lime is unpalatable to many pests and thus it also acts as a repellent (Smyth, 1991). Much research has demonstrated the efficacy of these methods, as in the case of cowpeas weevils, which cause 90% of insect damage, mainly in storage, where infestation may reach 100% (Boeke *et al.*, 2004). In northern Cameroon, farmers thresh cowpeas, mix them with ash and place the mixture into a mud granary or clay jar, compress it and then cover it with an additional ash layer, which effectively controls weevils (Wolfson *et al.*, 1991).

Traditional communities in Ghana provide examples of short-term tuber storage methods. Highly perishable cassava roots are generally stored in the field and harvested when needed but, when they must be harvested and stored short-term, they are immersed in hot water and then buried in damp soil. The soil moisture content is 'maintained at a level that is not high enough to accelerate rotting and not low enough to cause hydration', keeping roots fresh for about 2 weeks (Pace *et al.*, 1989, p. 5). In the south-east, roots are dried or immersed in hot water and then stored in baskets or bags, or kept in a water-filled bucket or in a basket covered with jute that is sprinkled regularly to keep the roots moist until they can be further processed (Dei, 1990). Yams are less perishable than cassava and Dei (1990) found that, while yams can be maintained in the field for up to 1½ years, some varieties are harvested and stored for a few weeks to 5 months using special structures or techniques. In one mode, all soil is removed to keep yams from growing and from developing diseases; rotten yams are removed from stores to prevent disease spread. They may then be tied with twine to wooden stakes and a cross-bar, which also prevents the spread of disease, provides ventilation, and allows inspection, or they may be stored in small heaps on the ground covered with yam vines, soil, or sticks. Or they may be stored in pits lined with yam vines and covered with soil, or placed on a low platform made of stakes and palm fronds located under a tree, which provides ventilation and shade to avoid drying and cracking. They may be packed into gaps at the base of a tree buttress, 'leaving yam tubers in the soil after the stems have been cut ... packing [them] in ashes and covering them with soil' (Dei, 1990, p. 14).

Highly perishable crops can also be stored for longer periods in pits, as in Fiji (Aalbersberg *et al.*, 1988), where women stored breadfruit, cassava, taro, plantain and giant swamp taro. Men dug pits in a shaded area with good drainage, near a fresh water supply. Pits were lined with dried banana leaves and then with green banana leaves that were folded and overlapped in several layers to avoid soil contamination, leaving some leaves to extend over the top as a cover. Washed fruit or roots were stored in these pits and covered with dried leaves weighted with stones. When the food was well fermented and soft (3–4 weeks), it was removed and replaced with fresh stores. The pH in the storage pit decreased over time, preventing spoilage. Fijians also inoculated the stored material with an aroid plant, *Amorphophallus campaculatus*, to aid fermentation.

In Bangladesh, women usually store rice seed in baskets prepared by coating them inside and out with cow dung and mud, which keeps the basket from splitting and protects against rodents. Rice-filled containers are covered with straw, a clay pot, or coconut shell, and sealed with mud (Abdullah, 1985). 'The application of airtight seals, which requires considerable expertise, is essential for controlling humidity and temperature ... Seed varieties must be kept in separate containers and farmers need to be able to identify the varieties held in each' (Oakley and Momsen, 2007, p. 92). Parrish (1995) reported that, in an Egyptian oasis, grains were stored in holes dug in and covered with sand. Ashes, lime, or crushed red peppers were mixed with the grain, acting as insect irritants and desiccants.

Traditional storage ITK and technologies not only largely protect plant materials for future use; they can even enhance these materials and their performance. Modi (2004, 2007) showed that, in South Africa, traditional taro corm storage increases the viability of taro propagules. Taro corms are susceptible to rot when stored in the field. Farmers thus maintain taro germplasm *in situ* using pre-planting storage methods. Propagules are layered in dry pits separated with straw to prevent water infiltration. Using indigenous methods, 'corm rot can be minimized and propagules are enhanced through shoot and root development' (Modi, 2007, p. 218). The high quality propagules allow for successful crop stand establishment and the planting season is also prolonged. Farmers also put corms into pits for 3 months, which allows fresh seed corm peels to dry and harden, accelerating sprouting and minimizing corm losses from pathogens (Modi, 2004).

Chemical control of storage pests: botanicals and fumigation

Most smallholders use plants and their products to control storage pests chemically. Many botanical pesticides and repellents are also used as medicinals and flavourings, since these tend to possess insecticidal properties. Plants produce phytochemicals and secondary metabolites that defend them from pests. Plants and secondary plant products such as edible oils, distilled essential oils, soaps and powdered roots, stems, fruits and leaves are mixed with stored plant materials (Golob *et al.*, 1999; Walia *et al.*, 2014) or rubbed on the sides of storage structures to make use of these properties (Parrish, 1995). Their use depends on locally transmitted ethnobotanical knowledge to identify and manage them (wild or cultivated) and to process, apply and store them effectively.

Botanicals act against insects through six mechanisms: through respiration, as with a fumigant; through contact or digestion; by preventing reproduction or causing sterilization; by inhibiting normal feeding behaviour (anti-feedant); by repelling or altering insect behaviour; or by a combination of these actions. Scientific knowledge about these effects is quite limited: 'for most of the botanicals that have been tested ... the only data available concern strictly efficacy, while the mode of action for many of them remains unknown' (Athanassiou *et al.*, 2014, pp. 131–132). Hundreds of species have been tested for efficacy against major storage insects and yet, 'there is a dearth of information concerning actual use of plants by farmers ... very many plants used as grain protectants by rural communities ... have yet to be identified ... Very little information has been acquired which describes how farmers apply plant protectants' (Golob *et al.*, 1999, p. 7).[5]

A few African examples suffice to illustrate the use of different plant species and plant parts as pesticides. In Senegal, one of the few ethnobotanical studies found (Toumnou *et al.*, 2012) surveyed 184 farmers and documented a total of 35 species in 21 families used as storage insecticides. Factorial analysis revealed two groups of plants: one for conserving peanuts, corn and millet; and the other to conserve beans, sorghum, fonio and rice. The most commonly used plants were of the *Cesalpiniaceae*, *Meliaceae*, *Piperaceae* and *Combretaceae* families, and the parts most commonly used were leaves, barks and fruit. In Benin, 33 plant species are used as insecticides or repellents against weevils in fresh, dried or processed form (Boeke *et al.*, 2004). In Ghana's Ashanti region, a survey of 500 farmers identified 26 plant species used to protect stored grain; more than 90 other plant species were reported that had insecticidal properties. A quarter of the farmers used plant protectants, but few used them exclusively; the most common means of protecting stored maize was smoking (Golob *et al.*, 1999).

The use of smoke and fumigation are among the most effective traditional methods for protecting stored plant materials from insect infestation. Fumigants diffuse throughout a storage area, reaching interstitial areas in grain and killing insects (Athanassiou *et al.*, 2014). Fumigants used for storage worldwide include spices, herbs, 'vegetable oil, inert dusts, plant extracts like essential oils, lectins, proteins, and leaf powders, which have insecticidal and antimicrobial activity' (Shaaya *et al.*, 1997, p. 7). Burning causes biogases and carbon dioxide to permeate the storage space. The concentration of carbon dioxide suffocates, dehydrates and poisons insects (Shaaya *et al.*, 1997). Modi (2004, p. 17) reported: 'In the past two decades smoke has been shown to improve seed germination of almost 200 species from more than 40 families.'

Synthetic versus botanical pesticides

Traditional smallholders often also deploy synthetic pesticides. Writing on eastern and southern

Africa, Stathers *et al.* (2013, p. 366) reported that the decisions to use synthetic or natural insecticides or repellents in storage was related to 'cost, knowledge, information networks, experience, availability of the protectant, the planned storage duration of the grain and objective of storage'. One of the few ethnographic studies on post-harvest systems, among the Embu in Kenya (Nathani, 1996), discussed recent changes in storage pest control. According to one woman, 'what I use depends on the crop. For instance, for maize I use marigold and for beans and potatoes ashes,' while another woman stated, 'most of the time I use [synthetic] pesticide, it is only when there is no money that I use these natural preservatives' (Nathani, 1996, p. 162). Of 177 women surveyed, 25 used synthetic pesticides alone, 132 used natural pesticides and 20 used nothing. One woman compared synthetic versus botanical pest control: 'the indigenous practices are time consuming and tedious ... the good thing ... is that they are free ... during the time of scarcity, people turn to these herbs to preserve their food, but they do not talk about it because they are embarrassed' (Nathani, 1996, p. 163). This embarrassment, discussed below, is due to the stigma associated with indigenous practices in an era when using 'modern' farming techniques confers higher social status and government support.

Many researchers note that, certainly in smallholder agriculture, local botanical products are superior to synthetic pesticides, which pose health hazards for humans and animals during storage, application or food consumption. Synthetic pesticides bring more generalized environmental risks, are often difficult to access and prohibitively expensive, and are largely imported, implying a burden of debt and subjection to international market fluctuations. Especially, pesticides that are commonly used in storage lead to insect resistance (see e.g. Poswal and Akpa, 1991; Shaaya *et al.*, 1997; Golob *et al.*, 1999; Pandey *et al.*, 2014). Women are very likely to be more exposed than men to health risks because of their multiple interactions with the chemicals in and outside the home – mixing, applying, and disposing of pesticides; contact during sowing, weeding, thinning, harvesting and collecting residues; and applying them against domestic and storage pests. They are also subject to drift into the domestic environment, food and water contamination (including through washing and reuse of pesticide containers), and by washing contaminated clothing (London *et al.*, 2002). Many studies attest that smallholders' knowledge about synthetic pesticide application and access to protective gear is very limited, so the risks posed are very high. On the other hand, smallholders collect or produce botanical insecticides and repellents themselves and so complex distribution networks are not required. Their use is generally sustainable – they are renewable and biodegradable, with no negative impact on the environment (Golob *et al.*, 1999).

Storage ITK Change and Loss

> ... serious [on-farm storage] losses do sometimes occur and these may have resulted from agricultural developments for which the farmer is not pre-adapted. These include the introduction of high yielding cereal varieties that are more susceptible to pest damage, additional cropping seasons that result in the need for harvesting and drying when weather is damp or cloudy, or farmers producing significant surplus grain which, because it is to be marketed rather than ... [consumed] is less well tended ... new pests can be a problem, as in the case of the larger grain borer (LGB – *Prostephanus truncatus*) in Africa.
>
> (Hodges, 2012, pp. 2–3)

Major changes in traditional storage systems result from more generalized changes in farming systems, mostly intensification. Especially important are increases in storage volumes and storage pest incidence, the diffusion of HYVs that often have poorer storage characteristics compared with traditional varieties, and the use of synthetic pesticides. A virtually undocumented source of erosion is likely to be changes in household labour supply and demand due to phenomena such as the feminization of agriculture and male out-migration, HIV–AIDS, increasing female wage labour employment and participation of children in formal education, all of which may impact especially women's ability to manage what are often labour-intensive and time-sensitive processes. In some places, the expertise associated with traditional agriculture and storage systems is considered as an emblem of backwardness instead of a source of prestige. Traditional storage knowledge is likely to be no different than other forms of ITK, where 'a major threat to the sustainability of natural resources is the erosion of people's knowledge, and the basic reason for this erosion is the low value attached to it' (Hoppers, 2002, p. 7).

With increased yields and conversion to monoculture, crops must be stored in bulk, possibly exceeding the capacity of traditional storage structures and of households to process and manage the quantities involved. Pesticides have been widely adopted in areas where crop production has intensified although, as is the case with the Embu people in Kenya, many smallholders combine their use with traditional storage methods. Intensification and extra-regional trade have led to new insect pests for which no traditional control methods exist. This compels innovation: either storage methods change and synthetic pesticides are used, as in the Egyptian case below, or new botanicals are introduced, as in Nigeria. In some cases, as in Nepal, innovation is lacking.

Other forces implicated in the erosion of agricultural ITK are also affecting post-harvest TDK. In Egypt's Western Desert (Parrish, 1995), yields increased when wells were introduced some 50 years ago. Farmers expanded storage facilities and changed methods and now contend with increased losses due to new weevil pests that they say arrived when the Ministry of Agriculture began to offer extension services. Some crops were traditionally stored in pits and sand was used to fill the interstices, but this is no longer possible as increased irrigation caused ground saturation and sand dunes shifted, so sand is not available near storage areas. Sun drying has ended. Indoor granaries and mud silos became too small. By the early 1990s, 80% of farmers used only synthetic chemicals in storage, while 26% mixed old and new technologies. The higher the education of the farmer and the greater the area farmed, the more likely it was that synthetic pesticides and rodenticides would be applied and traditional practices would be abandoned. Many now work off-farm and synthetic pesticides are labour-saving; wages permit purchases of synthetic inputs. Nearly 70% of farmers learned modern pest control methods from extensionists; those who learned traditional methods learned them from family members and neighbours.

Among the Tharu in Nepal, 70–90% of all grains produced are stored on-farm (Björnsen Gurung, 2002). There is little traditional pest control knowledge since pest pressure was low. Farmers moved both fields and villages, which probably suppressed pests; rice was cropped only once per year and pest-resistant landraces were preferred. Then, farmers became sedentary, and pests became common when rice HYVs were introduced in the 1990s. Throughout Nepal, the need for storage pest management is associated with the cultivation of pest-susceptible crops and varieties. Traditional pest-resistant crops (millet, buckwheat, and barley) did not require sophisticated control practices, but these are now mainly minor crops. Björnsen Gurung (2002, p. 102) found little innovation in crop storage practices, even with the increase in pests, and argued that this was because crop losses were more related to natural calamities than to storage pests: 'Entomologists and extension workers commonly attribute higher importance to pests and their control than farmers do.'

Although the literature dealing with change in traditional storage systems does not consider traditional knowledge, the use of synthetic pesticides can certainly lead to its loss, as evidenced in Mali, where traditional storage methods

> ... formed an integral part of the informal education of youths ... Particularly in the last two decades, chemicals have been introduced at an ever-increasing rate into peasant farming. Profiting sometimes from the label of modernism, they were often accepted without comparing their performance (preservation effects, environmental impact) with that of traditional techniques. Now, young people are no longer being taught how to use the local resources.
>
> (Koné, 1993, p. 1)

Embu women said that, before World War II, there were no weevils or worms to destroy stored food. People thought that weevils and moths were introduced by European soldiers or crops. One woman said, '... through trial and error we learnt that various leaves could be used as repellents ... Marigold and others became very important ... once we got to know that they could be used to preserve food' (Nathani, 1996, p. 166). By the 1990s, many women in their 20s knew no traditional storage pest control methods and used only synthetic pesticides, in part because they believed that it was 'backward and outdated' to use indigenous storage practices, or they feared that people would perceive them as poor. 'What would my neighbours say if they found out that I use marigold or ashes to preserve food? They would laugh at me and think I am old fashioned' (Nathani, 1996, p. 163). Land clearing for agriculture had reduced the availability of wild plants used as botanical repellents and insecticides, and women spent less time harvesting them: '... many of these repellents are not easily available ... in the olden days, these plants used to grow even behind our houses' (Nathani, 1996, pp. 164–165).

It can be presumed that, in addition to the factors discussed above, women's time constraints play an important role in the loss of traditional storage knowledge. In gender and rural development literature, it is widely recognized that rural women's time is often very constrained. Some literature that discusses technical aspects of traditional storage also stresses the associated labour burdens and promotes more modern technologies as a means to free women's time. This is certainly true in the case of other post-harvest processes, such as in Swaziland, where the decline in the cultivation of indigenous food crops is attributed in part to labour constraints in processing (Malaza, 2003). When younger women attend school or engage in wage labour, they are less likely to have time available and are more likely to reject traditional foods and agricultural activities (see e.g. Howard, 2003, 2006).

One study documented the potential threat of over-harvesting. In Ghana, 16 plant species used as stored product pesticides were researched across four different geographical zones. Ecological studies determined that natural regeneration of one of these species was low due to repeated wild fires, and four others 'face threat of extinction in the likely event of being extensively used in grain storage' and thus would require intensive propagation. With the exception of three species, the rest had very low seed germination rates (Belmain, 2002, p. 34).

Conclusions and Needs for Further Research

There is a substantial and largely uninformed effort under way to change smallholder knowledge, technologies and practices based on a set of often erroneous or unsubstantiated assumptions about the irrationality, inefficiency and backwardness of traditional post-harvest storage systems, which lead to high estimates of food loss. However, several studies report that improvements to traditional storage systems are seldom proffered and, when they are, they are often not adopted or effective (Greeley, 1991; World Bank *et al.*, 2011; Affognon *et al.*, 2015). Modi (2007, p. 214) summarized the reasons: there is '(i) poor understanding of their agroecosystem realities, in that introduced technologies do not match the complex nature of smallholder agriculture, and (ii) underestimation or poor esteem of traditional knowledge'.

There is also recognition among some well-informed scientists and practitioners that traditional systems are in fact more sustainable, appropriate and resilient, since they are based on local renewable resources, including botanical pesticides, that are not polluting or dangerous, nor do they generate insect resistance. Traditional systems are well adapted to local crops and environments and serve to reinforce demand for landraces. From an economic standpoint, they are generally quite superior, as they do not rely on expensive inputs, volatile input markets, foreign exchange, or on often non-existent formal distribution networks, or external knowledge and extension advice. However, even those who extol the multiple virtues of traditional storage and post-harvest systems lack in-depth understanding of the spatial, temporal and material interrelationships entailed in storage systems (Smyth, 1991), or the complex knowledge, skills, decision-making, strategies, labour processes and social relations entailed. Social scientists have barely been involved in such research, and those who do engage usually do not focus on contemporary agroecological systems. Even gender researchers concerned with agriculture tend to avoid what is considered to be the 'reproductive' domestic sphere, overlooking the rich, complex, dynamic and diverse sets of knowledge and skills, as well as the high labour demand and bargaining processes entailed that in part define women's status and welfare.

Storage contributes decisively to food security at both household and higher levels, and to extensive networks of inter-community exchange – both monetary and non-monetary – that interlock socio-ecological systems for people's mutual benefit and resilience. Given the importance of post-harvest systems generally and especially of on-farm storage for livelihoods and food security, particularly in regions where smallholder agriculture predominates, there is an urgent need to advance a multi-purpose research agenda. One requirement is for in-depth ethnobiological /anthropological and politico-ecological research that can illuminate the nature and complexity of on-farm post-harvest systems, including knowledge and skills, their relations with technology (choices, options and change), socio-cultural (especially gender and other power) relationships, agroecological environments, and the larger economic and political context. This would allow researchers who need to understand the systems that they propose to change to at least be more capable of recognizing the authentic needs of smallholders and women, and of better predicting outcomes of change processes. Research is certainly required that questions whether and how technological innovations undermine or improve smallholders' (and especially women's) decision-making capacities, knowledge acquisition and transmission, status and welfare. Research is also urgently needed that deals with the ways in which climate and other types of environmental change (e.g. in agrobiodiversity), market expansion and agricultural intensification affects the resilience of on-farm post-harvest processes and storage systems.

Notes

[1] Referring in all likelihood to peasant agriculture (Brock, 2005).

[2] Here the focus is on indigenous technical knowledge (ITK) relating to stored plant materials rather than livestock or fish, which merit their own reviews. I also exclude the ITK related specifically to storage structures when these are built (e.g. constructed of wood, bamboo, maize cobs, etc.) which also merit their own review, as they may be highly specialized. The ITK involved in post-harvest processes relating to livestock and captured game and fish are part of the overall ITK held by subsistence and traditional peoples, who do not depend on plant materials alone for nutrition.

[3] Greeley (1982, p. 51) citing Lester Brown, 1970: '... according to one calculation, based on local reports, 50 per cent of the grain crop of India was lost to rodents, 15 per cent was lost during milling and processing, 15 per cent was lost to cows, birds and monkeys, 10 per cent was lost to insects and 15 per cent was lost during storage and transit – a grand total of 105 per cent.'

[4] Even this literature is very restricted: 'Our poor understanding of storage behaviour results from the inability of archaeologists to unambiguously recognize remains associated with past storage strategies ... particularly at the domestic level of analysis. This methodological shortcoming has ... contributed to the significant but generally overlooked role of storage in the development of complex societies' (Smyth, 1991: 1). Ethnoarchaeological research on contemporary on-farm storage systems could potentially overcome such limitations, but it is extremely rare.

[5] Published databases of plants used for storage protection include: Golob and Webley, 1980; Rees *et al.*, 1993; Dales, 1996; Golob *et al.*, 1999.

References

Aalbersberg, W.G.L., Lovelace, C.E.A., Madhoji, K. and Parkinson, S.V. (1988) Davuke, the traditional Fijan method of pit preservation of staple carbohydrate foods. *Ecology of Food and Nutrition* 21(3), 173–180. doi: 10.1080/03670244.1988.9991030.

Abdullah, T. (1985) Women in rice farming systems in Bangladesh and how technology programs can reach them. In: *Women in Rice Farming. Proceedings of a Conference on Women in Rice Farming Systems.* The International Rice Research Institute (IRRI), Manila, 26–30 September 1983. Gower Publishing Company Ltd., Hants, UK, pp. 209–220. Available at: http://pdf.usaid.gov/pdf_docs/PNAAV812.pdf (accessed 17 July 2016).

Affognon, H., Mutungi, C., Sanginga, P. and Borgemeister, C. (2015) Unpacking postharvest losses in Sub-Saharan Africa: a meta-analysis. *World Development* 66, 49–68. doi: 10.1016/j.worlddev.2014.08.002.

Altieri, M.A., and Koohafkan, P. (2008) *Enduring farms: climate change, smallholders and traditional farming communities*. Third World Network Environment and Development Series 6. Available at: http://www.fao.org/nr/water/docs/enduring_farms.pdf (accessed 17 July 2016).

Athanassiou, C.G., Rani, P.U. and Kavallieratos, N.G. (2014) The use of plant extracts for stored product protection. In: Singh, D. (ed.) *Advances in Plant Biopesticides.* Springer, New Delhi, pp. 131–147.

Balbo, A.L. (2015) Storage: introduction to the special issue. *Environmental Archaeology* 20(4), 305–313. doi: 10.1179/1749631415Y.0000000021.

Belmain, S. (2002) *Optimising the indigenous use of botanicals in Ghana. Final Technical Report.* Natural Resources Institute, UK; Royal Botanic Gardens, Kew, UK; Ministry of Food and Agriculture, Ghana, Forestry Research Institute, Ghana, University of Ghana. Available at: http://r4d.dfid.gov.uk/PDF/Outputs/CropPostHarvest/R7373_FTR.pdf (accessed 4 September 2016).

Björnsen Gurung, A. (2002) Indigenous knowledge of storage pest management in Nepal. PhD Dissertation, Diss. ETH No. 14784. Swiss Federal Institute of Technology, Zurich, Switzerland. Available at: http://e-collection.library.ethz.ch/eserv/eth:25881/eth-25881-01.pdf (accessed 23 August 2016).

Boeke, S.J., Baumgart, I.R., van Loon, J.J.A., van Huis, A., Dicke, M. and Kossou, D.K. (2004) Toxicity and repellence of African plants traditionally used for the protection of stored cowpea against *Callosobruchus maculatus. Journal of Stored Products Research* 40, 423–438. doi: 10.1016/S0022-474X(03)00046-8.

Boxall, R. (1989) Storing grain on the farm. In: Carr, M. (ed.) *Women and the Food Cycle.* Intermediate Technology Publications, London, pp. 27–36.

Brock, J.F. (2005) *Four Surveyors of the Gods: In the XVIII Dynasty of Egypt – New Kingdom c. 1400 BC.* International Federation of Surveyors Article of the Month, March 2005. Available at: https://www.fig.net/resources/proceedings/fig_proceedings/cairo/papers/wshs_02/wshs02_01_brock.pdf (accessed 5 September 2016).

Campbell, J.R. (2015) Development, global change and traditional food security in Pacific Island communities. *Regional Environmental Change* 15, 1313–1324. doi: 10.1007/s10113-014-0697-6.

Cunningham, P. (2011) Caching your savings: the use of small-scale storage in European prehistory. *Journal of Anthropological Archaeology* 30, 135–144. doi: 10.1016/j.jaa.2010.12.005.

Dales, M.J. (1996) *A Review of Plant Materials Used for Controlling Insect Pests of Stored Products.* NRI Bulletin 65. Natural Resources Institute, Chatham, UK.

Dei, G. (1990) Indigenous knowledge and economic production: the food crop cultivation, preservation and storage methods of a West African community. *Ecology of Food and Nutrition* 24(1), 1–20. doi: 10.1080/03670244.1990.9991115.

Ellen, R. (2016) *Environmental threats to sustainable living in low-lying reef systems: biodiversity preservation and Moluccan trading networks between 1600 and the present*. Paper presented at the Royal Anthropological Institute and British Museum Conference 'Anthropology, Weather and Climate Change', London, 27–29 May.

Golob, P. (1997) Current status and future perspectives for inert dusts for the control of stored product pests. *Journal of Stored Products Research* 33, 69–79. doi: 10.1016/S0022-474X(96)00031-8.

Golob, P. and Webley, D.J. (1980) *The Use of Plants and Minerals as Traditional Protectants of Stored Products*. G138. Natural Resources Institute, Chatham, UK.

Golob, P., Moss, C., Dales, M., FIdgen, A., Evans, J. and Gudrups, I. (1999) *The Use of Spices and Medicinals as Bioactive Protectants for Grains*. FAO Agricultural Services Bulletin No. 137. Food and Agriculture Organization of the United Nations (FAO), Rome. Available at: http://www.fao.org/docrep/x2230e/x2230e00.HTM (accessed 15 July 2016).

Greeley, M. (1982) Farm-level post-harvest food losses: the myth of the soft third option. *IDS Bulletin* 13(3), 51–60. doi: 10.1111/j.1759-5436.1982.mp13003007.x.

Greeley, M. (1991) *Postharvest Technologies. Implications for Food Policy Analysis.* EDI Development Policy Case Series. Analytical Case Studies Number 7. The World Bank, Washington, DC. Available at: http://documents.worldbank.org/curated/en/171111468764392177/pdf/multi-page.pdf (accessed 20 June 2016).

Gustavsson, J., Cederberg, C., Sonesson, U., van Otterdijk, R. and Meybeck, A. (2011) *Global Food Losses and Food Waste. Extent, Causes and Prevention.* Food and Agriculture Organization of the United Nations (FAO), Rome. Available at: http://www.fao.org/docrep/014/mb060e/mb060e.pdf (accessed 2 June 2016).

Haines, C.P. (1995) Grain storage in the tropics. In: Jayas, D.S., White, N.D. and Muir, W.E. (eds) *Stored-Grain Ecosystems.* Marcel Dekker Inc., New York, pp. 55–100.

Hamilton, R. (2004) *The Art of Rice: Spirit and Sustenance in Asia*. UCLA Fowler Museum of Cultural History, Los Angeles, California.

Hodges, R.J. (2012) *Postharvest weight losses of cereal grains in Sub-Saharan Africa*. APHLIS – African Postsharvest Losses Information System, September. Available at: http://www.erails.net/FARA/aphlis/aphlis/weight-losses-review/. (accessed 2 June 2016).

Hoppers, O. (2002) Indigenous knowledge and the integration of knowledge systems. In: Odora, C. (ed.) *Indigenous Knowledge and the Integration of Knowledge Systems.* New Africa Books, Claremont, South Africa, pp. 2–22.

Howard, P. (2003) Women and plants: an exploration. In: Howard, P. (ed.) *Women and Plants: Gender Relations in Biodiversity Management and Conservation.* Zed Press and Palgrave MacMillan, London and New York, pp. 1–48.

Howard, P. (2006) Gender and social dynamics in swidden and homegardens in Latin America. In: Kumar, B.M. and Nair, P.K. (eds) *Tropical Homegardens: A Time-tested Example of Sustainable Agroforestry.* Springer, the Netherlands, pp. 159–182. doi: 10.1007/978-1-4020-4948-4_10.

IRRI (1985) *Women in Rice Farming. Proceedings of a Conference on Women in Rice Farming Systems.* The International Rice Research Institute (IRRI), Manila, Philippines, 26–30 September 1983. Gower Publishing Company, Aldershot, UK. Available at: http://pdf.usaid.gov/pdf_docs/PNAAV812.pdf (accessed 13 September 2016)

Koné, S. (1993) Traditional storage of farm products. ILEIA Newsletter 9(3): 12–13. Available at: http://edepot.wur.nl/397869 (accessed 24 August 2017).

Kroma, M.M. (2002) The gender division of labor in rice post-harvest processing in Sierra Leone: implications for extension and technology development. *The Journal of Agricultural Education and Extension* 8(4), 181–194. doi: 10.1080/13892240285300221.

London, L., Grosbois, S. de., Wesseling, C., Kisting, S., Rother, H.A. and Mergler, D. (2002) Pesticide usage and health consequences for women in developing countries: out of sight, out of mind? *International Journal of Occupational and Environmental Health* 8, 46–59. doi: 10.1179/oeh.2002.8.1.46.

Malaza, M. (2003) Modernization and gender dynamics in the loss of agrobiodiversity in Swaziland's food system. In: Howard, P. (ed.) *Women and Plants: Gender Relations in Biodiversity Management and Conservation*. Zed Press and Palgrave MacMillan, London and New York, pp. 243–257.

Manda, J. and Mvumi, B.M. (2010) Gender relations in household grain storage management and marketing: the case of Binga District, Zimbabwe. *Agriculture and Human Values* 27, 85–103. doi: 10.1007/s10460-008-9171-8.

Modi, A.T. (2004) Short-term preservation of maize landrace seed and taro propagules using indigenous storage methods. *South African Journal of Botany* 70(1), 16–23. doi: 10.1016/S0254-6299(15)30263-5.

Modi, A.T. (2007) Effect of indigenous storage method on performance of taro [*Colocasia esculenta* (L.) Schott] under field conditions in a warm subtropical area. *South African Journal of Plant and Soil* 24(4), 214–218. doi: 10.1080/02571862.2007.10634812.

Nathani, N. (1996) Sustainable development: indigenous forms of food processing technologies. PhD Dissertation, University of Toronto, University Microfilms International, Ann Arbor, Michigan.

Oakley, E. and Momsen, J.H. (2007) Women and seed management: a study of two villages in Bangladesh. *Singapore Journal of Tropical Geography* 28, 90–106. doi: 10.1111/j.1467-9493.2006.00278.x.

Pace, R.D., Plahar, W.A. and Lu, J.Y. (1989) Status of traditional food preservation methods for selected Ghanaian foods. *Food Reviews International* 5(1), 1–12. doi: 10.1080/87559128909540842.

Panagiotakopulu, E., Buckland, P.C., Day, P.M. and Doumas, C. (1995) Natural insecticides and insect repellents in antiquity: a review of the evidence. *Journal of Archaeological Science* 22, 705–710. doi: 10.1016/S0305-4403(95)80156-1.

Pandey, A.K., Singh, P., Palni, U.T. and Tripathi, N.N. (2014) In vivo evaluation of two essential oil based botanical formulations (EOBBFs) for the use against stored product pathogens and pests, *Aspergillus* species and *Callosobruchus* species (Coleoptera: Bruchidae). *Journal of Stored Product Research* 59, 285–291. doi: 10.1016/j.jspr.2014.09.001.

Parrish, A.M. (1995) 'There were no *sus* in the old days:' post-harvest pest management in an Egyptian oasis village. *Human Organization* 54(2), 195–204. doi: 10.17730/humo.54.2.k3x0438510517467.

Poswal, M.A.T. and Akpa, A.D. (1991) Current trends in the use of traditional and organic methods for the control of crop pests and diseases in Nigeria. *Tropical Pest Management* 37(4), 329–333. doi: 10.1080/09670879109371609.

Proctor, D.L. (1994) *Grain Storage Techniques. Evolution and Trends in Developing Countries*. FAO Agricultural Services Bulletin No. 109. Food and Agriculture Organization of the United Nations (FAO), Rome. Available at: http://www.fao.org/docrep/t1838e/t1838e00.htm (accessed 13 August 2016).

Rees, D.P., Dales, M.J. and Golob, P. (1993) *Alternative Methods for the Control of Stored-product Insect Pests: a Bibliographic Database*. Natural Resources Institute, Chatham, UK.

Rockefeller Foundation (2015). *Perspectives to reducing post-harvest losses of agricultural products in Africa*. Background Paper, African Development Bank Conference Feeding Africa, 21–23 October 2015. AFDB, Dakar, Senegal. Available at: http://www.afdb.org/fileadmin/uploads/afdb/Documents/Events/DakAgri2015/Agriculture_Industrialization_and_post-harvest_losses.pdf. (accessed 30 August 2016).

Shaaya, E., Kostjukovski, M., Eilberg, J. and Sukprakarn, C. (1997) Plant oils as fumigants and contact insecticides for the control of stored-product insects. *Journal of Stored Products Research* 33(1), 7–15/ doi: 10.1016/S0022-474X(96)00032-X.

Sinha, R.N. (1995) The stored-grain ecosystem. In: Jayas, D.S., White, N.D. and Muir, W.E. (eds) *Stored-Grain Ecosystems.* Marcel Dekker Inc., New York, pp. 1–31.

Situmeang, V.S.N. (2013) Ume Kbubu: Household granary and food security in Timor Tengah Selatan. Master's Thesis, University of Oregon, University Microfilms International, Ann Arbor, Michigan.

Smyth, M.P. (1991) *Modern Maya Storage Behaviour. Ethnoarchaeological Case Examples from the Puuc Region of Yucatan*. University of Pittsburgh Memoirs in Latin American Archaeology No. 3. Department of Anthropology, University of Pittsburgh, Pennsylvania.

Stathers, T., Lamboll, R. and Mvumi, B.M. (2013) Postharvest agriculture in changing climates: its importance to African smallholders. *Food Security* 5, 361–392. doi: 10.1007/s12571-013-0262-z.

Tapia, M. and de la Torre, A. (1998) *Women Farmers and Andean Seeds*. Food and Agriculture Organization of the United Nations (FAO) and International Board for Plant Genetic Resources (IPGRI), Rome. Available at: http://www.bioversityinternational.org/uploads/tx_news/Women_farmers_and_Andean_seeds_308.pdf (accessed 23 September 2016).

Testart, A. (1982) The significance of food storage among hunter-gatherers: residence patterns, population densities, and social inequalities. *Current Anthropology* 23(5), 523–537.

Toledo, A. and Burlingame, B. (2006) Biodiversity and nutrition: a common path toward global food security and sustainable development. *Journal of Food Composition and Analysis* 19(6–7), 477–483. doi: 10.1016/j.jfca.2006.05.001.

Toumnou, T., Seck, D., Thiaw, C. and Cisse, N. (2012) Farmers' pesticidal plant use in the protection of stored cereal and legume grains: ethnobotanical surveys in some rural communities in Senegal. *International Journal of Science and Advanced Technology* 2(3), 25–33. Available at: http://www.ijsat.com/view.php?id=2012:March:Volume%202%20Issue%203 (accessed 20 June 2016).

Van Liere, M.J., Brouwer, I.D. and den Hartog, A.P. (1996) A food ethnography of the Otammari in north-western Benin: a systematic approach. *Ecology of Food and Nutrition* 34, 293–310. doi: 10.1080/03670244.1996.9991467.

Walia, S., Saha, S. and Rana, V.S. (2014) Phytochemical pesticides. In: Singh, D. (ed.) *Advances in Plant Biopesticides*. Springer, New Delhi, pp. 295–322.

Winniefridah, M. and Manuku, M. (2013) Traditional science of seed and crop yield preservation: exploring the contribution of women to indigenous knowledge systems in Zimbabwe. *International Journal of Humanities and Social Science* 3(4), 234–244. Available at: http://www.ijhssnet.com/journals/Vol_3_No_4_Special_Issue_February_2013/23.pdf (accessed 27 June 2016).

Wolfson, J.L., Shade, R.E., Mentzer, P.E. and Murdock, L.L. (1991) Efficacy of ash for controlling infestations of *Callosobruchus maculatus* (F.) (Coleopetera: Bruchidae) in stored cowpeas. *Journal of Stored Product Research* 27(4), 239–243. doi: 10.1016/0022-474X(91)90006-X.

World Bank (2009) *Gender in Agriculture Sourcebook*. The World Bank, Washington DC. Available at: http://siteresources.worldbank.org/INTGENAGRLIVSOUBOOK/Resources/CompleteBook.pdf (accessed 15 June 2016).

World Bank, Natural Resources Institute, and Food and Agriculture Organization of the United Nations (2011) *Missing Food: The Case of Postharvest Grain Losses in Sub-Saharan Africa.* Economic and Sector Work Report No. 60371-AFR. The World Bank, Washington DC. Available at: https://openknowledge.worldbank.org/handle/10986/2824 (accessed 15 June 2016).

15 The Local Wisdom of Balinese *Subaks*

Wayan Windia, Gede Sedana, Thérèse de Vet and J. Stephen Lansing*

Since the 1970s, Balinese *subaks* (local water-user associations) have played a prominent role in development plans for the island. But they are now perceived by many Balinese to be in a crisis of sustainability. Ironically, the origins of the current crisis can be traced to the success of two earlier development programmes that began in the 1970s: the Green Revolution in agriculture; and the expansion of tourism. At the onset of the post-colonial era, the government of Indonesia began to mobilize the *subaks* to support the goal of self-sufficiency in rice (Sedana, 2012). To avoid conflict with traditional agriculture, new facilities for the tourist industry were to be ring-fenced in coastal areas. But as both tourism and rice agriculture expanded, the *subaks* took on new, often contradictory roles in subsequent development plans. In the 1990s, policy makers came to see the *subaks* as advantageous for both rice production and tourism. Aspects of the traditions of the *subaks* were re-interpreted in light of ever-changing plans for economic development. These contradictions came into sharp focus in 2012, with the creation of the United Nations Educational, Scientific and Cultural Organization (UNESCO) World Heritage designation: *Cultural Landscape of Bali: the Subak System as a manifestation of the Tri Hita Karana Philosophy*. This recognition by UNESCO came at a time when the very survival of the *subaks* had begun to be debated in Balinese newspapers.

The chapter begins with an overview of this history, from the Green Revolution and master plans for tourism of the 1970s to the present controversies surrounding the UNESCO World Heritage. From the outset, development planners struggled to comprehend the functional role of the *subaks* in the management of Bali's rice terraces. The second part of the chapter reviews what has been learned about that question.

The Green Revolution

In Bali, a fragile system of cooperative management has sustained an equally fragile infrastructure of terraces, tunnels and aqueducts for many centuries. The landscape of Bali is dominated by two active volcanoes, with steep slopes reaching almost to the sea. According to Balinese legend, these symmetrical peaks are fragments of the cosmic mountain that were brought to the island by the Hindu gods. The flanks of the volcanoes are deeply incised by ravines containing fast-flowing rivers and streams, with small diversionary dams or weirs. In ancient times, as tunnels and canals proliferated on the slopes of the volcanoes,

* Corresponding author; e-mail: jlansing@ntu.edu.sg

the irrigation systems began to link communities, like melons on a vine. When that occurred, the task of enforcing equitable rules for irrigation schedules would have required cooperation at two levels: among the farmers on each terraced hillside; and between entire communities sharing the same irrigation system. References to a specialized institution for this purpose begin to appear in Balinese inscriptions in the 11th century. This institution, called *subak*, is not identical to the village. Instead the membership of a *subak* consists of all the farmers who own land watered by a common source, like a spring or tertiary canal (Lansing, 2006).

In 1971, French consultants prepared a master plan for tourism development in Bali for the United Nations Development Programme (UNDP). The island's population was then 2.2 million and the economy was still dominated by traditional agriculture. At that time, tourism played a minor role in Bali's economy; the annual number of foreign visitors only reached 300,000 in 1973. The UNDP's master plan for tourism foresaw 'concentrating future hotel development in an area where the potentially negative effects of tourism on Bali's traditional life can be minimized'. 'Cultural tourism' was emphasized in official plans, along with protection for the environment and for Bali's traditional society. UNDP argued that the expansion of tourism should have 'high developmental priority', because nearly all available agricultural land was already under cultivation (Picard, 1990).

As these plans were taking shape at UNDP and the World Bank, a massive intervention in Balinese rice cultivation was about to begin, spearheaded by the Asian Development Bank (ADB). The 'Green Revolution' promised to enable the government of Indonesia to meet its goal of self-sufficiency in rice production. By the mid-1970s, plans were under way for both the expansion of tourism and a Green Revolution in Balinese agriculture. Around the southern beaches, the ADB's plan for tourism development encouraged the proliferation of new hotels and tourist facilities. Meanwhile, in the countryside, the Green Revolution was rapidly replacing native Balinese rice varieties with hybrid rice, which promised higher yields (Poffenberger and Zurbuchen, 1980). With the goal of providing the extra irrigation water needed to support this intensification of rice agriculture, the ADB provided loans to finance a costly overhaul of the irrigation works in more than 50 *subaks*. By 1977, 70% of the *subaks* in southern Bali were planting the new rice (Sutawan and Pitana, 1993).

Meanwhile, the expansion of tourism outran even the most optimistic scenarios of the ADB. Indonesia's National Statistics Board reports that 3.41 million foreign tourists visited Bali in 2014, prompting complaints that the island was in danger of being 'loved to death'. Continuing support for cultural tourism encouraged visitors to spill out of the beachside hotels into the countryside, where the attractions included the beauty of the rice terraces and *subak* temples. Ironically, cultural tourism succeeded so well that the original master plan (intended to confine the expansion of tourist infrastructure to the beaches) soon became moot. Hotels, restaurants, gift shops and private residences proliferated in much of Bali, especially in the vicinity of the most attractive views. Real estate developers bid up the price of attractive building sites, enriching a few farmers but triggering higher taxes for their neighbours.

While this was going on, the Green Revolution was running into unexpected problems in the countryside. Ironically, the same landscape features that attracted the tourists – the traditional rice terraces and temples – were perceived by consultants to be potential obstacles to the modernization of Balinese agriculture. This story is well known from many publications, so the summary here will be brief (for an overview see Lansing and de Vet, 2012). In Bali, the Green Revolution meant the replacement of slow-growing native rice with fast-growing high-yielding varieties (HYVs). HYV rice was bred to make efficient use of chemical fertilizer and promised to double rice yields if sufficient fertilizer was applied. To achieve these gains, it was also necessary to speed up the cropping cycles. Before the advent of the Green Revolution, irrigation flows were pegged to calendrical cycles, which provided a template for annual cycles of festivals and rituals linking the villages to their fields and to the perceived cyclical rhythms of the natural world. The Balinese calendar is famous among scholars for its mathematical precision: like the Mayan calendar, it tracks time not as a single linear sequence of days and months, but rather by multiple interlocking cycles of different-sized weeks, which are linked to the lunar month and solar year (Liefrinck, 1969a). To obtain maximum yields from HYV rice, it was necessary to set this calendar aside, to permit continuous (unscheduled) rice cropping. Consultants to the Green Revolution programmes suggested that the traditional calendar could still be used to schedule rituals in the temples and fields, but the link to the actual cycles of water flows and rice growth had to be broken.

This policy inadvertently triggered an ecological crisis. Continuous rice cropping provided rice pests with an uninterrupted food supply and by 1985 Balinese agricultural experts reported that their numbers had exploded (Lansing, 2007). When irrigation flows stopped being timed by calendrical cycles, they became unpredictable. It soon became apparent that even massive doses of pesticides could not control the pests, whereupon nearly all *subaks* quietly returned to the traditional calendrical timing of irrigation, adjusted to fit the growth cycles of the new rice. For agricultural experts and planners, this return to traditional farming practices looked like conservative resistance to agricultural modernization. As the American engineer in charge of the ADB irrigation project remarked in 1985, 'They don't need a high priest, they need a hydrologist.'

The view of *subaks* as obstacles to agricultural modernization was soon to change. A 1988 World Bank study of the environmental impact of agrochemicals in Bali was highly critical (Machbub *et al.*, 1988). In 1992 the post-evaluation office of the ADB faulted their Bali Irrigation Project for bypassing the *subak* system, noting: 'The cost of lack of appreciation of the merits of the traditional regime has been high.' The report concluded that 'the issue has to be raised whether the Government should continue to be involved in the irrigation sector in Bali or whether the farmers would benefit more without government involvement' (Anonymous, 1992).

While planners and consultants debated the lessons to be learned from the Green Revolution in Bali, the problems of the *subaks* were also becoming a growing concern for the provincial government. In 2004 the national government required all provincial governments to create 20-year development plans. Plans for assistance to the *subaks* took on a significant role in the *Long-term Development Plan for Bali 2005–2025*, which began to be implemented in 2006. This plan proposed direct subsidies for *subaks* from both the provincial (Bali) and regional (district or *Kabupaten*) governments. The initial annual amounts for each *subak* were targeted at US$1550 from the provincial government and US$260 from the district governments. By 2012 these subsidies had risen to US$2575 and US$310, respectively. To be eligible for these funds, *subaks* had to submit annual proposals and account for their expenditures. One-quarter of these funds was allocated to 'creative economic' initiatives; the rest was designated for the support of *Tri Hita Karana* (Pedersen and Darmiasih, 2015).

Indigenous Knowledge: *Tri Hita Karana* and the *Subaks*

Tri Hita Karana (THK) is a Sanskrit religious formula, translated literally as 'three causes of prosperity (or goodness)'. Since 1966, THK has frequently been cited by Balinese planners and religious authorities to signify the interdependence of humans (*pawongan*), nature (*palemahan*) and spirit (*parhyangan*) (Pedersen and Darmiasih, 2015). The *Long-term Development Plan for Bali* stated that THK is exemplified by the *subaks*. This had far-reaching consequences. Firstly, by advocating direct government support for *subak* temples and temple rituals, it signalled an end to the argument that these were an obstacle to development. Secondly, it upheld THK, as exemplified by the *subak* system, as a model for the whole of Balinese society. Thirdly, it implied that the well-being of the *subaks* was important for the survival of the Balinese way of life. Thus the *Long-term Development Plan* affirmed the importance of the *subak* system and offered a roadmap for government support.

For the *subaks*, THK is more than a general philosophical principle or goal. Instead it connects specific social groups with specific features of the natural landscape, by means of specific temples and ritual activities. The responsibilities of each *subak* to sustain THK are connected to the management of its lands and waters. Each *subak* has its own water sources, irrigation works and temple networks, and well-defined relationships with other *subaks* and villages. To belong to a *subak* is to accept specific rights and responsibilities, which vary between *subaks* and are often encoded in traditional law books (*awig-awig*). As the Green Revolution showed, the resulting coordinated system of management has immediate practical benefits for efficient irrigation flows and pest control. And for the reasons discussed below, it is based on the capacity of the *subaks* for self-governance.

Traditional autonomy of the *subaks*

Subaks are autonomous water-user groups that include both physical infrastructure (irrigation structures including weirs, canals and diversions) and temples and shrines. The physical, social and religious aspects of the *subaks* are inseparably connected (Windia, 2012). *Subaks* are not under the formal control of villages and their physical boundaries are seldom identical with those of villages. Instead *subak* boundaries are based on hydrology and may extend across more than one sub-district. This occurs when (as is usually the case) the source of irrigation water is located far enough upstream from the terraces that the canals must pass through neighbouring villages (Fig. 15.1). The size and location of *subaks* depend on the availability of water. In contrast, villages are residential communities. On the other hand, *subaks* and villages are not entirely separate and they often share responsibility for particular activities (Windia, 2012). Membership in either a *subak* or a village brings responsibilities to care for their physical well-being as a sacred responsibility (*pangempon*). Village lands are usually owned collectively and passed on by inheritance. Village temples define a sacred topography and families are responsible for contributions to the annual cycle of village temple rituals. Similarly, ownership of rice fields in a *subak* brings responsibilities for the performance of rituals in the *subak* temples, as well as the upkeep of the irrigation works. Not infrequently, some *subak* rituals are performed in village temples, but the distinction between *subak* and village membership (and responsibilities) is always clear.

Most *subaks* also coordinate both irrigation and ritual activities with other *subaks*, which enables them to manage the ecology of the rice terraces on a larger scale. These practices also sustain the administrative separation of *subaks* and villages. Typically, groups of *subaks* that share water from the same sources belong to a larger congregation (*subak gede*, or greater *subak*), which coordinates water management and rituals at the temple constructed near the weir. At a higher level, several *subak gede* may cooperate in water management and temple rituals at the river or watershed scale.

The existence of *subaks* as autonomous water-user groups is evident from some of the earliest known historical records from Bali, which were royal inscriptions issued by Balinese kings. The oldest Balinese text (Royal inscription Sukawana A 1, dated AD 882) mentioned irrigated rice fields, and the third inscription (Bebetin A 1, dated AD 896) mentioned irrigation tunnel

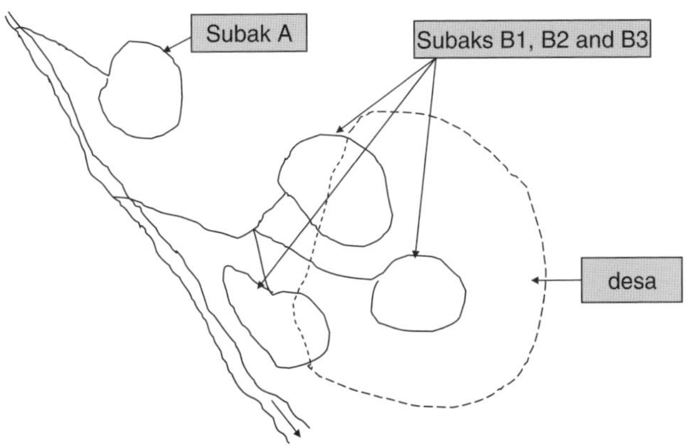

Fig. 15.1. Why *subak* and village (*desa*) boundaries are not identical.

builders. The first appearance of the term *subak* occured in the Pandak Bandung inscription of AD 1071. The following year, the Klungkung C inscription included a royal order calling for the re-measurement of the rice fields of the *subak* of Rawas, and listed the irrigated areas that belonged to this *subak*, which were located in at least 18 communities (Lansing *et al.*, 2009). As archaeologist Jan Wisseman Christie noted, 'the spatial distribution of Balinese irrigation canals, which by their nature cross community boundaries, made it impossible for irrigation to be handled at a purely community level' (Christie, 1992). After the conquest of north Bali by the Dutch in the 19th century, colonial officials also emphasized the autonomy of the *subaks*. After surveying the conquered kingdoms in the 1880s, a senior Dutch colonial official F.A. Liefrinck concluded: 'The explanation of the amazingly high standard of rice cultivation in Bali is to be found in Montesquieu's conclusion that "the yield of the soil depends less on its richness than on the degree of freedom enjoyed by those who till it"' (Liefrinck, 1969b, translated from 1887 report in Dutch).

This view was later echoed by Bali's great ethnographer, V.E. Korn, who wrote in 1932: 'Even an author like Dr. Julius Jacobs (who visited Bali in the 1880s), who has hardly any good words for Bali, feels obliged to express his admiration for the functioning of irrigation on the island. He writes: "the irrigation works are excellent and the manner for sharing the water among different owners of the sawahs is organized in an exemplary way; each subdivision of these works is under the immediate supervision of the *klijan soebak* [*subak* head], who cares for the regular sharing of the water, while a *sedahan* takes in the taxes on the water".' Korn (1932) continued: 'The care for these excellent irrigation works and the exemplary regular sharing of the water across the sawahs of the various parties who are entitled to it, is almost entirely in the hands of the irrigation societies.'

Collective management: 'The voice of the *subak* is the voice of God'

The ancient tradition of autonomous self-governance by the *subaks* continues today. *Subaks* meet regularly to make decisions ranging from water management and pest control to the organization of water temple rituals. Although each *subak* functions independently, subaks also routinely engage in voluntary cooperation with other *subaks*. As E.J. van Naerssen, a Dutch irrigation engineer, reported a century ago, 'if due to lack of water not all areas can get water, then they create a turn-taking which is decided upon during the monthly meetings'. Van Naerssen (1918) also noted that downstream *subaks* routinely 'borrow water' from upstream *subaks*, a practice that continues today. For example, the head of *subak* Gadon 1 in Tabanan remembered borrowing water from distant upstream neighbours in 1988/9, quoting the proverb that 'come better or worse we're together'. This entailed an agreement extending upstream from the Gadon dam to the dams of Tapesan, Sungsan and Cangi. The following year, the Gadon *subaks* moved up their planting schedule to allow their upstream neighbours to obtain more water during the dry season and so repay the loan (Lansing and deVet, 2012).

Ritual plays a vital role in the activities of the *subaks*, but not as a simple expression of faith in THK or other religious principles. Instead *subak* rituals organize the cycles of collective labour in the fields. These include not only rites performed by individual farmers like *Nuasen* (planting), but also large rituals that bring together multi-*subak* congregations, like *Mapag Toyo* (greeting/opening the water) and *Nangluk Merana* (controlling pests). *Subaks* elect their own leaders, whose duties include collecting fines from farmers who have not met their obligations. Leaders are usually chosen by unanimous consent, and once the *subak* has made its choice, this honour is not easily refused, because 'the voice of the *subak* is the voice of God'. Theft is regarded as a religious offence and is subject to religious sanctions. A farmer or *subak* leader who has been accused of theft may be asked to swear this oath in a temple: 'If I speak falsely, may a curse fall on my family for seven generations.' Unsurprisingly, theft is rare. The perennial rites of the *subaks* are intended to 'make the foundations strong' (*negteg linggih*) and are considered to be beneficial for the whole of the Balinese world.

Tri Hita Karana as a World Cultural Heritage

In June 2012, the relationship between THK and the *subaks* was elevated to become the guiding principle for a UNESCO World Heritage Cultural Landscape: 'The *Subak* system as a manifestation of the *Tri Hita Karana* Philosophy'. UNESCO cited two key reasons for endorsing the new World Heritage designation: (i) celebrating the 'outstanding universal value' of the *subaks*; and (ii) helping to conserve them in the face of imminent threats to their sustainability (UNESCO, 2012). As is common in many UNESCO sites, these two goals were in some respects in conflict. National governments actively compete to secure World Heritage recognition from UNESCO, which almost invariably increases tourism. But the expansion of tourism is often one of the major threats to the sustainability of the chosen sites. This was particularly true in the case of the *subaks*.

In Bali, taxes on agricultural land are based on the sale price of nearby parcels; consequently the sale of even small parcels to developers can trigger skyrocketing land taxes for their neighbours, creating a domino effect of land sales as the neighbours find that the value of their rice harvests is no longer sufficient to pay their taxes. *Subaks* that escape this problem are not immune, because they must compete for water with the hotels and other tourist facilities, and the tourism-fuelled urbanization of Bali. Finally, tourism offers service industry jobs that usually pay better than agricultural labour, and the price of Green Revolution rice is kept low as a matter of national policy.

The most direct indication of the effect of these threats to the sustainability of the *subaks* is the rate of their disappearance. According to the Bureau of Statistics in Bali, during 2005–2009 approximately 1000 ha of *subak* rice terraces were converted to other uses every year. Since then, agricultural extension staff estimate that this rate has approximately doubled. In 2012, at the time of the World Heritage inscription, the total area of irrigated rice fields in Bali was approximately 80,000 ha, down from about 86,000 ha in 2000. Three *subaks* located near the capital city of Denpasar recently ceased to exist.

The Management Plan for the World Heritage landscape that was approved by UNESCO sought to address this crisis of sustainability by establishing a Governing Assembly composed of the elected heads of the *subaks* and villages inside the World Heritage landscape. With assistance from a professional Secretariat, the Governing Assembly was designed to empower the *subaks* to address the crisis of sustainability, as explained in the Management Plan (UNESCO, 2012):

> The structure of the Governing Assembly includes the professional Secretariat, headed by an appointed Secretary who oversees the work of three units: Planning; Monitoring and Evaluation; and Finance and Human Resources. Each of these units includes at least one full-time professional staff person. To ensure effective liaison with Provincial and Regency-level government departments, part-time staff from relevant departments are also appointed to two of these units. . . .
>
> The Governing Assembly is designed to function as a learning institution, with the capacity to mobilize, synthesize and make decisions based on different knowledge and operational systems, ranging from the traditional management systems of the subak to recent and successful work by Bali's Department of Agriculture to promote organic rice farming and monitor the social and ecological outcomes of these efforts. . . .
>
> Strategic priorities for implementation are:
>
> - Ongoing support for the role of subaks in sustaining Tri Hita Karana;
> - Livelihood protection and enhancement
> - Conservation and promotion of ecosystem services
> - Conservation of material culture
> - Appropriate development of cultural tourism and education
> - Infrastructure and facility development.

However, this management plan was not implemented. Although various committees were formed to oversee the World Heritage landscape, none of them included the elected representatives of the *subaks*. There was also a problem with funding. The World Heritage landscape was officially listed by UNESCO on 29 June, 2012, but the government budget for 2012 had already been allocated and the budget for fiscal 2013 was finalized in 2012. Consequently, funding was not available until 2014. With no governance structure in place, the rate of land conversion in the World Heritage *subaks* increased as developers sought to capitalize on the World Heritage status. In 2014, the UNESCO World Heritage Committee responded to these issues, stating that it:

1. Notes with concern that the vulnerabilities of the cultural landscape that were acknowledged at the time of inscription, and the need to support the traditional practices of the subak communities through their engagement in the management of the property, have not been addressed clearly;
2. Regrets that the laudable governance structures and Management Plan developed with the nomination have not been fully put in place and implemented, and that incentives and subsidies to support prosperous rural livelihoods and strong subak institutions, and land use regulations to prohibit inappropriate development within the property, have so far not been delivered as envisaged;
3. Urges the State Party to operationalize the Governing Assembly which incorporates the traditional practices underpinning the property as envisaged in the Decree of 2010, as soon as possible, and include in its membership representatives of the subak communities;
4. Also urges the State Party to allow the Governing Assembly to implement the approved Management Plan, as set out at the time of inscription, in order that the various multi-disciplinary Action Plans based on agreed Strategic Priorities can be delivered.

Indonesia responded with a State of Conservation Report in 2014. The report noted that a Management Coordination Forum (MCF) for the World Heritage landscape was created by the Governor of Bali in 2014 (Gubernatorial Decree No. 11/03-H/HK/2014). However, the MCF did not include the elected representatives of the *subaks*. Instead, it was made up of government departments as well as academic experts, institutes and museums. *Subak* leaders expressed their disappointment with the MCF to an evaluation team from UNESCO, and in 2015 UNESCO's World Heritage Committee reiterated its concerns (UNESCO, 2015):

> The World Heritage Committee notes with concern that the pressure for land conversion remains significant, creating a considerable vulnerability that is challenging the ability of the authorities to sustain Outstanding Universal Value and that, although it was envisaged at the time of inscription that there would be full engagement of the subak farming communities with the Governing Council for the effective implementation of the management plan, this seems not to have been effectively implemented.

In summary, the problems that now threaten the *subaks* originated in the 1970s with the Green Revolution and the expansion of tourism. Although both tourism and Green Revolution rice had positive benefits for the economy of Bali, their success also created unexpected difficulties. The goal of the ADB's Bali Irrigation Project in the 1980s was to increase rice production and farm incomes. In the end, the ADB's final evaluation team concluded, 'The cost of lack of appreciation of the merits of the traditional regime has been high,' noting that 'Bali irrigators and their *subak* organizations attach the utmost importance to equitable water sharing and to participatory decision-making' (Anonymous, 1992).

In 2012, Bali's UNESCO World Heritage cultural site had been created to help preserve the *subaks* and their 'indigenous knowledge', codified as *Tri Hita Karana*. After UNESCO approved the nomination, planners concluded that these goals could best be achieved by bypassing the traditional governance system of the *subaks*. The need for 'capacity building' in the *subaks* was cited by foreign consultants promoting the substitution of the Management Coordination Forum for the Governing Assembly of *subak* leaders. However, when asked, *subak* leaders consistently express a preference for doing their own capacity building.

Modelling the Functional Role of the *Subaks*

The Green Revolution inadvertently exposed the vital functional role of the *subaks*, by temporarily halting their management of irrigation in order to speed up cultivation cycles. The *subaks* were in effect taken offline for several years, an experiment that soon revealed the basis of their emphasis on the timing of irrigation flows. Paddies must be flooded and drained to deliver nutrients and promote plant growth, while also controlling weeds and rice pests. Synchronized harvests can reduce pest populations by removing their habitat, but for this to work the geographical extent of the fallow period must be large enough to prevent the pests from migrating to fields that are still in cultivation. The larger the area that is encompassed by the post-harvest flooding, the fewer are the pests, but if too many *subaks* try to flood their fields at the same time, there will not be enough water. Thus the timing of irrigation involves a trade-off between preventing water shortages versus pest infestations.

Lansing and Miller (2005) developed a formal model to try to capture the dynamics of this trade-off in a two-player game, where the players may be considered to be either individual farmers or whole *subaks*. One *subak* is located upstream of the other and so controls the flow of water. The *subaks* can adopt one of two possible cropping patterns, A or B (for example, A could fix planting dates for 1 January and 1 May, while B plants on 1 February and 1 June). The water supply is assumed to be adequate for both *subaks* if they stagger their cropping pattern, but if both plant at the same time, the downstream *subak* will experience water stress and its harvests will be somewhat reduced. Assume further that pest damage will be higher if plantings are staggered (because the pests can migrate from one field to the next) and lower if plantings are synchronized. Let p ($0 < p < 1$) represent the damage caused by the diffusion of pests between the fields, and w ($0 < w < 1$) represent the damage caused by water shortage. Given these assumptions, the payoff matrix is as in Table 15.1, where U and D designate the actions of the upstream and downstream *subaks* respectively.

Here the first number in each cell is the payoff for the upstream *subak* and the second is the payoff for the downstream *subak*. For example, if both plant on schedule A, the payoff for the upstream *subak* is 1, but it is $1-w$ for the downstream *subak* because of insufficient irrigation water.

Several conclusions follow from this simple model. The upstream *subaks* are never affected by water stress, but their downstream neighbours may be. (This is known to rural sociologists as the 'tail-ender' problem: the farmers at the 'tail end' of an irrigation system are at the mercy of their neighbours upstream, who control the irrigation flow.) However, the upstream farmers do care about pest damage, because pests, unlike water, can often move upstream. So a strategy of synchronized cropping patterns to control pests will always produce higher yields for the upstream *subaks*. When $p > w$, the downstream player will also achieve higher yields by synchronizing. Note that if he does so the aggregate harvest is higher (i.e. the total harvest for both farmers goes up). If $p < w$, the upstream farmer does better by staggered planting, which eliminates his water shortage. Interestingly, adding more pests to the fields until $p > w$ actually increases the aggregate harvest for the pair of *subaks*, because it encourages the upstream farmer to cooperate in a synchronized schedule (even though he must give up some water). But if the farmers are not worried about pests, the upstream player has no incentive to give up some of his water.

Table 15.1. Payoffs for synchronized or unsynchronized irrigation schedules for upstream (U) and downstream (D) *subaks*.

	DA	DB
UA	$1, 1-w$	$1-p, 1-p$
UB	$1-p, 1-p$	$1, 1-w$

Based on this logic, behaviour in accordance with the model may be predicted. In general, the downstreamers should prefer greater offsets in irrigation schedules and be willing to accept higher losses from pests as a result, up to $p > w$. The upstreamers, meanwhile, should be willing to give up some of their water to enable the downstreamers to synchronize their irrigation schedule. Both then benefit from a coordinated fallow period and consequently fewer pests. Put another way, the presence of pests in the ecosystem gives the downstream farmers a bargaining lever they can use to persuade their upstream neighbours to give them the water they need to avoid shortages.

We tested these predictions in two surveys, by asking farmers: 'Which is worse, pest damage or water shortages?' The results from the first survey performed in 1998 are shown in Table 15.2. In 2010 we repeated this survey in a different part of Bali, with similar results: upstream farmers tend to worry about pests, while downstream farmers are more concerned about water shortages ($P = 0.032$) (Lansing *et al.*, 2014). Thus the threat of increased pest damage from downstream neighbours provides an incentive for upstream farmers to synchronize their irrigation

Table 15.2. Responses of 117 farmers in 10 *subaks* in the Gianyar Regency to the question, 'Which is worse, pest damage or water shortages?', according to the location of their fields within the *subak* (Pearson's chi-squared 14.083, $P < 0.001$).

Location of farm	Pest damage	Water shortage
Upper	20	18
Middle	8	29
Lower	7	35

schedules. But if the synchronized patches grow too large, water stress will increase. How successful are the *subaks* at solving this problem?

A watershed-scale model

Using empirical data on the location, size and field conditions of 172 *subaks* in the watershed of the Oos and Petanu rivers in southern Bali in 1987/8, we modelled changes in the flow of irrigation water and the growth of rice and pests as *subaks* decided whether to cooperate with their neighbours (Fig. 15.2). In this model, each *subak* behaves as an adaptive agent that seeks to improve its harvest by imitating the cropping pattern of more successful neighbours. The model simulates the flow of water from the headwaters of the two rivers to the sea, at monthly intervals. The amount of water available for any given *subak* depends on seasonal patterns of rainfall and groundwater flow, and the amount of water diverted by upstream *subaks* for their own needs. As a new year begins, each of the 172 *subaks* is given a planting schedule that determines which crops it will grow and when they will be planted. As the months go by, water flows, crops grow and pests migrate across the landscape. When a *subak* harvests its crop, losses due to water shortages or pests are tabulated. At the end of the year, aggregate harvest yields are calculated for the *subaks*. Subsequently, each *subak* checks to see whether any of its closest neighbours got higher yields. If so, the target *subak* copies the cropping schedule of its (best) neighbour. If none of the neighbours got better yields, the target *subak* retains its existing schedule.

When all the *subaks* have made their decisions, the model cycles through another year. These simulations begin with a random distribution of cropping patterns. After a year the *subaks* in the model begin to aggregate into patches following identical cropping patterns, which helps to reduce pest losses. As time goes on these patches grow until they overshoot and cause water stress, causing patch sizes to become smaller. Yields fluctuate but gradually rise. The programme continues until most *subaks* have discovered an optimal cropping pattern, meaning that they cannot do better by imitating one of their neighbours.

Experiments with this model indicate that the entire collection of *subaks* quickly settles down into a stable pattern of synchronized cropping schedules that optimizes the trade-off between pest control and water sharing. In the model, as patterns of coordination resembling the water temple networks emerge, both the mean harvest yield and the highest yield increase, while variance in yield across *subaks* declines (Fig. 15.3). In other words, after just a few years of local experimentation, yields rise for everyone and variation in yields declines. Subsequent simulations showed that if the environment is perturbed, either by decreasing rainfall or by increasing the virulence of pests, a few *subaks* change their cropping patterns, but within a few years a new equilibrium is achieved.

To validate the model, we undertook a field survey and obtained 2 years of data on hydrology, actual planting schedules and harvest yields from August to December 1988 for 43 of the 172 *subaks* included in the model. We calibrated the model with this data and compared the simulated and reported rice harvests for two harvests for 72 *subaks* in Bali in 1989 (Fig. 15.4). The accuracy of the model improved from the first crop of the year ($r = 0.85$) to subsequent harvests ($r = 0.96$), because as time proceeded the accuracy of the pest sub-model increased (when the simulation starts, pests are at background levels, and it takes time for the effects of synchronized cropping to affect population levels). Considering the simplicity of the model, yields per hectare were also well correlated with $r = 0.5$. To eliminate the possibility that the model results were simply not responsive to variations in cropping plans, we ran additional simulations in which we disrupted the local coordination implicit in the planting schedules followed by the *subaks* in 1989. When the cropping patterns were randomized but the actual crops

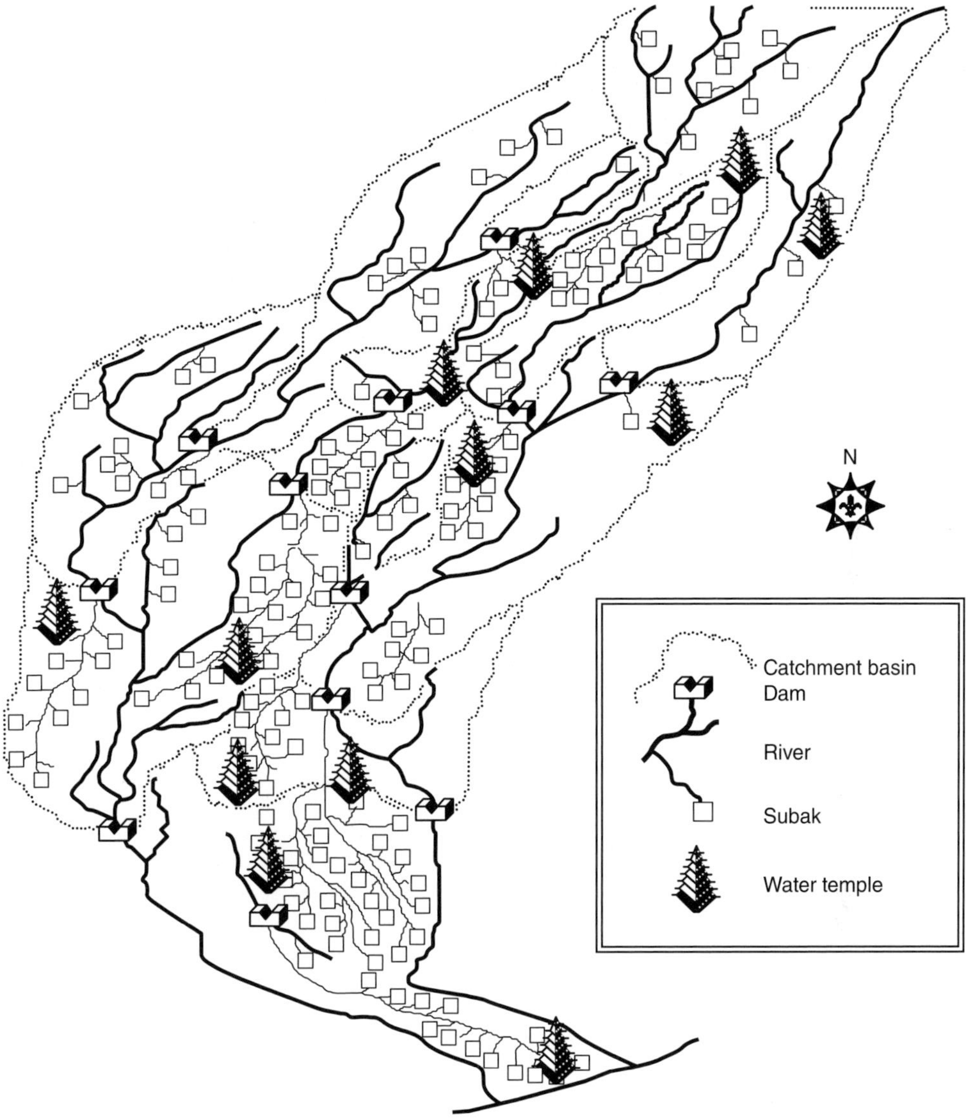

Fig. 15.2. *Subaks*, rivers and irrigation systems along the Oos and Petanu rivers of southern Bali. Traditionally, each *subak* is free to choose its own irrigation schedule. By synchronizing irrigation with different-sized clusters of neighbouring *subaks*, crop losses due to pests or water shortages can be avoided. Map is not to scale, *subaks* are not rectangular, and many more water temples exist than are depicted here. (From Lansing, 2007, p. 119. A free version of this model can be downloaded from https://github.com/mars0i/bali)

planted remained the same, the correlation for the second crop in 1989 dropped from 0.50 to 0.01 (Lansing and Kremer, 1993).

This model captures an evolving feedback relationship between the decisions of the *subaks* and the responses of the environment. Simple trial-and-error at the local level produces a patchwork of synchronized irrigation schedules which over time improves harvests and also reduces variance in harvests. The reduction in variance is potentially significant, because large differences in harvests could discourage

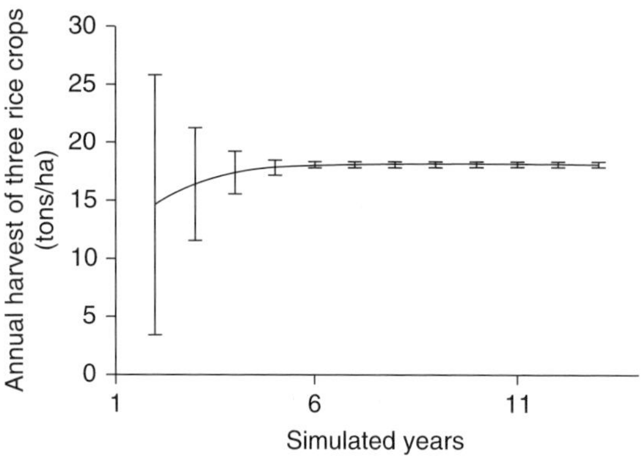

Fig. 15.3. Reduction in variance of harvest yields as cooperation spreads in the simulation model of the Oos and Petanu watersheds.

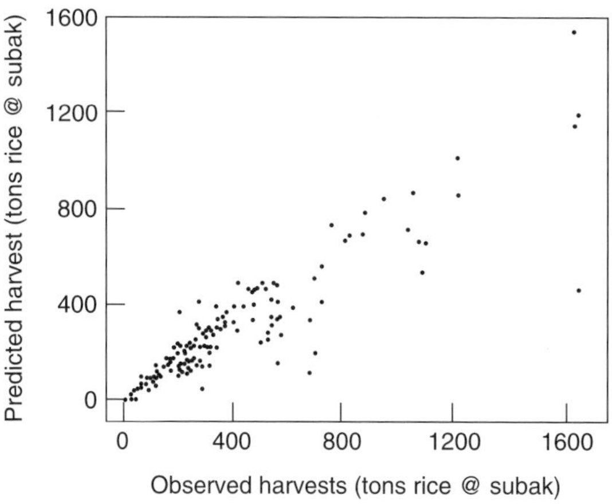

Fig. 15.4. Comparison of model predictions with actual rice harvests by *subak* in 1988–1989 for 72 (of 172) *subaks* in the Oos–Petanu watershed.

cooperation by farmers with suboptimal harvests. These simulated results are supported by responses from farmers to another question in the survey: 97% stated that their own harvest is about the same as that of the other farmers in their *subak* (Lansing, 2006).

Conclusion: *Tri Hita Karana*

In the models described above, the beliefs of the *subaks* and *Tri Hita Karana* play no role. Instead, adaptive agents merely imitate their neighbours with the goal of improving harvests. In reality, more is required. Balinese irrigation systems consist of physically fragile tunnels, canals and aqueducts, which often extend for several kilometres, require constant maintenance and are vulnerable to water theft. While the models provide some insight into the functional structure of this system, they do not account for the high levels of cooperation, planning and social investment that are required to sustain it.

Part of the answer lies in the secular institutions of the *subaks*. *Subaks* are self-governing assemblies of farmers, which hold regular meetings and assess fines on members who do not abide by their decisions. However, in surveys, farmers report that punitive fines and sanctions are seldom needed. From their perspective, the most important responsibility of the *subaks* is the performance of calendrical rites in water temples. By encouraging the farmers' awareness of their shared dependence on one another and the gods, these rites clearly have functional significance. In the model, the key to the emergence of global functional structure is the ability of each *subak* to respond to local ecological feedback involving just two parameters: pests and water. This mathematical analysis might seem suspiciously simple, given the obvious complexity of the Balinese landscape, but it received a measure of empirical support when government agricultural policies severed the local feedback channels, resulting in the almost instantaneous collapse of rice harvests.

A functional perspective is less useful in understanding the principal tool used by the farmers to manage the ecology of the rice terraces: the agricultural calendar. This hybrid device grafts a permutational calendar of 10 concurrent weeks, which vary in duration from 1 to 10 days, to the ancient Indian luni-solar Icaka calendar. It enables groups of *subaks* to organize complex interlocking irrigation schedules, involving different combinations of water turns and planting schedules, and provides a template for irrigation management. Over the centuries the uses of this calendar have expanded to encompass many other phenomena besides irrigation, including musical notation and cosmology. The historical development of this elaborate concept of time and its successive application to many aspects of the phenomenal world is not well captured by a functional perspective. Instead, it appears to reflect what G.W.F. Hegel (1770–1831) described as the desire of Reason to make the world congruent to itself. *Tri Hita Karana* is more than an expression of coupling between social and environmental processes; it maps them on to an abstract concept of divine order created by interlocking cyclical patterns (Lansing *et al.*, 2011). As the greatest Roman pastoral poet observed, '*felix qui potuit cognoscere causas ... fortunatus et ille deos qui nouit agrestis* (it is well for one to understand causes ... fortunate also to comprehend the gods of the countryside)' (Virgil, *Georgics* 2:490).

Acknowledgments

An earlier draft of this chapter was presented by Dr Wayan Windia at the seminar on The Spiritual Dimensions of Rice Culture in Southeast Asia, Srinakharinwirot University, Bangkok, Thailand 11–14 May, 2015. This chapter also draws on the results of several decades of collaborative research involving dozens of researchers, who are acknowledged in the cited references. Permission for research in Indonesia was granted by the Indonesian Institute of Science, the Indonesian Ministry for Research and Technology, and the Ministry of Agriculture. These studies were supported by grants from the National Science Foundation and the James S. McDonnell Foundation.

References

Anonymous (1992) Project Performance Audit Report, Bali Irrigation Sector Project. PE-241 L-552-INO. Manila. Page 13 para 88. Available at: http://www.adb.org/sites/default/files/evaluation-document/35372/files/14014-ino-ppar.pdf (accessed 9 April 2017).

Christie, J.W. (1992) Water from the ancestors: irrigation in early Java and Bali. In: Rigg, J. (ed.) *The Gift of Water: Water Management, Cosmology and the State in Early Southeast Asia.* School of Oriental and African Studies, London, pp. 7–25.

Korn, V. (1932) *Het Adatrecht van Bali.* 2nd edn (1st edn 1924). G. Naeff, The Hague.

Lansing, J. (2006) *Perfect Order: Recognizing Complexity in Bali.* Princeton University Press, Princeton, New Jersey.

Lansing, J. (2007) *Priests and Programmers: Technologies of Power in the Engineered Landscape of Bali.* Revised 2nd edn (1st edn 1991). Princeton University Press, Princeton, New Jersey.

Lansing, J.S. and Kremer, J.N. (1993) Emergent properties of Balinese water temples. *American Anthropologist* 95, 97–114.

Lansing, J.S. and Miller, J.H. (2005) Cooperation games and ecological feedback: some insights from Bali. *Current Anthropology* 46(2), 328–334.

Lansing, J. and de Vet, T. (2012) The functional significance of Balinese water temples: a reply to critics. *Human Ecology* 40(3), 453–467.

Lansing, J., Downey, S., Jannsen, M. and Schoenfelder, J. (2009) A robust budding model of Balinese water temple networks. *World Archaeology* 41(1), 112–133.

Lansing, J., Fox, S. and Fox, K. (2011) Niche construction on Bali: the gods of the countryside. *Philosophical Transactions of the Royal Society B* 366, 927–934.

Lansing, J.S., Cheong, S.A., Chew, L.Y., Cox, M.P., Ho, M.R. and Arthawiguna, W.A. (2014) Regime shifts in Balinese Subaks. *Current Anthropology* 46 (2), 232–239.

Liefrinck, F. (1969a) The Balinese Calendar. In: Swellengrebel, J.L. (ed.) *Bali: Further Studies in Life, Thought and Ritual.* W. van Hoeve, The Hague.

Liefrinck, F. (1969b) Rice Cultivation in Northern Bali. In: Swellengrebel, J.L. (ed.) *Bali: Further Studies in Life, Thought and Ritual*. W. van Hoeve, The Hague.

Machbub, B., Ludwig, H. and Gunaratnam, D. (1988) Environmental impact from agrochemicals in Bali (Indonesia). *Environmental Monitoring and Assessment* 11, 1–23.

Van Naerssen, E. (1918) Irrigatie en Waterverdeeling volgens de Opvatting der Baliërs. *Adatrechtbundel* 15, 27–39.

Pedersen, L. and Dharmiasih, W. (2015) The enchantment of agriculture: state decentering and irrigated rice production in Bali. *The Asia Pacific Journal of Anthropology* 16(2), 141–156. doi: 10.1080/14442213.2014.992458.

Picard, M. (1990) 'Cultural Tourism' in Bali: cultural performances as tourist attraction. *Indonesia* 49, 37–74.

Poffenberger, M. and Zurbuchen, M. (1980) The economics of village Bali: three perspectives. *Economic Development and Cultural Change* 29(1), 91–133.

Sedana, I.G. (2012) *Modal Sosial Dalam Pengembangan Agribisnis Pada Subak* [Social capital in the development of agribusiness in subaks]. Dissertation. Fakultas Sastra, Universitas Udayana, Denpasar, Indonesia.

Sutawan, N. and Pitana, G. (1993) *Subak, Sistem Irigasi Tradisional di Bali*. Upada Sastra, Denpasar, Bali.

UNESCO (2012) Documents on Cultural Landscape of Bali Province: the *Subak* System as a Manifestation of the *Tri Hita Karana* Philosophy. Available at: http://whc.unesco.org/en/list/1194/documents (accessed 9 April 2017).

UNESCO (2015a) SOC Report 2015. Available at: http://whc.unesco.org/en/list/1194/documents (accessed 17 April 2017).

UNESCO (2015b) Draft Decision: 39 COM 7B.66. Available at: http://whc.unesco.org/en/soc/3281 (accessed 17 April 2017).

UNESCO (2015) Decision: 39 COM 7B.66. *Cultural Landscape of Bali Province: the Subak System as a Manifestation of the Tri Hita Karana Philosophy (Indonesia) (C 1194rev)*. Available at: http://whc.unesco.org/en/decisions/6321 (accessed 9 April 2017)

Windia, W. (2012) 'Intangible Heritage of Subak Irrigation System in Bali'. Hong Kong: Nordic Digital Excellence in Design conference on Future Culture: [In]tangible Heritage. NODEM Digital Heritage Repository. Available at: http://repo.nodem.org/?objectId=48.

16 Indigenous Agriculture and the Politics of Knowledge

Alder Keleman-Saxena,* Samara Brock, Luisa Cortesi, Chris Hebdon, Amy Johnson, Francis M. Ludlow and Michael R. Dove

In agriculture, as in other arenas, the 20th century was a period of sweeping technological change. This period saw the introduction and widespread adoption of Green Revolution technologies, including synthetic chemical fertilizers and pesticides, hybrid seeds and, in the late 20th century, genetically modified crops. In parallel, national governments and international organizations undertook major infrastructural projects, including the construction of dams and irrigation systems, which both changed the possibilities of agriculture and had major environmental and social impacts. These trends have underpinned paradoxical outcomes: while agricultural output has vastly increased (Evenson and Gollin, 2003), dietary quality has not necessarily improved apace. Human diets have undergone significant homogenization, now consisting primarily of three major staple crops (Khoury *et al.*, 2014). Critics of the Green Revolution point out that it created the conditions for unprecedented environmental damage and social marginalization (Gupta, 1998; Tsing and Greenough, 2003). In the early 21st century, the contradictions sparked by these trends persist, with agriculture threatening damage to the environment, social relationships and human health, just as it offers opportunities for more socially and ecologically sustainable outcomes.

The push for greater integration of 'traditional' or 'indigenous' agricultural practices and forms of knowledge into agricultural development has received renewed attention in the context of current calls to make agriculture more environmentally and socially sustainable. With the advance of climate change, formal-sector agricultural technologists increasingly recognize that the defining features of future agricultural production and natural resource management are likely to be variability and unpredictability, both between annual cycles and within years. As such, the current push is to develop agricultural technologies, like 'sustainable intensification' or 'climate smart agriculture', which make better use of scarce land, water and other resources and respond dynamically to changing conditions. In many ways, these new initiatives mimic traditional agricultural systems, a key function of which is the buffering of farm households from risks posed by weather, the market, or other unforeseen events (cf. Firth, 1959; Bellon, 1996). Historically, such systems seldom functioned on production-maximizing logics, but were rather embedded in social systems promoting 'moral ecologies', balancing the immediate food harvests with longer-term sustainability (Yapa, 1993; Dove and Kammen, 1997).

* Corresponding author; e-mail: alder.keleman@gmail.com

In the face of current environmental threats, many developers of agricultural technology are now trying to adopt principles of risk-buffering and sustainable increase. However, as argued in this chapter, integrating these spheres of knowledge and practice requires more than a simple mechanical overlapping of techniques and technologies. Researchers and practitioners interested in actively supporting new approaches to agriculture co-developed with traditional or indigenous agricultural knowledge (TIAK) must pay close attention to the politics of knowledge. Below, we provide a brief background on how political currents have historically been reflected in the development of agricultural technologies, and we draw on case studies to demonstrate the political aspects of the recognition, valuation and implementation of knowledges that might be defined as 'traditional' or 'indigenous'. These case studies draw on the fieldwork and analysis of Ludlow (Ireland), Dove (Indonesia), Cortesi (India) and Hebdon (Ecuador).

Read together, these cases underscore key points relevant to understanding the politics of TIAK. They highlight the fact that not all successful agricultural practices are easily visible, nor are all forms of innovation equally recognized. They point out that when agricultural technologies and practices circulate, they do not always circulate unchanged, and they highlight that agricultural knowledge and practice do not operate in isolation, but rather are mediated by state, non-governmental organization (NGO) and market actors. We expand on these common threads in the final section, and conclude by offering observations about the implications of these cases for new and ongoing research in agricultural development and natural resource conservation.

The Politics of Indigenous Knowledge in Agricultural Technology Development: Conceptual Framework

In the social sciences, the concept of politics goes beyond the limited sphere of the formal competition among political parties for citizens' votes. Rather, social science theorists have broadly understood 'politics' to be a question of the competition for – and disposition of – power between and among social groups and actors.[1] A range of definitions – from the 'macropolitics' of institutions and bureaucracies, to the 'micropolitics' of, for example, the policing of one's own or one's peers' behaviour – have utility for understanding the relationship of indigenous knowledge and practices to 20th century agricultural development.

Agriculture, and the management of natural resources associated with agricultural production, represents a concrete form of power. Agriculture generates material resources that contribute to subsistence, that underpin the collection of state surpluses via taxation and that are the material object of trade relationships among groups. Further, the practices of agriculture may generate key forms of symbolic power, expressed in relationships between individuals and within groups. Symbolic power may be enacted discursively and socially, for example, through the moral condemnation of a given farmer's 'poor practices' by another farmer, another ethnic group, or the lens of formal science.

In agriculture practised within the purview of state bureaucracies, these discursive designations have material effects, functioning in a feedback loop with policy. While the forms of agriculture that are noticed and approved of receive research investment and material support through subsidies or enabling regulations, those forms of agricultural practice that remain unacknowledged, or are tacitly disapproved of, are subject to neglect or active discouragement through policy mechanisms. From fine-grained aspects of the care of plants and land to aggregate patterns of influence and policy making, agricultural practices, technologies and power politics are intertwined.

Within this context, the characterization, categorization and implementation of 'indigenous' or 'traditional' forms of agricultural knowledge are highly political. In agricultural research and development, forms of knowledge that receive the designations of 'indigenous' or 'traditional' tend to be those that are responsive or reactive to locally specific conditions and that derive from place, context and lived experience. These knowledges often build inductively from experience to abstraction. Meanwhile, 'formal' knowledge is usually defined by deduction and hypothesis-testing, and the application of abstract concepts to specific circumstances.[2] Due to these differences, formal science, often characterizes TIAK as 'pre-scientific' or 'unreliable.' Further, as a function of their association with specific histories of place and ethnic groups, knowledges that are 'indigenous' are by definition judged not to be cosmopolitan, malleable, or responsive to change. As such, the characterization of a certain body of knowledge or practices as 'indigenous' is something of a double-edged sword: while on the one hand, this characterization may confer legitimacy, on the other, it may also position that knowledge and its holders as backwards, static, or antiquated, placing them at a disadvantage in relation to formal science.

Precisely *whose* knowledge is recognized as TIAK is also a political question. In multi-ethnic contexts, although many groups may be able to make similar claims to indigeneity, one group may be better able than another to access the benefits associated with this designation, as Li (2000) described in detail in the comparative cases of the Lindu and Lauje groups in Indonesia. In Li's case, the claim and recognition of indigeneity conferred benefits on the more affluent, lake-based Lindu group, who were able to advocate on their own behalf against a major hydropower project. However, in other contexts, the recognition of indigeneity may also constrain the forms of action available, as in Shah's description of the 'eco-incarceration' of *adivasi* groups in the state of Jharkand, India (Shah, 2010).[3] Notably, even the distinction between 'formal' scientific and 'informal' indigenous knowledge may be problematic. Agrawal (1995) pointed out that, when viewed up close, the categories of knowledge described as 'indigenous' and 'scientific' are not internally coherent, but rather are highly diverse. At the level of specific knowledge systems, these designations may hold more similarities across categories than within them.

The effects of these political currents can be seen in the development of 20th century agricultural technologies. For example, in the 1980s and 1990s, various authors criticized the role of state-run experimental stations in generating information that was useful and relevant within only a certain set of knowledge parameters (Ceccarelli, 1989; Lipton and Longhurst, 1989; Tripp, 1997). Experimental stations, sited in favourable lands and managed under the best possible conditions, often represent an 'ideal type' of what, in the eyes of formal science, agriculture *should* look like; but observers note that the 'wide adaptation' in formally bred crop varieties that these stations seek to underpin may not encompass the necessities of poor farmers cultivating land in marginal environments (Ceccarelli, 1989). Further, the strict basing of regulations for the labelling and sale of 'formally' improved seed on results from these experimental stations may make it impossible for less common varieties – or varieties that underperform on experimental stations but do well in other environments – to be recognized as saleable, reliable seed (Tripp, 1997). Such confluences of technology development and legal regulatory frameworks effectively marginalize technologies, knowledges and genetic materials emerging from place-based agricultural practices and instead favour the technologies generated by 'formal' science and its associated actors.

As in the case of the testing and registration of formal seed, when indigenous knowledge and 'formal' knowledge are juxtaposed, their merits are often judged on narrow technical terms, or in terms of market value, with little reference to historical, environmental or social context. However, agricultural technologies and practices are always implemented by people who exist within local, governmental or international institutions, as well as many other forms of social hierarchy, which may include gender, caste, ethnicity or age-group relationships. As Scott (1998) demonstrated in his discussion of the relationship of informal knowledge (*metis*) to formalized technical knowledge (*techne*) in *Seeing Like a State*, the relative valuations and privileges accorded to knowledge and technology are the result of long-term processes, in which hegemonic or autocratic actors may use the resources afforded by formal science to advance their own influence. While Scott's argument rests on historical observations, similar patterns can be observed in the production and valuation of agricultural knowledge in the present, through international markets, regulatory frameworks (such as standards for organic agriculture, or the Agreement on Trade-Related Aspects of International Property Rights (TRIPS)) and public investment in agricultural technology. The case studies below present specific examples of these trends.

Case 1: The Misperception of Irish Agriculture under English Settler Colonialism

The long span of history furnishes numerous examples of the indivisibility of agricultural knowledge, politics and culture; of the adaptations necessary to subsist successfully in new or changing environments; and of the political and cultural factors that hinder adaptation. These factors are often most transparent in colonial contexts in which established agricultural practices have been deemed technologically or culturally inferior and thus ignored or replaced in favour of more familiar practices, often commercially driven and export-oriented, in which subsistence is not the sole or even primary goal. Even where the material technology of the colonizer is nominally more advanced, its application may prove unsustainable when insufficient awareness of the surprises (e.g. extreme weather) and variability of newly colonized environments is compounded by a cultural imperialism that is blind to insights offered by established agricultural knowledge, which may embody more nuanced environmental understandings.

The case of Medieval and Early Modern Irish history allows us to examine successive waves of colonization from the island of Great Britain (i.e. England, Wales and Scotland), including how the inherited culture, politics and environmental perceptions of the settlers affected the fortunes of the colonial enterprise. While having nominally similar environments and climates, agriculture with a different pastoral/arable balance often held a comparative economic advantage in Britain, with a greater focus on arable agriculture particularly in the south-east of England, a populous region with a more continental climate (Britnell, 2004). In a famous description of Ireland, based upon his travels with Prince John of England in 1185, the priest-scholar Giraldus Cambrensis (of Welsh–Norman lineage) noted the remarkable physical and cultural features of the island. Of climate and agriculture, he highlighted, 'The grass is green in the fields in winter, just the same as in summer. Consequently the meadows are not cut for fodder, nor do they ever build stalls for their beasts' (O'Meara, 1982, p. 53).

Giraldus here described the climatic characteristics that made the practice of booleying (transhumance) not simply feasible but economically advantageous in Ireland. This involved the removal of cattle to higher or less fertile pastures during summer months of strongest grass growth and their return to more sheltered or fertile lower-lying pastures set aside for winter grazing (Kelly, 1997). Giraldus's reference to a lack of stalls and fodder also suggests the gamble inherent in maximizing benefit from the Irish grass-growing environment. Unhoused cattle, without stored fodder, relied on sufficient availability of winter grass and suffered from sporadically severely cold winters, as made repeatedly clear in Gaelic Irish chronicles. The *Annals of Inisfallen* furnish an example for 1280, reporting 'Very bad weather ... heavy snow in the first week of March ... so that innumerable cows died and live-stock of all kinds largely perished ...' (Mac Airt, 1944, p. 375).

There is little doubt that the Gaelic Irish knew of haymaking and fodder storage, given the proximity to Britain and the strong contacts between both islands. Making hay is labour intensive and the practice of booleying can be seen, at least partly, as a conscious and sophisticated agricultural strategy playing the odds that cold winters would be sufficiently rare so that intermittent losses would be outweighed by longer-term gains from avoiding haymaking. Giraldus, however, writing with propagandistic intent to support the Anglo-Norman colonization of Ireland (Smith, 2008), asserted that the Gaelic had:

> ... not progressed at all from the primitive habits of pastoral living. While man usually progresses from the woods to the fields, and from the fields to settlements and communities of citizens, this people despises work on the land ... and desires neither to abandon, nor lose respect for, the life which it has been accustomed to lead in the woods and countryside ... The fields cultivated are so few because of the neglect of those who should cultivate them ...
>
> (O'Meara, 1982, pp. 101–102).

This belies archaeological evidence for arable agriculture of considerable scope since the early medieval period (Monk, 1991; Monk and Power, 2012), the quick adoption of new types of grain, and the spread of agricultural technologies, such as the watermill, in the same period and the Gaelic legal texts in which the conduct of agriculture and natural resource management is treated in extraordinary detail, including procedures for dealing with neglected land (Kelly, 1997). But in needing to argue that Gaelic culture was inferior and the land underutilized (Smith, 2008), there was little room for Giraldus to acknowledge (or perhaps even privately perceive) that Gaelic agriculture was an evolved adaptation to Ireland's environmental setting.

The Anglo-Norman invasion and colonization of east and south-east Ireland began with the invasion of Ireland in 1169 by Cambro-Normans (from Wales) and the subsequent arrival of English (Anglo-Norman) colonists. This created a feudalized economy and landscape focusing heavily on arable agriculture, modelled after the English seigniorial system structured around manorial centres of feudal lords. For at least the first 100 years, the colony fared well, choosing with precision the best soils for arable agriculture in Ireland, but deteriorated in the 14th century (Britnell, 2004). One reason was soil exhaustion by the export-oriented colonial economy during an arable 'boom' period in the middle to late 13th century, supplying growing markets when England was urbanizing and conducting internal and foreign wars, leading to demand for grain that could not always be met internally. Profits made serving this demand guided decision making regarding the intensification of cereal production (over 90% on some manor demesne lands (Hennessy, 2004)) at the expense of longer-term sustainability. Meanwhile a failure to fully integrate the Gaelic Irish into the manorial system and a continued misperception of the agricultural potential of the uplands into which the Gaelic had often been driven (Fig. 16.1) likely militated against the colony's success.

Yet as the colony declined into the 14th century, so did the differences between the Anglo-Normans and the Gaelic Irish, through processes of intermarriage and acculturation (Booker, 2010, 2013). English commentators

Fig. 16.1. Art Mór Mac Murchadha Caomhánach rides out from what is depicted as a wild and undeveloped area to meet orderly Anglo-Norman knights on more civilized terrain. This illustration comes from the early 15th century text *Histoire du roy d'Angleterre Richard II* by Jean Creton, who accompanied King Richard II to Ireland in 1399. (BL, Harleian MS. 1319, reproduced by permission of the British Library.)

viewed the adoption of Gaelic cultural and agricultural practices with hostility, representing the degeneracy of the colony and the English therein (Palmer, 2001). Attempts were made to legislate against intercultural contact, as with the 35 Acts of the Statutes of Kilkenny in 1366 (Fitzsimons, 2001). The acculturation of Anglo-Normans in Ireland can, however, be seen as a successful adaption to the Gaelic cultural and natural environment, including the marked climatic changes experienced through the 14th century, often now considered part of the first phase of the Little Ice Age. These spurred hybrid agricultural practices in former colonial areas involving an increased emphasis on cattle rearing but with hay still grown for fodder (Nicholls, 2004).

Despite this experience, the lessons of history were hard learned; in efforts by the English to re-colonize Ireland in the 16th and 17th centuries and their introduction of official 'plantations', 'planter' settlers from Britain (mainly English but also Welsh and Scottish) were again encouraged to introduce English farming practices. However, faced with Irish environmental realities and a series of harvest failures in the early 17th century against a background of further rapid climatic change as the Little Ice Age progressed, many new settlers finally pursued the comparative economic advantage afforded by grazing and cattle-rearing as Ireland was further incorporated into the globalizing economy of the period (Smyth, 2006; Ludlow and Crampsie, in press). In this case, then, the politics of globalization were overshadowed by the politics of opening markets, creating a space for the re-evaluation of Irish traditional agricultural knowledge.

Case 2: Technology Development for Smallholder Oil Palm in Indonesia

One of the most controversial agricultural developments today is the vast expansion over the past generation of oil palm (*Elaeis guineensis* Jacq.) cultivation in the Indo-Malay region (Carlson *et al.*, 2012). Many scholars are studying this development, some interpreting it in terms of concepts like 'land grabbing' or 'accumulation by dispossession', which suggests a process of forceful assault on the traditional rights of local peoples (White *et al.*, 2012). There has been less attention paid to the ontological mechanisms – the history and politics of knowledge of export crop development – that make this possible (but see Bissonnette, 2013). We may ask: how has the system of knowledge that underpins these traditional rights been so undermined as to make the abrogation of local rights not seem like an abrogation at all?

The region's most important export crop in the 20th century was natural rubber, which made it the centre of global production. The South-east Asian rubber industry was based on cultivation of an exotic species, *Hevea brasiliensis* (Müll. Arg.), which was introduced to the region late in the 19th century from South America. The system of exploitation in its native habitat in the Amazon did not transfer with the tree. Henry Nicholas Ridley, director of Singapore's Botanical Gardens, developed *de novo* a system of cultivation, tapping and processing the latex. Native smallholders in British Malaya and the Dutch East Indies spontaneously adopted much of Ridley's system, but they also experimented and made their own innovations, notably by intercropping rubber in their rice swiddens and swidden fallows, such that the rubber grew up in the middle of a semi-natural forest.

The smallholder system proved to be highly successful; their production overtook that of estates early in the 20th century and their dominance continues to this day. The colonial estate sector eventually adopted some of the smallholders' innovations, notably by abandoning the early faith in clean weeding around the trees, but it never acknowledged this debt, instead insisting that smallholder practices were profligate and dangerous and, on these grounds, enacted punitive anti-smallholder legislation across the region in the 1920s and 1930s. This hostile stance towards smallholders has persisted to the present day. Almost all government research and development in Indonesia, and to a lesser extent Malaysia, has focused on estate cultivation of rubber; and the minor fraction going to smallholders has consisted not of an effort to develop smallholder technology, but of efforts to transfer estate technology to smallholdings.[4]

This is the historical context for the contemporary development of oil palm in the region. An exotic from West Africa, oil palm (*Elaeis guineensis*) was first introduced to the East Indies in the second half of the 19th century. Cultivation in the Indo-Malay region exploded in the late 20th and early 21st centuries. As with rubber, the native system of smallholder exploitation in its homeland did not transfer along with the plant. Oil palm was developed by the parastatal sector in Indonesia and Malaysia strictly as an estate crop. Smallholder cultivation of oil palm has been supported only when attached to estates, as in the so-called 'nucleus estate' schemes, in which smallholdings surround an inner core of estate cultivation and processing facilities. The fact that 80% of production comes from smallholders in Thailand, the region's third largest producer, suggests that an alternative development path, based on smallholdings, is viable (Byerlee, 2014). Smallholder development, independent of government schemes, is also accelerating in Sarawak, East Malaysia. Many studies across Malaysian and Indonesian Borneo show a similar desire among smallholders to adopt oil palm (Cramb and Sujang, 2013).

Most of the burgeoning literature on oil palm ignores the relevance of estate and smallholder histories in the region, with a few astute exceptions. Cramb (2011) suggested that the contemporary decision to develop oil palm along estate lines represents a reinvention of the colonial thesis of 'dualism', a belief in distinct native and European logics of production, which renders export crops unsuited to traditional native agriculture. Byerlee (2014, p. 591) attributed the dominance of the estate model today to factors similar to those that initially favoured the estate model early in the 20th century: high commodity prices; a convergence of state and investor interests; and a high modern belief in the virtues of agribusiness. Taking these arguments a step further, we suggest that the export crop history of the region made it impossible to think of cultivating oil palm other than in a centralized capital-intensive para-statal model.

The post-humanist turn in anthropology suggests that we cannot ignore the plants at the centre of these industries. It is no accident that oil palm, like rubber, is an exotic in South-east Asia. Ives (2014a, b) has drawn attention to the significance of indigenous versus non-indigenous vegetation. The alienness of oil palm, its removal from its native socio-ecology, as was also the case with rubber, creates a *tabula rasa* that de-privileges time-worn place-specific knowledge. This alienness made it easier to imagine the estate-based model of cultivation, because there was no competing, alternative model to imagine in its place.

Bonneuil (2000) suggested that there is an epistemic imperialism to development schemes, such that schemes like estates privilege metropolitan versus local knowledge. But the history of smallholder rubber shows the other side of coin. The estate model is almost by definition, as Bonneuil has shown, an insular one. Dependent upon the valorization of its own imported system of knowledge, it is highly vulnerable to competitive innovation by local smallholders, especially during economic downturns when its heavy capitalization becomes a liability instead of an asset. By virtue of this same dependency, it is ontologically vulnerable as well: smallholder development is an unexpected and literally incomprehensible development from the perspective of the estate sector.

The trajectory of modern oil palm development in South-east Asia has been viewed by critics of its social and environmental impacts with unfolding surprise. Given the export crop history of the region, it should have been a surprise only if oil palm development had followed a different path. However, this same history shows that the estate sector's self-privileging often co-develops with an irresistible economic force of smallholder agriculture. There are signs that something like this may happen with oil palm, as it gets spontaneously adopted by smallholders across the region (Byerlee, 2014).

Case 3: System of Rice Intensification (SRI) in India: Moral and Organic or More of the Same?

The System of Rice Intensification (SRI) is a methodology of rice cultivation developed by a French Jesuit in Madagascar in the 1980s and spread around the world from the late 1990s by Cornell Professor Dr Norman Uphoff. SRI is employed by millions of farmers in Asia and Central and East Africa, because it requires fewer inputs and yet yields greater output (Rabenandrasana, 1999; Uphoff, 2001; Uphoff and Fernandes, 2002; Vermeulen, 2009). In India, for more than a decade, it has been defined as 'indigenous' and 'traditional' by enthusiastic farmers and NGOs in the country, but also as 'innovative' and 'global'. Paradoxically, this designation as traditional is at odds with its novelty, just as the categorization of indigeneity contrasts with its exogenous, decontextualized and global character.

Similarly, while not necessarily averse to, and most often implemented with, synthetic pesticides and hybrid seeds, SRI is often named 'organic' and has been considered 'the' solution to troubles induced by the Green Revolution. These incongruous definitions refer to two apparently contrasting logics/chains of assumptions (Cortesi, 2013), which are less at odds when examined through the logic of practice. The success of SRI in rural India is crucially linked to this practical resolution of dichotomies, but its failures are to be associated to another oppositional understanding of agriculture, more political than economic.

The official package of SRI centres on practices rather than inputs. For example, it relies on the use of moisture and drainage instead of flooding; the preparation and management of the seeds and the nursery; precise timings and modality of transplanting the seedlings; and timely and frequent intercultivation with a special manual weeder. These techniques are drawn from the best agricultural practices observed over several decades in Madagascar by Father Henri de Laulanié, and in effect, SRI practices require the farmer to be closely attentive to the life of the crop. For example, the seedlings must be transplanted at the exact moment in which two leaves open from the sprout, not earlier or later. This is understood to be the moment when the plant is at its peak of germination, strong enough to be transplanted, but not so established as to perish from the uprooting. However, SRI prescribes that each one of the seedlings must be accompanied out of the nursery and into the field with its own soil, in order to prevent breakage of even the smallest root.

SRI is divergent from, although not opposed to, the Green Revolution agriculture practised in India today. It does not assume an increase, but in many cases allows for a decrease, of external inputs. More profoundly, and differing from Green Revolution agriculture, its practices are considered 'moral', because their discourse refers to understanding and respecting the plant as a living being. Yet SRI is not totally dissimilar from Green Revolution technologies,[5] because it derives and reproduces the same core values of productivity, intensification, monoculture and efficiency (see Yapa, 1993; Dove and Kammen, 1997). SRI is often informally defined as organic not because it does not use fertilizers, pesticides and hybrid varieties, but in response to the excessive 'in-organicity' of the Green Revolution. As a package of inputs, technologies and practices, Green Revolution approaches are applied by the large majority of farmers, often without alternative, yet internally criticized because, as farmers often repeat, these 'medicines' (read as the recurrent and indispensable use of pesticides and fertilizers) 'poison our land and then us'.

In this context, SRI is considered, at least in farmers' fields if less often on sympathizers' websites and manuals, to be 'organic,' 'traditional' and 'indigenous'. SRI refers to values of morality conceptually related to indigeneity, and to the ethics of living in harmony with nature, often believed to be intrinsically traditional and indigenous. In this moral and ethical perspective, SRI is more traditional and indigenous than other current agricultural practices, by virtue of referring to the mythical time in which humans started observing nature, farmed in tune with it and transformed it into their habitat and livelihood. Hence, SRI symbolically rewinds to the moment when the rupture between humans and nature occurred and agriculture became a source of poison and disease. In this way, SRI heralds a synthesis of dichotomies; in it, innovation can be traditional, indigeneity can be global and high productivity can be organic.

SRI implicitly assumes, and demonstrates, that farmers' alternative cultivation practices are neither efficient, nor moral, nor healthy. Farmers accept the critique, recognizing the divergence of their practices from what they consider morally acceptable. Nevertheless, the same farmers are unwilling to sacrifice high yields for higher morality. SRI's triumphant proposal to combine ecological morality and high yields is hence worth the extra labour required (Bijker, 2007), and the effort of learning the new practices – at least for the farmers who can afford it.

Anti-conventional, yet easily institutionalized, SRI has been included in many NGO programmes and governmental policies in the country. Boasted as the reason for Tamil Nadu's boom harvest in 2014 (Vidal, 2014), SRI has become a source of 'hope' for many Indian states, particularly for those gifted with high soil fertility and labour availability, but cursed with less ample or subsidized water resources and with lower output. In a context where agriculture is embedded in feudal landownership and manorialism and where the state has not been able to perform a much-advocated land reform or even a land census, the institutionalization of SRI via state support has transformed it into a site for the reproduction of patronage relationships of governance and citizenship. As a result, the support of the state or its subsidiaries (NGOs) is perceived as indispensable for SRI to be successful.

However, it remains to be seen whether SRI's success would persist without such government and NGO support. In many cases, when subsidies are removed, farmers return to their conventional Green Revolution technologies. While the economic motive can still be in favor of SRI vis-à-vis conventional agriculture, its importance as a symbolic arena, in which political relationships are negotiated and solidified, may outweigh its economic convenience. Institutions wishing to promote SRI as a realistic and beneficial practice may need to pay greater attention to these epistemological and political relationships and processes, if they hope to support the continued implementation of this technology.

Case 4: *Sumak Kawsay* and Socialism: Traditional Indigenous Agricultural Knowledge and State Policy in Ecuador

Can people have knowledge of different 'worlds' rather than just holding different 'worldviews' (Carrithers *et al.*, 2010) and, if so, in what sense (Salmond, 2014)? This question resonates with an issue that has long preoccupied scholars interested in indigenous knowledge – the problem of indigenous knowledge being ignored or not taken seriously (Ellen, 2003).

In the 1990s and 2000s, indigenous scholars in Ecuador's Amazonian region, having witnessed state- and corporate-driven incursions and disasters, began to question not only the validity of different kinds of development but the presumed universality of development itself (Sabin, 1998). One of the results was the creation of a new concept by these indigenous scholars that was meant to represent and strengthen an alternative to development based on living in and coexisting with the forest. In particular it was Runa (Quichua-speaking) scholars who coined the term *sumak kawsay* – from *sumak* ('good,' 'beautiful,' 'delicious' or 'plentiful') and *kawsay* ('life'). They coined this term with an eye to 'self-representation' rather than using a non-local analytics of 'development'. In the years since its coining, this concept has generated intense interest because of its 'ontological' ambition to offer a vision of what development is based in 'traditional' Amazonian Runa agricultural and ecological knowledge and practice, especially swidden agriculture.

One of the first texts to define *sumak kawsay* was an ethnography by Runa anthropologist Carlos Viteri about his home village of Sarayaku: *Súmak Kausai: A Viable Answer to Development* (Viteri, 2003). This text and other key texts were written by Sarayaku Runa authors about attaining and maintaining 'good living' or *sumak kawsay* in Sarayaku, a community of 1500 people with a 135,000 ha Amazonian forest territory in Pastaza, Ecuador. Sarayaku's economy centres around manioc swidden gardening (*chagra*), hunting, fishing and wage labor. Sarayaku, like many other communities in the region, had taken an organized stance against oil drilling on their lands since the 1970s. *Sumak kawsay* emerged out of this conflictual context as a way to counter the notion that indigenous Amazonians – hunting and practising swidden agriculture out in the forest – lacked development. In their descriptions of their territory and ways of living in it, Sarayaku authors emphasized how well they already lived and through what practices (such as swidden gardening and hunting) and through what knowledges (such as the secrets of forest living taught by elders and shamans). Further outlining the concept, Carlos Viteri (2003, p. iv) wrote that:

> In contrast with *súmak káusai*, development is conceived of only in regard to lack and problems, and consequentially it sets out a behind state of underdevelopment in order to appear like the 'medicine' or formula for overcoming this behind state through a lineal transit. *Súmak káusai* on the other hand functions as a social practice oriented precisely to *avoiding* a fall into aberrant conditions of existence.

Sumak kawsay is defined here as a way of 'avoiding a fall', of maintaining security, or not losing a degree of advancement already achieved. Runa ways of life were presented in Viteri's ethnography as being highly adequate and, in a sense, 'already developed'. As one community member noted, 'The forest is already developed. ... What petroleum industries do is destroy what is already developed' (Viteri, 2011). *Sumak kawsay*, as Viteri (2000, 2003) insisted, was not a 'kind of development' but an alternative concept that started from an assumption of community well-being rather than lack. Viteri emphasized that the cornerstone of this already achieved good life is swidden agriculture. In particular, Sarayaku's swidden economy affords them high levels of economic security and political autonomy. It depends on a 'living forest', and the more forest, the better (Sarayaku, 2015). By contrast, the oil economy produces manifold risks for the forest and its indigenous inhabitants.

In 2008, after 15 years of indigenous-led protests and the destabilization of more than ten presidential regimes, Ecuadorians passed a new constitution at a critical moment when the government had a mandate to address social movement demands (Becker, 2012). One of the demands that was heeded was to incorporate *sumak kawsay* into law as a national alternative concept to development. In the constitution, however, *sumak kawsay* was defined quite differently than it had been by Sarayaku authors. The constitution's Preamble calls for a 'new form of public coexistence, in diversity and in harmony with nature, to achieve the good way of living, the *sumak kawsay*'. Notably here the term was presented as a goal to be achieved out on the horizon, fitting what came to be imagined by government intellectuals as a future 'Socialism of *Sumak Kawsay*' (Ramírez-Gallegos, 2012a, b, 2014). Crucially as well, the constitutional definition of *sumak kawsay* made no reference to the original indigenous writings about it, nor to the defence of swidden agriculture and forest ways of life that spurred it to be coined.

Freed from these historical connections, *sumak kawsay* became a highly portable and flexible term. It came to be used by the government, for instance, to justify virtually all of its public policies, including, most ironically, oil drilling on indigenous lands in the Amazon.

While traditional agricultural and ecological knowledge may seem irrelevant to modern 'development' and 'progress', the reluctance of the Ecuadorian state to recognize the full historical context and meaning of *sumak kawsay* hints at how politically consequential struggles between different understandings of key concepts like development can be. It also highlights the limits of nominal recognition of others' knowledge (Povinelli, 2002). Although *sumak kawsay* was nominally recognized, it was recognized only as more or less the same thing as conventional 'development', as little more than an 'indigenous' take on an already discovered idea (socialism). This reflected a state presumption that Runa had only a different worldview concerning an unchangeable concept of development, rather than a different concept based on assumptions drawn from a different framing of life (Salmond, 2014; Kohn, 2013, 2015; Stevenson 2014), in this case one that values the security and autonomy afforded by swidden gardening and Runa culture more generally.

Discussion: Agricultural 'Development' and the Politics of TIAK

The four case studies presented above demonstrate the politics that come into play when knowledge and practices implemented by a particular ethnic or social group become a site in which the distribution of political, material, or symbolic power is disputed among groups. In English-colonized Ireland, the symbolic inferiority of local practices of livestock management was of such political utility that it rendered English colonizers unable to consider adopting ecologically appropriate practices, until climatic imperatives and economic incentives shifted to support the development of a 'hybrid' system. In the case of oil palm in Indonesia, the politics of resource distribution are centred on who has (or is perceived as able to have) important knowledge about the use and management of oil palm, worthy of recognition and support by the state. While estate production is privileged, smallholder production is not acknowledged, in parallel to earlier patterns of non-recognition of the Indonesian smallholder rubber-tapping sector.

Our case study from India explores the curious juxtaposition of a technology that is externally generated but locally categorized and adopted as 'indigenous', despite the fact that it has greater parallels to Green Revolution technologies than to the 'local', 'organic' or 'sustainable' technologies with which it is equated. This is explained by the extent to which the moral content of SRI parallels ethical principles that underscore respect for non-human life. The moral valence of agricultural development is also at play in the example of *sumak kawsay* from Ecuador. In this case, however, while the initial moral concept embedded in *sumak kawsay* is drawn directly from the principles of swidden agriculture, these moral valences are stripped and replaced when *sumak kawsay* is reappropriated as an 'indigenous' platform for Ecuador's national development.

The commonalities among these cases demonstrate four principles of broader relevance for understanding the politics of TIAK. Firstly, they underline that not all successful agricultural practices are equally visible. Practices implemented by powerful groups, like Indonesian estate holders or English colonists, may be more likely to be acknowledged than practices implemented by less powerful groups, like smallholder oil palm producers and transhumant Irish peasants. Secondly, and relatedly, these cases demonstrate that not all forms of innovation are equally recognized. Often, innovation stemming from an exotic locale, external to the place where it is implemented, is granted more legitimacy than innovation that comes from the grassroots. This can be seen in the moral legitimacy afforded to SRI in India, as well as in the non-recognition of the ontological distinctions between *sumak kawsay* and 'development' discussed in the Ecuadorian case.

Thirdly, these examples show that ideas and technologies do not circulate unchanged. Rather, they are taken up by specific people in specific places and for specific purposes, which may entail significant reconfigurations of their initial content and intent. This is demonstrated in the Ecuadorian case by the appropriation of the term *sumak kawsay* and the development of indigenous smallholder technology for the cultivation of both exotic oil palm and rubber in Indonesia. In the case of SRI, such reconfigurations are evidenced not in the technological sphere but rather in the political sphere, with the attachment of SRI's suggested practices not only to larger symbolic narratives but also to a concrete politics of state- and NGO-sponsored patronage.

Fourthly, and taken together, the examples discussed here demonstrate that agricultural knowledge and practice do not operate in a vacuum. They are instead mediated by state, NGO and market actors, all of whom may be driven by specific political or material interests. Whether embodied by the state bureaucracies in Indonesia and Ecuador, the extension-oriented NGOs and government agencies in India, or the colonial English government in 14th century Ireland, such institutional actors play key roles in defining the symbolic and material possibilities associated with the implementation of particular 'indigenous' or 'exotic' technologies.

With reference to this volume's theme, 'optimizing the contribution of indigenous knowledge to agriculture', these broad principles have implications for practitioners seeking to co-develop agricultural technologies with indigenous groups and practices. First and foremost, they highlight the importance of considering the nuances of history, context and ethnic identities. Understanding why and how a particular technology or practice has come to be categorized as 'indigenous', and by whom, may be just as important as understanding the content of the practice itself.

Further, for researchers and practitioners alike, the choice to identify a research topic as 'indigenous' or 'traditional' should be a deliberate designation, as opposed to a romanticized description. Some researchers may choose this term as a way of signalling solidarity with ongoing political movements, or out of respect for collaborators' own language of self-identification. Others may choose to focus less on these broad relational labels and more on specific technologies or practices, which may be adopted regardless of one's self-identification or political positioning. The latter stance may be particularly appropriate in contexts where two or more groups are in competition for material and symbolic power and where using the term 'indigenous' to describe one group would effectively contribute to de-legitimizing the rights and knowledge of the other.

Research on 'indigenous' technologies, like research on poverty alleviation and development, is sometimes assumed to offer, by prior definition, an absolute moral good. However, this view often masks the nuanced politics that underlie the characterization of a particular form of knowledge as indigenous. Those engaging in agricultural development projects with indigenous communities or indigenous knowledges, then, should acknowledge that their engagement of the term represents an entry into a slippery political landscape, in which 'science' has no option to remain neutral. This may place the researcher in a 'double bind' (Ludlow *et al.*, 2016): well-intentioned advocacy on behalf of a group or a concept may well be subsequently rejected by the very people who a researcher seeks to support, precisely due to their own positioning within local currents of politics.

Despite the challenges inherent in navigating the politics of TIAK, there may yet be value in reflecting and articulating technology development with the body of practices and knowledge that the term connotes. Firstly, many formally trained scientists, including plant breeders, geneticists and others, subscribe to understandings of their own knowledge projects that correspond closely with the hypothesis-testing model. To the extent that 'indigenous knowledge' is the 'other' of hypothesis-driven science, an emphasis on integrating this knowledge may provide opportunities to integrate other forms of alterity (otherness), including ethnicity, race and onto-epistemological perspectives, into larger processes of technology development.

Secondly, and perhaps representing the other side of the coin, agriculture is in fact an arena in which 'indigenous' and 'scientific' knowledge have, historically, been somewhat successfully brought into dialogue. Examples include the farming systems research of the 1980s, or participatory research exploring the role of local agricultural experimentation in development (Prain *et al.*, 1999; Cleveland and Soleri, 2002). As compared with other arenas where 'traditional' and 'modern' knowledges are directly at odds, such as medical practice or conservation science, the gap in how farmers and plant breeders perceive their object and space of work may be less insurmountable. As such, agricultural technology development may represent a fertile arena for the synthesis of new knowledges and practices drawing from the full spectrum and diversity of 'indigenous' and 'scientific' knowledge.

Notes

[1] Theorizations of the political have taken a range of forms, for which the German theorist Max Weber (1864–1920) and the French philosopher Michel Foucault (1926–1984) are useful referents. Weber defined politics as 'striving to share power or striving to influence the distribution of power either among states or among groups within a state' (Weber, 1991, p. 78). Weber's writings focused on politics writ large, within the formal contexts of state governing bodies and bureaucracies. Meanwhile, Foucault's writings elucidated the enactment of the political through diffuse, pervasive and discursive forms of power which, particularly in his later writings, he characterized as being manifested through human relationships and the expression of individual agency (Sluga, 2011).

[2] See, for example, Agrawal (1995) and Latour (1993) for useful overviews of how 'traditional' and 'modern' forms of knowledge have been conceived of and implemented. Cleveland and Soleri (2002) and Prain *et al.* (1999) similarly discussed the relationship between the 'formal' knowledge of plant breeders and farmers' knowledge and experimentation.

[3] See also Heatherington's (2010) discussion of conservation politics in Sardinia.

[4] This paragraph drew on Dove (2011).

[5] In this way, it contrasts with the One Straw revolution, the Fukuoka method, or other methodologies of natural farming.

References

Agrawal, A. (1995) Dismantling the divide between indigenous and scientific knowledge. *Development and Change* 26, 413–439. doi: 10.1111/j.1467-7660.1995.tb00560.x.

Becker, M. (2012) Building a plurinational Ecuador: complications and contradictions. *Socialism and Democracy* 26(3), 72–92.

Bellon, M.R. (1996) The dynamics of crop insfraspecific diversity: a conceptual framework at the farmer level. *Economic Botany* 50(1), 26–39.
Bijker, W.E. (2007) Dikes and dams, thick with politics. *Isis* 98(1), 109–123.
Bissonnette, J.F. (2013) Development through large-scale oil palm agribusiness schemes: representations of possibilities and the experience of limits in West Kalimantan. *SOJOURN: Journal of Social Issues in Southeast Asia* 28(3), 485–511.
Bonneuil, C. (2000) Development as experiment: science and state building in late colonial and postcolonial Africa, 1930–1970. *Osiris* 15, 258–281.
Booker, S. (2010) Ashamed of their very English names? Identity and the use of Irish names by the English of late medieval Ireland. In: Signori, G. and Rolker, C. (eds) *Umstrittene Zugehörigkeit(en). Spätmittelalterliche Praktiken der Namengebung im europäischen.* Vergleich, Konstanz, Germany, pp. 131–148.
Booker, S. (2013) Intermarriage in fifteenth-century Ireland: the English and Irish in the 'four obedient shires' in Ireland. *Proceedings of the Royal Irish Academy* 113C, 219–250.
Britnell, R. (2004) Britain and Ireland, 1050–1530. *Economy and Society*, Oxford University Press, Oxford, UK.
Byerlee, D. (2014) The fall and rise again of plantations in tropical asia: history repeated? *Land* 3(3), 574–597.
Carlson, K.M., Curran, L.M., Ratnasari, D., Pittman,A.M., Soares-Filho, B.S. *et al.* (2012) Committed carbon emissions, deforestation, and community land conversion from oil palm plantation expansion in West Kalimantan, Indonesia. *Proceedings of the National Academy of Sciences* 109(19), 7559–7564.
Carrithers, M., Candea, M., Sykes, K., Holbraad, M. and Venkatesan, S. (2010) Debate: Ontology is just another word for culture. *Critique of Anthropology* 30(2), 152–200.
Ceccarelli, S. (1989) Wide adaptation: how wide? *Euphytica* 40, 197–205.
Cleveland, D.A. and Soleri, D. (2002) Introduction: farmers, scientists and plant breeding: knowledge, practice, and the possibilities for collaboration. In: Cleveland, D.A. and Soleri, D. (eds) *Farmers, Scientists, and Plant Breeding: Integrating Knowledge and Practice.* CAB International, Wallingford, UK, pp. 1–18.
Cortesi, L. (2013) Nature is hard to know: conflicts over floods as struggles of knowledge. In: Prasad, E., Joy, K.J., Paranjape, S. and Vispute, S. (eds) *Agony of Floods: Flood Induced Water Conflicts in India.* Forum for Policy Dialogue on Water Conflicts in India, Pune, India, pp. 124–138.
Cramb, R.A. (2011) Re-inventing dualism: policy narratives and modes of oil palm expansion in Sarawak, Malaysia. *The Journal of Development Studies* 47(2), 274–293.
Cramb, R.A. and Sujang, P.S. (2013) The mouse deer and the crocodile: oil palm smallholders and livelihood strategies in Sarawak, Malaysia. *The Journal of Peasant Studies* 40(1), 129–154.
Dove, M.R. (2011) *The Banana Tree at the Gate: The History of Marginal Peoples and Global Markets in Borneo.* Yale University Press, New Haven, Connecticut.
Dove, M.R. and Kammen, D.M. (1997) The epistemology of sustainable resource use: managing forest products, swiddens, and high-yielding variety crops. *Human Organization* 56, 91–101.
Ellen, R. (2003) 'Deja vu' all over again: reinvention and progress in applying local knowledge to development. In: Sillitoe, P., Bicker, A. and Pottier, J. (eds) *Participating in Development: Approaches to Indigenous Knowledge.* Routledge, London and New York, pp. 235–258.
Evenson, R.E. and Gollin, D.E. (2003) Assessing the impact of the green revolution, 1960 to 2000. *Science*, 300(5620), 758–762. doi: 10.1126/science.1078710.
Firth, R. (1959). Social change in Tikopia: re-study of a Polynesian community after a generation. *Critical Pressures on Food Supply and Their Economic Effects.* Allen and Unwin, London, pp. 51–76.
Fitzsimons, F. (2001) Fosterage and gossipred in late medieval Ireland: some new evidence. In: Duffy, P.J., Edwards, D. and Fitzpatrick, E. (eds) *Gaelic Ireland, c.1250–c.1650: Land, Lordship and Settlement.* Four Courts Press, Dublin, pp. 138–149.
Gualinga Montalvo, J. and Pueblo Originario Quichua de Sarayaku (2005) Propuesta Declaratoria de Kausak Sacha – Selva Viviente (Territorio Sagrado, Patrimonio de la Biodiversidad y Cultura Kichwa en Ecuador). *Ecología Política: Cuadernos de debate internacional* 29, 168–175.
Gupta, A. (1998) *Postcolonial Developments: Agriculture in the Making of Modern India.* Duke University Press, Durham, North Carolina.
Heatherington, T. (2010) *Wild Sardinia: Indigeneity and the Global Dreamtimes of Environmentalism.* University of Washington Press, Seattle, Washington.
Hennessy, M. (2004) Manorial agriculture and settlement in early fourteenth-century co. Tipperary. In: Clarke, H.B., Prunty, J. and Hennessy, M. (eds) *Surveying Ireland's Past.* Geography Publications, Dublin, pp. 99–118.
Ives, S. (2014a) Farming the South African 'Bush': ecologies of belonging and exclusion in rooibos tea. *American Ethnologist* 41(4), 698–713.

Ives, S. (2014b) Uprooting 'indigeneity' in South Africa's Western Cape: the plant that moves. *American Anthropologist* 116(2), 310–323.

Kelly, F. (1997) *Early Irish Farming*. Dublin Institute for Advanced Studies, Dublin.

Khoury, C.K., Bjorkman, A.D., Dempewolf, H., Ramirez-villegas, J., Guarino, L., Rieseberg, L. and Struick, P. (2014) Increasing homogeneity in global food supplies and the implications for food security. *Proceedings of the National Academy of Sciences of the United States of America*. doi: 10.1073/pnas.1313490111.

Kohn, E. (2013) *How Forests Think: Toward an Anthropology Beyond the Human.* University of California Press, Berkeley, California.

Kohn, E. (2015) Anthropology of ontologies. *Annual Review of Anthropology* 44, 311–327.

Latour, B. (1993) *We Have Never Been Modern*. Harvard University Press, Cambridge, Massachusetts.

Li, T. (2000) Articulating indigenous identity in Indonesia: resource politics and the tribal slot. *Comparative Studies in Society and History* 42(1), 149–179.

Lipton, M. and Longhurst, R. (1989) New seeds and poor people. Johns Hopkins University Press, Baltimore, Maryland.

Ludlow, F. and Crampsie, A. (in press) Environmental history of Ireland, 1550–1730. In: Ohlmeyer, J. (ed.) *Cambridge History of Ireland: Volume 2, Early Modern Ireland, 1550–1730*. Cambridge University Press, Cambridge, UK.

Ludlow, F., Baker, L., Brock, S., Hebdon, C. and Dove, M.R. (2016) The double binds of indigeneity and indigenous resistance. *Humanities* 5(3), 53. doi: 10.3390/h5030053.

Mac Airt, S. (ed.) (1944) *The Annals of Inisfallen.* MS Rawlinson B 503. Dublin Institute for Advanced Studies, Dublin.

Monk, M. (1991) The archaeobotanical evidence for field crop plants in early historic Ireland, In: Renfrew, J.M. (ed.) *New Light on Early Farming: Recent Developments in Palaeoethnobotany*. Edinburgh University Press, Edinburgh, pp. 315–328.

Monk, M. and Power, O. (2012) More than a grain of truth emerges from a rash of corn-drying kilns? *Archaeology Ireland* 26, 38–41.

Nicholls, K. (2004) *Gaelic and Gaelicized Ireland in the Middle Ages* (revised edn). Lilliput, London.

O'Meara, J.J. (ed.) (1982) *The History and Topography of Ireland* (revised edn). The Dolmen Press, Portlaoise, Ireland.

Palmer, P. (2001) *Language and Conquest in Early Modern Ireland: English Renaissance Literature and Elizabethan Imperial Expansion*. Cambridge University Press, Cambridge, UK.

Povinelli, E. (2002) *The Cunning of Recognition: Indigenous Alterities and the Making of Australian Multiculturalism*. Duke University Press, Durham, North Carolina.

Prain, G., Fujisaka, S. and Warren, M.D. (eds) (1999) *Biological and Cultural Diversity: The Role of Indigenous Agricultural Experimentation in Development*. Practical Action, Bourton on Dunsmore, UK.

Rabenandrasana, J. (1999) Revolution in rice intensification in Madagascar. *LEISA Magazine* 15(3/4), 48–49.

Ramírez Gallegos, R. (2012a) *La Vida (Buena) Como Riqueza de Los Pueblos: Hacia Una Socioecología Del Tiempo*. Instituto de Altos Estudios Nacionales, Quito, Ecuador.

Ramírez Gallegos, R. (2012b) *Socialismo Del Sumak Kawsay, O Biosocialismo Republicano*. Don Bosco, Quito, Ecuador.

Ramírez Gallegos, R. (2014) *El Socialismo Ya No Es Lo Que Era. Y Su Nuevo Rostro ¿Se Perfila Más Latino?* Available at: http://reneramirez.ec/el-socialismo-ya-no-es-lo-que-era-y-su-nuevo-rostro-se-perfila-mas-lation (accessed 7 December 2016)

Sabin, P. (1998) Searching for middle ground. *Environmental History* 3(2), 144–168.

Salmond, A. (2014) Transforming translations: addressing ontological alterity. *HAU: Journal of Ethnographic Theory* 4(1), 155–187.

Sarayaku, Pueblo de (2003) *Sarayaku Sumak Kawsayta Katina Killka*. Pueblo Originario de Sarayaku, Puyo, Ecuador.

Sarayaku, Pueblo de (2015) *Kawsak Sacha – Selva Viviente:Propuesta de los Pueblos Originarios frente al cambio climático*. Available at: http://sarayaku.org/kawsak-sacha-selva-viviente-propuesta-de-los-pueblos-originarios-frente-al-cambio-climatico/ (accessed 30 June 2017).

Scott, J.C. (1998). *Seeing Like a State: How Certain Schemes to Improve the Human Condition Have Failed*. Yale University Press, New Haven, Connecticut.

Shah, A. (2010) *In the Shadows of the State: Indigenous Politics, Environmentalism, and Insurgency in Jharkand, India*. Duke University Press, Durham, North Carolina.

Sluga, H. (2011) 'Could you define the sense you give the word "political"?' Michel Foucault as a political philosopher. *History of the Human Sciences* 24(4), 69–79.

Smith, A. (2008) Written off the map: cleared landscapes of medieval Ireland, In: Smith, A. and Gazin-Schwartz, A. (eds) *Landscapes of Clearance: Archaeological and Anthropological Perspectives.* Left Coast Press, Walnut Creek, California, pp. 49–70.

Smyth, W.J. (2006) *Map-Making, Landscapes and Memory: a Geography of Colonial and Early Modern Ireland c.1530–1750*. Cork University Press, Cork, Ireland.

Stevenson, L. (2014) *Life Beside Itself: Imagining Care in the Canadian Arctic*. University of California Press, Berkeley, California.

Tripp, R. (1997) *New Seed, Old Laws: Regulatory Reform and the Diversification of National Seed Systems.* Intermediate Technology Publications on behalf of the Overseas Development Institute, London.

Tsing, A. and Greenough, P. (2003) Introduction. In: Tsing, A. and P. Greenough, P. (eds) *Nature in the Global South: Environmental Projects in South and Southeast Asia*. Duke University Press, Durham, North Carolina.

Uphoff, N. (2001) The system of rice intensification – agroecological opportunities for small farmers? *LEISA* 17(4), 15–16.

Uphoff, N. and Fernandes, E. (2002) System of rice intensification gains momentum. *LEISA* 18(3), 24–29.

Vermeulen, M. (2009) More from less, from less to more: dissemination of a rice cultivation technique. *Farming Matters* 24(4), 22–24.

Vidal, J. (2014) Miracle grow: Indian rice farmer uses controversial method for record crop. *The Guardian,* 13 May 2014. Available at: http://www.theguardian.com/global-development/2014/may/13/miracle-grow-indian-rice-farmer-sri-system-rice-intensification-record-crop (accessed 17 April 2017).

Viteri Gualinga, C. (2000) Visión indígena del desarrollo en la Amazonía. *Polis* 3. Available at: http://polis.revues.org/7678 (accessed 6 December 2016),

Viteri Gualinga, C. (2003) Sumak Kausai: Una Respuesta Viable Al Desarollo. Master's Thesis. Universidad Politecnica Salesiana, Quito, Ecuador.

Viteri Gualinga, F.T. (2011) 'Franco Viteri'. Available at: https://youtu.be/yp3ELrU1c8E?t=1m53s (accessed 1 February 2016).

Weber, M. (1991) Politics as a vocation. In: Gerth, H., Weber, M. and Wright Mills, C. (eds) *Max Weber: Essays in Sociology*. Routledge, New York, pp. 77–128.

White, B., Borras Jr, S.M., Hall, R., Scoones, I. and Wolford, W. (2012) The new enclosures: critical perspectives on corporate land deals. *Journal of Peasant Studies* 39(3–4), 619–647.

Yapa, L. (1993) What are improved seeds? an epistemology of the green revolution. *Economic Geography* 69(3), 254–273.

Index

Page numbers in **bold** type refer to figures, tables and boxed text.